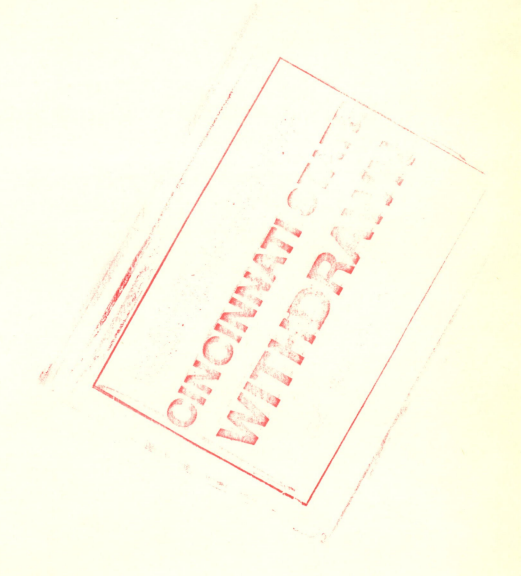

Electronic instruments
and measurement techniques

Electronic instruments and measurement techniques

F.F. MAZDA

Chief Engineer, Multiplexor Systems
STC Telecommunications Ltd

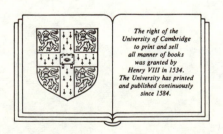

The right of the
University of Cambridge
to print and sell
all manner of books
was granted by
Henry VIII in 1534.
The University has printed
and published continuously
since 1584.

CAMBRIDGE UNIVERSITY PRESS

Cambridge

New York   New Rochelle   Melbourne   Sydney

Published by the Press Syndicate of the University of Cambridge
The Pitt Building, Trumpington Street, Cambridge CB2 1RP
32 East 57th Street, New York, NY 10022, USA
10 Stamford Road, Oakleigh, Melbourne 3166, Australia

First published 1987

Printed in Great Britain at
the University Press, Cambridge

British Library cataloguing in publication data

Mazda, F.F
Electronic instruments and measurement
techniques
1. Electronic measurements   2. Electronic
instruments
I. Title
621.3815'48   TK7878.4

Library of Congress cataloguing in publication data

Mazda, F.F.
Electronic instruments and measurements techniques.
Bibliography
Includes index.
1. Electronic instruments.   2. Electronic
measurements. I. Title.
TK7878.4.M36   1987   621.381   86–18863

ISBN 0   521   26873   7

TM

Contents

Preface

The field of electronic measurements and instrumentation is vast, and it would be foolish if I claimed to have covered it all in this book! Instead I have tried to provide an understanding of the principles involved in electronic measurements, and to introduce the instruments which are most commonly used in electronics laboratories. The fundamental aspects of the instruments and their basic operations are described; detailed procedures are usually covered in the manuals associated with the instruments.

Having read this book the student should be in a good position to decide on the type of instrument for any application, the basic procedures to be used, and the techniques for obtaining the best results.

The book is divided into three parts. The first part covers measurement fundamentals such as measurement standards; transducers which form the front ends of instruments; instrument buses which interconnect different equipments in a large measurement system; and the statistics of measurement errors, and how they can be minimised.

The second part of the book describes general purpose instruments, which are used in a wide variety of applications. These include voltmeters, ammeters and ohmmeters in both analogue and digital form; bridges for measuring components and parameters such as frequency; power measuring instruments; signal generators and signal analysis instruments such as counters, timers, wave analysers and spectrum analysers; and the oscilloscope, which is justly called the workhorse of the electronics industry!

Part 3 describes measurement techniques used for specialised application areas, and their instruments. Electronic component testing covers a wide field from resistors through to complex integrated circuits. The emphasis here is on testers which are used in the laboratory, rather than in a factory environment or by component manufacturers. Audio and video circuit tests, such as in radio and television, are then described, followed by a chapter on testing transmission systems such as antennas and transmission lines. With the growth in the use of microprocessors, digital circuit analysis is an increasingly important field. A chapter is devoted to this, and covers topics such as signature analysers, logic analysers, and microprocessor development systems, which play such an important role in software development.

The book concludes with a chapter on optoelectronic measurements. Although optical communication is playing such a vital role within the electronics industry the optical terminology used is often confusing to electronics engineers. These are first introduced, followed by a description of optical units, optical measurement techniques, and fibre optic systems.

I am grateful to the many, many authors who have enriched the field of electronic measurements with their writings, and on whose works I have drawn freely. I am also grateful to Dr Peter Spreadbury of the University Engineering Department, at the University of Cambridge, for reading through the text of my book, and for providing many useful suggestions.

F.F. MAZDA

Bishop's Stortford 1987

To Roshan

For twenty-eight very happy years

Part 1: Fundamentals

1. Units and constants

1.1 Introduction

The measurement of quantities requires a system of units which is accurate, reliable, and easy to understand and to use. Mechanics led the field in scientific units, and the second was well established by the time the metric system of length and mass was adopted in France in 1799. There were several variations on these early units, such as metre–gram–second and millimetre–milligram–second, but gradually the centimetre–gram–second (cgs) unit became universally accepted. This developed into the MKS unit (metre–kilogram–second).

Workers in the electrical field soon realised that they too needed a system of universal units. In 1832 Karl Gauss measured the strength of the earth's magnetic field in terms of length, force, mass and time, and established the first basic set of units. Friedrich Kohlrausch measured resistance in the above units in 1849. Wilhelm Weber introduced the first complete system of electrical units in 1851. These were based on mechanical units, and they form the basis of our present system of electrical units.

In a coherent system of units the relationships between the various units do not involve any conversion factors. The system of units is made up of fundamental units and derived units. Fundamental units consist of a minimum set of independent base units, whilst derived units consist of a combination of the base units. Sometimes, if the derived units are in common use, they are given a name of their own.

Apart from units, the names for units and their symbols are also standardised. This enables scientists to have a universal language, where scientific equations can be understood worldwide.

1.2 Units and symbols

In 1960 the 11th Conférence Générale des Poids et Mesures adopted the Système International d'Unités as an international system of units (SI units). These SI units include the MKS system of units (mechanical units) and the MKSA or Georgi system of units (electrical units). Six basic quantities were proposed for the SI units: (i) length in metres (ii) mass in kilograms (iii) time in seconds (iv) temperature in degrees Kelvin (v) current in amperes (vi) luminous intensity in candelas. The 1967 Conference changed the name of the standard unit of temperature from the degree Kelvin to the kelvin. In 1971 a seventh basic unit was adopted, that for the amount of substance, the mole.

From the basic SI units other units can be derived. Tables 1.1 to 1.5 give the fundamental and derived SI units. The symbol shown for the quantity is that most commonly used in equations, and this should not be confused with the abbreviation used for the unit. For example a current of 10 amperes is often written as $I = 10\,\text{A}$.

Table 1.6 gives the terms used for multiples of a unit. Special care must be taken not to confuse the symbol for the decimal prefix with that for

Table 1.1. *Fundamental units*

Quantity	Symbol	Unit name	Unit symbol	Dimensions
length	l	metre	m	L
mass	m	kilogram	kg	M
time	t	second	s	T
electric current	I	ampere	A	QT^{-1}
temperature	T, θ	kelvin	K	θ
luminous intensity	I_v	candela	cd	I_v

Table 1.2. *Derived mechanical units*

Quantity	Symbol	Unit name	Unit symbol	Dimensions
plane angle	$\alpha, \beta, \theta, \phi$	radian	rad	—
solid angle	ω	steradian	sr	—
area	A	square metre	m^2	L^2
volume	V	cubic metre	m^3	L^3
frequency	f	hertz	Hz	T^{-1}
angular frequency	ω	radian/second	rad/s	T^{-1}
velocity	v	metre/second	m/s	LT^{-1}
acceleration	a	metre/second²	m/s^2	LT^{-2}
angular velocity	ω	radian/second	rad/s	T^{-1}
angular acceleration	α	radian/second²	rad/s^2	T^{-2}
wavelength	λ	metre	m	L
rotational frequency	n	revolutions/second	r/s	T^{-1}
volume density	ρ	kilogram/metre³	kg/m^3	ML^{-3}
weight	W	newton	N	MLT^{-2}
weight density	γ	newton/cubic metre	N/m^3	$M^{-2}LT^{-2}$
force	F	newton	$N(kg \cdot m/s^2)$	MLT^{-2}
pressure	p	newton/metre²	N/m^2	$ML^{-1}T^{-2}$
moment of force	M	newton metre	$N \cdot m$	ML^2T^{-2}
torque	T	newton metre	$N \cdot m$	ML^2T^{-2}
work	w	joule	$J(N \cdot m)$	ML^2T^{-2}
power	P	watt	$W(J/s)$	ML^2T^{-3}
energy	E	joule	J	ML^2T^{-2}
efficiency	η	—	—	—
gravitational field strength	g	newton/kilogram	N/kg	LT^{-2}

Table 1.3. *Derived electrical units*

Quantity	Symbol	Unit name	Unit symbol	Dimensions
quantity of electricity	Q	coulomb	$C(A \cdot s)$	Q
electrical dipole moment	p	coulomb metre	$C \cdot m$	LQ
polarization	P	coulomb/metre²	C/m^2	$L^{-2}Q$
permittivity of free space	ε_0	farad/metre	F/m	$M^{-1}L^{-3}T^2Q^2$
electric field intensity	E	volt/metre	V/m	$MLT^{-2}Q^{-1}$
electric flux	Ψ, Φ_e	coulomb	C	Q
electric flux density	D	coulomb/metre²	C/m^2	$L^{-2}Q$
electric potential	V	volt	V	$ML^2T^{-2}Q^{-1}$

(Contd)

Table 1.3. (Cont.)

Quantity	Symbol	Unit name	Unit symbol	Dimensions
capacitance	C	farad	F(A·s/V)	$M^{-1}L^{-2}T^2Q^2$
resistance	R	ohm	Ω(V/A)	$ML^2T^{-1}Q^{-2}$
resistivity	ρ	ohm·metre	Ω·m	$ML^3T^{-1}Q^{-2}$
conductance	G	mho (siemen)	mho(S)	$M^{-1}L^{-2}TQ^2$
conductivity	γ	mho/metre (siemen/metre)	mho/m (S/m)	$M^{-1}L^{-3}TQ^2$
current density	J	ampere/metre²	A/m²	$L^{-2}T^{-1}Q$
electrical energy	U_e	joule	J	ML^2T^{-2}
electrical energy density	u_e	joule/cubic metre	J/m³	$ML^{-1}T^{-2}$
reactance	X	ohm	Ω	$ML^2T^{-1}Q^{-2}$
impedance	Z	ohm	Ω	$ML^2T^{-1}Q^{-2}$
admittance	Y	mho (siemen)	mho(S)	$M^{-1}L^{-2}TQ^2$
electrical power	P	watt	W	ML^2T^{-3}
electrical energy	W, U	joule	J	ML^2T^{-2}
characteristic impedance	Z_0	ohm	Ω	$ML^2T^{-1}Q^{-2}$
loss angle	δ	radian	rad	—

Table 1.4. Derived magnetic units

Quantity	Symbol	Unit name	Unit symbol	Dimensions
magnetic scalar potential	U_m	ampere	A	$T^{-1}Q$
magnetic field intensity	H	ampere/metre	A/m	$L^{-1}T^{-1}Q$
magnetic flux	Φ	weber	Wb	—
magnetic flux density	B	tesla (weber/metre²)	T(Wb/m²)	$MT^{-1}Q^{-1}$
magnetic vector potential	$\boldsymbol{A}$	weber/metre	Wb/m	$MLT^{-1}Q^{-1}$
magnetic flux linkage	A	ampere turn	A·t	$T^{-1}Q$
magnetic susceptibility	χ_m	—	—	—
magnetic permeability	μ	henry/metre	H/m	MLQ^{-2}
permeability of free space	μ_0	henry/metre	H/m	MLQ^{-2}
relative permeability	μ_r	—	—	—
electromagnetic dipole moment	m	ampere metre²	A·m²	$L^2T^{-1}Q$
self inductance	L	henry	H	ML^2Q^{-2}
mutual inductance	M	henry	H	ML^2Q^{-2}
reluctance	R	henry⁻¹	H^{-1}	$M^{-1}L^{-3}Q^2$
permeance	P	henry	H	ML^3Q^{-2}
magnetic polarization	J	tesla	T	$MT^{-1}Q^{-1}$
number of turns	N, n	—	—	—
hysteresis coefficient	k_H	—	—	—
eddy-current coefficient	k_e	—	—	—

Table 1.5. Derived optical units

Quantity	Symbol	Unit name	Unit symbol	Dimensions
radiant energy	W	joule	J	ML^2T^{-2}
radiant flux	P	watt	W	ML^2T^{-3}
radiant intensity	i_e	watt/steradian	W/sr	ML^2T^{-3}
radiance	L_e	watt/steradian metre²	W/sr·m²	$M^{-1}L^2T^{-3}$
radiant exitance	M_e	watt/metre²	W/m²	$M^{-1}L^2T^{-3}$

(Contd)

Table 1.5. (*Cont.*)

Quantity	Symbol	Unit name	Unit symbol	Dimensions
irradiance	E_o	watt/metre²	W/m²	$M^{-1}L^2T^{-3}$
quantity of light	Q_V	lumen second	lm·s	I_VT
luminous flux	Φ_V	lumen	lm	I_V
luminous exitance	M_V	lumen/metre²	lm/m²	I_VL^{-2}
illuminance	E_V	lux	lx	I_VL^{-1}
luminous intensity	I_V	candela	cd	I_V
luminance	l_V	candela/metre²	cd/m²(nit)	I_VL^{-2}
luminous efficacy	$K(\lambda)$	lumen/watt	lm/W	$I_VM^{-1}L^{-2}T^3$
total efficacy	K_t	lumen/watt	lm/W	$I_VM^{-1}L^{-2}T^3$

the unit. For example m·N is the product of a metre and a newton whereas mN is a millinewton. The Greek alphabet is commonly used to designate certain quantities. Table 1.7 lists the alphabet and gives some of the quantities for which they are used.

1.3 Conversion factors

The SI system of units is gradually replacing other systems. However, the English (or Imperial) system is still in widespread use in the UK, and in North America. Table 1.8 lists some of the commonly used units in the English system, and gives the multiplication factor needed to convert each to its SI equivalent. For example a magnetic flux density of 10 gauss (in the English system of units) is equivalent to $10 \times 10^{-4} = 10^{-3}$ tesla (in the SI system).

Table 1.6. *Decimal prefixes*

Multiple	Prefix	Symbol
10^{18}	exa	E
10^{15}	peta	P
10^{12}	tera	T
10^9	giga	g
10^6	mega	M
10^3	kilo	k
10^2	hecto	h
10	deca	da
10^{-1}	deci	d
10^{-2}	centi	c
10^{-3}	milli	m
10^{-6}	micro	μ
10^{-9}	nano	n
10^{-12}	pico	p
10^{-15}	femto	f
10^{-18}	atto	a

Table 1.7. *Greek alphabet*

Name	Capital	Small	Common designation
Alpha	A	α	angle, coefficient, attenuation, absorption factor
Beta	B	β	angle, coefficient, phase
Gamma	Γ	γ	specific gravity, conductivity, propagation constant
Delta	Δ	δ	density, angle, increment or decrement, permittivity
Epsilon	E	ε	natural logarithms, permittivity, dielectric constant
Zeta	Z	ζ	coordinate, coefficient, impedance
Eta	H	η	efficiency, field strength, hysteresis
Theta	Θ	θ	angle, reluctance, time constant
Iota	I	ι	inertia, unit vector
Kappa	K	κ	susceptibility, coupling coefficient
Lambda	Λ	λ	permeance, wavelength, attenuation
Mu	M	μ	coefficient of friction, permeability, amplification
Nu	N	ν	frequency, reluctivity, viscosity
Xi	Ξ	ξ	coordinates,
Omicron	O	o	—
Pi	Π	π	3.1416 (circumference ÷ diameter)
Rho	P	ρ	resistivity, coordinates, charge density

(*Contd*)

Table 1.7. (*Cont.*)

Name	Capital	Small	Common designation
Sigma	Σ	σ	summation, deviation, conductivity, leakage
Tau	T	τ	time constant, density, transmission factor
Upsilon	Y	υ	—
Phi	Φ	ϕ	angles, flux, phase
Chi	X	χ	reluctance, susceptibility, angles
Psi	Ψ	ψ	angles, phase, dielectric flux
Omega	Ω	ω	resistance (ohms), angular velocity

Table 1.8. *Conversion factors*

Quantities being converted From	To	Multiplication factor
inch	metre	2.54×10^{-2}
foot	metre	3.048×10^{-1}
yard	metre	9.144×10^{-1}
mile	kilometre	1.609
nautical mile	kilometre	1.852
mil	metre	2.54×10^{-5}
fathom	metre	1.8288
square inch	square metre	6.4516×10^{-4}
square foot	square metre	9.290×10^{-2}
square yard	square metre	8.361×10^{-1}
square mile	square kilometre	2.590
cubic inch	cubic metre	1.6387×10^{-5}
cubic foot	cubic metre	2.832×10^{-2}
cubic yard	cubic metre	7.645×10^{-1}
gallon (US)	cubic metre	3.785×10^{-3}
gallon (UK)	cubic metre	4.5×10^{-3}
pound	kilogram	4.536×10^{-1}
ton	kilogram	1.0161×10^{3}
pound/square foot	kilogram/square metre	4.882
pound/cubic foot	kilogram/cubic metre	1.60185×10
bar	newton/square metre	1×10^{5}
foot pound	kilogram metre	1.383×10^{-1}
torr	newton/square metre	1.3332×10^{2}
kilogram force	newton	9.8066
pound force	newton	4.4482
British thermal unit	joule	1.0548×10^{3}
calorie	joule	4.1868
horsepower	watt	7.457×10^{2}
kilowatt hour	joule	3.60×10^{6}
therm	joule	1.0551×10^{8}
mile/hour	kilometre/hour	1.609
knot	metre/second	5.15×10^{-1}
electron volt	joule	1.602×10^{-19}
gilbert	ampere	7.9577×10^{-1}
gauss	tesla	1.0×10^{-4}
maxwell	weber	1.0×10^{-8}
footcandle	lumen/square metre	1.0764×10
footlambert	candela/square metre	3.4263
lambert	candela/square metre	3.183×10^{3}

1.4 Constants

The values of commonly used physical constants are given in Table 1.9. The number in parentheses after each value is the standard deviation uncertainty in the last digit, calculated on the basis of internal consistency. In calculating additional quantities, using two or more of these constants, the general law of error propagation must be used.

Table 1.9. *Physical constants*

Quantity	Symbol	Value	Error p.p.m	Prefix	Unit
velocity of light	c	2.9979250(10)	0.33	10^8	$m \cdot s^{-1}$
electron charge	e	1.6021917(70)	4.4	10^{-19}	C
Planck's constant	h	6.626196(50)	7.6	10^{-34}	$J \cdot s$
	$\hbar(h/2\pi)$	1.0545919(80)	7.6	10^{-34}	$J \cdot s$
Faraday constant	F	9.648670(54)	5.5	10^7	$C \cdot kmol$
Rydberg constant	R_∞	1.09737312(11)	0.10	10^7	m^{-1}
gas constant	R_0	8.31434(35)	42	10^3	$J \cdot kmol^{-1} \cdot k^{-1}$
Boltzmann constant	k	1.380622(59)	43	10^{-23}	$J \cdot k^{-1}$
Stefan–Boltzmann constant	σ	5.66961(96)	170	10^{-8}	$W \cdot m^{-2} \cdot k^4$
gravitational constant	G	6.6732(31)	460	10^{-11}	$N \cdot m^2 \cdot kg^{-2}$
first radiation constant	c_1	3.741832(20)	7.6	10^{-16}	$W \cdot m^2$
second radiation constant	c_2	1.438833(61)	43	10^{-2}	$m \cdot K$
Avogadro number	N_A	6.022169(40)	6.6	10^{26}	$kmol^{-1}$
atomic mass unit	u	1.660531(11)	6.6	10^{-27}	kg
electron rest mass	m_e	9.109558(54)	6.0	10^{-31}	kg
	m_e^*	5.485930(34)	6.2	10^{-4}	u
proton rest mass	m_p	1.672614(11)	6.6	10^{-27}	kg
	m_p^*	1.00727661(8)	0.08	—	u
neutron rest mass	m_n	1.674920(11)	6.6	10^{-27}	kg
	m_n^*	1.00866520(10)	0.10	—	u
electron charge to mass ratio	l/m_e	1.7588028(54)	3.1	$10''$	$C \cdot kg^{-1}$
quantum of circulation	$h/2m_e$	3.636947(11)	3.1	10^{-4}	$J \cdot s \cdot kg^{-1}$
	h/m_e	7.273894(22)	3.1	10^{-4}	$J \cdot s \cdot kg^{-1}$
Bohr radius	α_0	5.2917715(81)	1.5	10^{-11}	m
classical electron radius	r_0	2.817939(13)	4.6	10^{-15}	m
Bohr magneton	μ_B	9.274096(65)	7.0	10^{-24}	$J \cdot T^{-1}$
nuclear magneton	μ_n	5.050951(50)	10	10^{-27}	$J \cdot T^{-1}$
electron magnetic moment	μ_e	9.284851(65)	7.0	10^{-24}	$J \cdot T^{-1}$
proton magnetic moment	μ_p	1.4106203(99)	7.0	10^{-26}	$J \cdot T^{-1}$

2. Measurement standards

2.1 Introduction

The units used in the measurement of mechanical and electrical quantities were introduced in Chapter 1. To ensure that these units are consistently applied a set of measurement standards is required, and these are described in this chapter. The philosophy between the various levels of standards is first introduced, followed by details of the standards used to measure mechanical, electrical, magnetic, thermal and optical quantities.

2.2 Levels of standards

There are in general four distinct levels of standards; working standards, secondary standards, primary standards and international standards.

The main measurement tools of an industrial laboratory are working standards. These standards are used to check and calibrate other laboratory instruments within that industry, and also to carry out certain precise measurements. Working standards are periodically checked against secondary standards.

Secondary standards are kept in various laboratories in industry. Their prime function is to check and calibrate working standards. Responsibility for maintenance of the secondary standard is with the industrial laboratory concerned, although periodically these standards may be sent to National Standards Laboratories for checking and calibration. The secondary standards are then returned with a certification from the National Laboratory, giving the measured value in terms of the primary standard.

Primary standards are kept in the National Laboratories of different countries. These standards are not available for use outside the National Laboratory, although they may be used

to calibrate secondary standards sent to that laboratory. Primary standards are themselves calibrated at the National Laboratories by making absolute measurements in terms of the fundamental units. These results, from the various National Laboratories, are compared to give an average value for the primary standard.

The International standard represents the unit of measurement concerned, to the closest possible accuracy which the measurement technology will allow. They are defined by international agreement, and are checked periodically by absolute measurement, in terms of the fundamental unit concerned. International standards are maintained at the International Bureau of Weights and Measures at Sèvres, near Paris. These standards are not available for general use or for calibration.

2.3 Mechanical standards

The three mechanical standards of mass, length and time are discussed and developed in this section.

2.3.1 Mass

The SI unit of mass is the kilogram, and in 1889 the 1st Conférence Générale des Poids et Mesures defined the standard kilogram as equal to the mass of the international prototype of the kilogram. This is a cylinder of platinum–iridium alloy which is kept in the International Bureau of Weights and Measures at Sèvres, France. Secondary standards of the kilogram, kept by the various industrial laboratories, have an accuracy of about 1 part per million, whereas working standards, which are available in a range of values, have an accuracy of about 5 parts per million.

The original standard for the Imperial pound

unit was a prototype made from platinum, although the pound has now been redefined as equal to 0.45359237 kilogram, exactly.

2.3.2 Length

The SI unit of length is the metre. The metre was originally defined as a ten-millionth part of the earth's meridian passing through Paris. This was represented by the distance between two lines, engraved on a platinum–iridium bar kept at the International Bureau of Weights and Measures. In 1960 the 11th Conférence Générale des Poids et Mesures redefined the metre more accurately, as equal to 1650763.73 wavelengths, in vacuum, of the orange–red radiation of the krypton-86 atom, in its transition between the levels $^2p_{10}$ and $^5d_5$.

Then in the 1983 Congress meeting the metre was again redefined, in terms of time which can be specified accurately using atomic standards. The metre is now the length travelled by light in vacuum in a time interval of 1/299792458 second. The speed of light is therefore made equal to 299792458 metres per second.

The yard was originally defined in terms of an imperial standard yard, but it is now defined as being equal to 0.9144 metre exactly.

Industrial working standards usually consist of steel gauge blocks, having two flat parallel surfaces set at a specified distance apart. These are relatively cheap, and have accuracies of the order of one part per million.

2.3.3 Time and frequency

Time and frequency are based on aspects of the same phenomenon, which can be measured with respect to some physical quantity. Having defined a precise time standard one can find frequency, by counting the number of cycles over the period of one second.

There are two types of time. The first is elapsed time, and is the time between the occurrence of two events. This time measurement is independent of a starting point. The second type of time defines when an event occurs in relation to some starting point. This is called Time-of-Day or Epoch. Measurement of Epoch occurs in terms of elapsed time standards, once the starting point has been defined.

This section describes time and frequency standards in greater detail, and outlines the techniques used to compare these around the world.

2.3.3.1 Atomic standards

Atomic standards provide the ability to measure the unit of time, the second, to a very high level of accuracy. When an atom moves between two energy states E_1 and E_2 it emits radiation, of frequency v defined by:

$$hv = E_2 - E_1 \qquad (2.1)$$

where h is Planck's constant. Provided there is no external interference, such as from electrical or magnetic fields, the frequency of emission is very stable and is determined by the internal structure of the atom.

In October 1967 the 13th General Conference of Weights and Measures defined the International Second as the duration of 9192631770 periods of radiation, corresponding to the transition between two hyperfine

Table 2.1. *Characteristics of frequency standards*

Characteristics	Caesium beam	Hydrogen maser	Rubidium vapour	Quartz
Reproducibility	$\pm 3 \times 10^{-12}$	$\pm 2 \times 10^{-12}$	NA	NA
Stability (1 s average)	5×10^{-12}	5×10^{-13}	5×10^{-12}	5×10^{-12}
Drift (ageing rate)	negligible	negligible	$\pm 1 \times 10^{-11}$/mon.	$\pm 5 \times 10^{-10}$/day
Resonance frequency (Hz)	9192631770	1420405751	6834682608	NA
Operating temperature	-20 to $+60\,°C$	0 to 50 °C	0 to 50 °C	0 to 50 °C
Weight (approximate)	30 kg	400 kg	15 kg	10 kg
Power demand (approximate)	40 W	200 W	40 W	20 W
Atomic interaction time	2.5×10^{-3} s	0.5 s	2×10^{-3} s	NA
Atomic resonance events/s	10^6	10^{12}	10^{12}	NA
Temperature of resonating atoms	360 K	300 K	330 K	NA

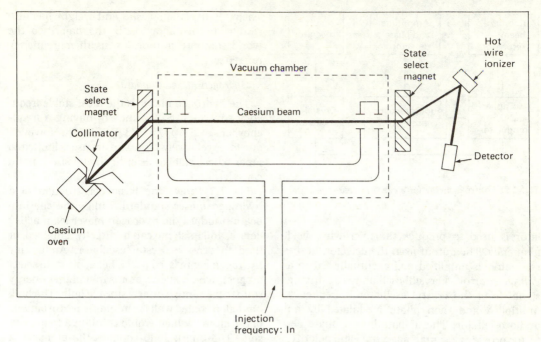

Fig. 2.1. Simplified construction of a caesium beam resonator.

levels ($F = 4$, $m_f = 0$ and $F = 3$, $m_f = 0$) of the ground state ($^2S_{1/2}$) of the caesium 133 atom.

Three main atomic standards have been used, caesium, hydrogen, and rubidium. The caesium beam and the hydrogen maser are primary or absolute standards, since they do not need to be referenced to any other standard for checking or calibration. The rubidium vapour standard, like a quartz oscillator, is a secondary standard, and it needs to be initially set up against a primary standard, and then periodically recalibrated.

Table 2.1 compares these four main types of frequency standards. Although the quartz oscillator is shown here as an independent standard, it is often used to provide feedback frequency to caesium or rubidium atomic standards, which are passive. In Table 2.1 'reproducibility' indicates how readily the standard can produce the same frequency, from one occasion to the next, without any recalibration. This parameter is only meaningful when applied to primary standards. Short term stability is a measure of the frequency variation of the standard, due to random noise in the oscillator. Drift, or long term stability, defines the changes with time of the absolute value of the frequency standard. The observation period

for this should be long enough to negate the effect of short term variations.

Caesium beam standard

Fig. 2.1 shows the basic construction of a caesium beam resonator. Caesium atoms are slowly evaporated in an oven and collimated into a beam. The state select magnet deflects each atom by an amount which depends on the energy state of that atom. Only those atoms in the $F = 4$, $m_f = 0$ state are allowed to form the caesium beam in the vacuum chamber. Here they pass through a low and uniform magnetic field (c-field) which is subjected to microwave excitation at a frequency of 9192.631770 MHz. This has exactly the correct energy to change the state of some of the atoms to $F = 3$, $m_f = 0$. The beam then passes a second state select magnet, identical to the first one. This deflects the atoms which have undergone the required state change onto a hot wire ionizer. The ionized caesium atoms are then filtered by a mass spectrometer, and are converted to electron current by the detector.

Fig. 2.2 shows a control scheme for the caesium beam resonator, where a crystal oscil-

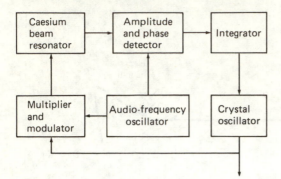

Fig. 2.2. Control scheme for a caesium beam resonator

lator is used to produce the microwave field frequency. The output from the detector, in the resonator, is amplified and eventually fed to a voltage controlled crystal oscillator. The driving signal from the crystal oscillator is frequency multiplied and then phase modulated by an audio oscillator. The amplitude and phase detector provides a signal magnitude and polarity, which corresponds to the difference of the oscillator frequency from the desired frequency. This signal is filtered by the integrator, to remove high-frequency components, and then used to drive the crystal oscillator, to the correct frequency required by the caesium beam resonator.

As indicated in Table 2.1, caesium beam standards are accurate to within 3 parts in 10^{12} and have good long term and short term stability.

Their relatively small size and weight has also resulted in their adoption, by the majority of the world standard authorities, as their frequency reference.

Hydrogen maser standard

The hydrogen maser is the most stable frequency source currently known, having a frequency of 1420405751.73 ± 0.03 Hz. However, due to its relatively large size, its use is limited to areas where stability is critical, and size is not a consideration.

Fig. 2.3 shows the basic construction of a hydrogen maser standard. Unlike the caesium beam standard the hydrogen maser is an active device, and its signal can be directly amplified, or used to drive a crystal oscillator. An atomic hydrogen beam is fed past a hexapole state select magnet, which allows atoms in a higher energy state to pass into a quartz storage bulb. The bulb has teflon coated walls to minimise perturbations of the atoms, which would result in a frequency shift. The quartz bulb confines the atoms to a uniform magnetic field, in a tuned microwave cavity. This is set to a frequency equal to the transition of the hydrogen atom, between the states $F = 1$, $m_f = 0$ and $F = 0$, $m_f = 0$. The hydrogen atoms undergo many reflections off the walls of the bulb, and interact with the microwave field. This relaxes the atoms, and they give up their energy to the microwave field within the tuned cavity. This stimulates more

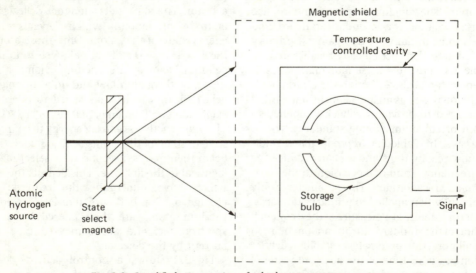

Fig. 2.3. Simplified construction of a hydrogen maser resonator.

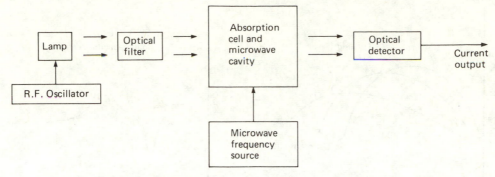

Fig. 2.4. Simplified representation of a rubidium vapour resonator.

atoms to radiate, until a maser operation is achieved.

Rubidium vapour standard

The rubidium vapour standard is based on the hyperfine transition in rubidium 87 gas, between the states $F = 2$ and $F = 1$. The transition frequency is 6834682608 Hz.

Fig. 2.4 gives the basis of the rubidium vapour resonator. Light from a spectral lamp is filtered, and then fed to the cell containing rubidium 87 vapour in a microwave cavity. An inert buffer gas is added to reduce Doppler broadening effects. The microwave signal is derived by multiplication of a quartz oscillator, operating in a closed loop, similar to that used for a caesium standard.

Optical pumping of the rubidium 87 gas causes the population of the $F = 2$ state to increase, at the expense of the $F = 1$ state. Application of microwave energy, close to the transition frequency of the rubidium vapour states, causes transitions from the $F = 2$ to the $F = 1$ level, so that more light is absorbed by the cell. At a frequency corresponding to 6834682608 Hz the photodetector output reaches a minimum.

The rubidium vapour standard has good short term stability, and small size, making it a useful portable instrument. It is not self calibrating, and must be calibrated against a primary standard. The standard exhibits drift, since the resonance frequency is dependent on the gas mixture and pressure in the cell. However its drift is over 100 times less than that of the quartz standard.

Quartz crystal standard

The quartz oscillator relies on the piezoelectric properties of quartz, to give a frequency standard which is used in a wide range of applications. The oscillator employs the high Q value of the quartz, when it is in a mechanical resonant mode. The frequency range is generally between 1 Hz–750 kHz and 1.5–200 MHz. Ranges between 750 KHz and 1.5 MHz are difficult to cover, due to strong secondary resonance. It is also preferable to use high frequency crystals, since the stability of low frequency crystals is less, and they have greater physical size.

The quartz crystal is aligned using X-rays, and then precision cut along one of several planes. It is then ground and lapped to exact dimensions, electrodes are vacuum deposited onto the sides and leads attached to them.

The quartz crystal has an atomic structure oriented about the x (electrical), y (mechanical), and z (optical) axes. Many different cuts can be made to it depending on the application. An X cut gives a mechanical change in thickness, in the x direction, for an electrical field applied in the y direction. It has a negative temperature coefficient of frequency. A Y cut gives mechanical change in length in the y direction for an electrical field in the x direction. It has a positive temperature coefficient of frequency. By making a cut rotated from the x and y axes, it is possible to combine the negative and positive temperature coefficients, to minimise the frequency shift in the crystal.

The two cuts most commonly used for quartz controlled oscillators are AT and BT. The AT cut has a lower variation of frequency with temperature, but the BT cut is preferred above 10 MHz, since it is less susceptible to drive levels and load capacitances. All crystals have a frequency–temperature characteristic similar to

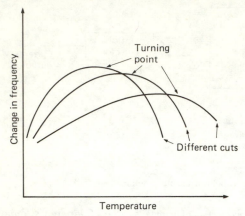

Fig. 2.5. Effects of crystal cut on its turning point.

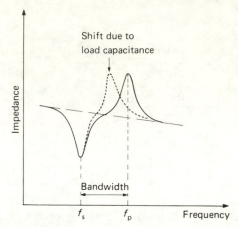

Fig. 2.7. Impedance characteristic of a quartz crystal oscillator.

those of Fig. 2.5. The position of the turning point is determined by the cut of the crystal, and can be chosen to occur in the range $-50\,°\text{C}$ to $+100\,°\text{C}$. Temperature controlled ovens are used to produce a quartz crystal standard having high stability and low drift.

In an oscillator the quartz is used as a stable resonant circuit, to control the frequency of the electronic system. Fig. 2.6 shows the equivalent

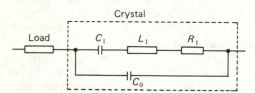

Fig. 2.6. Equivalent circuit of a quartz crystal.

circuit of a quartz crystal assembly. L_1 represents the vibrating mass of the crystal blank, R_1 the loading on the crystal due to ambient air and molecular friction, C_1 the elasticity of the quartz material, and C_0 the static capacitance between the electrodes plated on the blank, and the stray lead and holder capacitances.

A quartz crystal can operate in a series resonant mode or in a parallel antiresonant mode. Its characteristic is similar to that shown in Fig. 2.7 where f_s and f_p are the series resonant and parallel antiresonant frequencies respectively. The value of the series resonance frequency and of Q are given by:

$$f_s = \frac{1}{2\pi}(L_1 C_1)^{1/2} \tag{2.2}$$

$$Q = 2\pi f_s \frac{L_1}{R_1} \tag{2.3}$$

The ability to change the operating frequency of the crystal by varying the load capacitance, is called the 'pullability' of the crystal, and is used in applications such as voltage controlled and temperature controlled oscillators.

The ageing or frequency drift of the crystal is asymptotic with time. After an initially fast rate the ageing settles down to about 5 parts in 10^{10} per day, in a temperature controlled oven. If an oscillator, or its oven, is switched off for a few hours, then the crystal recovers, and the ageing process starts again at its initial rate; therefore a quartz standard should never be switched off. Ageing is caused by contamination in the crystal, fatigue in the mounting wires and solder, outgassing of material inside the sealed holder and leakage of the seals.

2.3.3.2 Time-of-day

Atomic standards give accurate and convenient time scales and intervals, but for navigation, satellite tracking, etc. correlation with the rotation of the earth is needed.

The earliest unit of time was obtained by observing the rotation of the earth about its axis, with respect to the sun. This is termed apparent solar time because the time is not a constant 24 hours for each rotation, as initially supposed. Since the earth's orbit is elliptical, and its orbital plane is at an angle of $23°\,27'$ to the plane of the equator, the apparent solar time deviates from the expected value. It has a maximum deviation of about 16 minutes, and this occurs in November.

To eliminate time variations, due to orbital eccentricity and the tilt of the earth's axis, an average can be taken of all the apparent solar days in a year. This is called a mean solar day, and a mean solar second is equal to a mean solar day divided by 86400. A solar year is equal to 365 days, 5 hours, 48 minutes and 45.5 seconds.

Universal time (UT), also known as Greenwich Mean Time, is referenced to the prime meridian, which passes through Greenwich, England. The units of UT are chosen such that on average local noon will occur when the sun is on the local meridian. UT is based on the rotation of the earth, which is known to be nonuniform, and when no corrections are made to it UT is equal to mean solar time, at Greenwich, and is denoted by UT_0.

Due to seasonal changes annual variations occur in the rotational speed of the earth. There are also semiannual variations, due to distortion of the shape of the earth, effected by tidal action of the sun. These cause the earth to be late in its rotation, by about 30 milliseconds near 1 June each year, and ahead by about the same amount near 1 October. When UT_0 is corrected for these periodic variations it is denoted by UT_1.

Unpredictable changes in the speed of rotation of the earth also occur, such as those due to turbulent motions in the core of the earth. Observations of these effects, from around the world, are made to the Bureau International de l'Heure, in Paris, which issues periodic corrections to UT_1 to give UT_2 time. This time is included in most radio time signals, and is used for celestial navigation and tracking of satellites.

Another time scale based on the rotation of the earth is sidereal time. However, this rotation is related to the vernal-equinox point in the sky, and not to the sun, as was done for the solar time. This means that sidereal time is not affected by errors due to orbital motion, although it is affected by other variations in the rotational speed. The mean sidereal day is equal to 23 hours, 56 minutes and 4.09 seconds.

In 1956 the International Committee of Weights and Measures defined the second as 1/31556925.9747 of the tropical year for January 0, 1900 at 12 hours. (January 0, 1900 =

December 31, 1899). This is known as an ephemeris second, and it is a constant, as defined. However it cannot be measured as accurately as its definition implies, and in practice it is determined by observing the motion of the moon, and then referring to lunar ephemeris tables.

2.3.3.3 Time zones

The world is divided into 24 time zones, based on longitude. The zones are established at 15° of longitude east and west of the Prime Meridian at Greenwich (0° longitude), and cover a belt $7\frac{1}{2}°$ on either side of these reference meridians. The time zones are therefore related to the mean solar time at Greenwich. Eastward from Greenwich the zones are numbered $+1$ to $+12$, to indicate the hour angle which must be added to universal time, to get the local 'standard' time. Westward from Greenwich the zones are numbered -1 to -12 to indicate that the hour angle must be subtracted. In practice the exact time boundaries in a country are modified from the 'standard' time for geographical or political reasons.

2.3.3.4 Time standard comparisons

Units of time, or frequency, cannot be kept in a safe for reference, like units of length or mass. Time units must be generated for each use, and so need regular comparison against recognised standards. This can be done by flying the standard, such as an atomic clock, to the various locations for comparison, or by transmission of time information by radio signals. The broadcasts are usually linked to a time standard, such as an atomic clock, at the standard authority.

VLF, LF and HF time transmission stations exist around the world. A few of these are listed in Table 2.2. The accuracy indicated in the table is that of the radio signal. Other errors are added to this due to propagation problems. High frequency propagation occurs between the ionosphere and the earth's surface, and variations of the ionosphere's characteristics and height cause the propagation time of the signal to vary continuously. For accurate readings it is therefore necessary to record time data over many days, and then to average them. Usually, with a signal accuracy of 0.2 parts in 10^{10}, as transmitted, the received accuracy is better than 1 part in 10^8.

Transmissions of low frequencies, below about 75 KHz, are more stable. These follow the earth's

Table 2.2. *Some time signal emissions in the UTC system*

Call sign	Place	Latitude and longitude	Authority	Accuracy (parts in 10^{10})	Carrier frequency (kHz)	Schedule	Type of signal
CHU	Ottawa Canada	45°18'N 75°45'W	National Research Council	0.05	3330 7335 14 670	continuous	Second pulses: 300 cycles of 1 kHz tone Minute pulses: 0.5 s long Voice announcement each minute DUT1: CCIR code by split pulses
DCF77	Mainflingen, West Germany	50°01'N 09°00'E	Physikalisch-Technische Bundesanstalt	0.1	77.5	continuous (except 4 h to 8 h on second Tuesday of each month)	0.1 s carrier interruptions at beginning of each second. Second 59 marker omitted DUT1: CCIR code by lengthening to 0.2 s
FFH	Chevannes, France	48°32'N 02°27'E	Centre National d'etudes des Telecommunications	0.2	2500	continuous from 8 h to 16 h (except Saturday and Sunday)	Second pulses: 5 cycles of 1 kHz tone Minute pulses: 0.5 s long DUT1: CCIR code by lengthening to 0.1 s
GBR	Rugby, UK	52°22'N 01°11'W	National Physical Laboratory	0.2	16	2 h 55 to 3 h 8 h 55 to 9 h 14 h 55 to 15 h 20 h 55 to 21 h	Second pulses: A1 type telegraphy signals DUT1: CCIR code by double pulse
IAM	Rome, Italy	41°52'N 12°27'E	Instituto Superiore Poste e Telecommunicazioni	0.5	5000	every 15 m from 7 h 30 to 8 h 30 (except Saturday PM and Sunday)	Second pulses: 5 cycles of 1 kHz tone Minute pulses: 20 cycles long
JJY	Tokyo, Japan	35°42'N 139°31'E	Ministry of Posts and Telecommunications	0.5	2500 5000 10 000 15 000	continuous (except between 25 and 34 minutes)	Second pulses: 8 cycles of 1.6 kHz tone Minute pulses: preceded by a 0.6 kHz tone DUT1: CCIR code by lengthening
LOL1	Buenos Aires, Argentina	34°37'S 58°21'W	Observatorio Naval	0.2	5000 10 000 15 000	11 h to 12 h 14 h to 15 h 17 h to 18 h 20 h to 21 h 23 h to 24 h	Second pulses: 5 cycles of 1 kHz tone Second 59 marker omitted Voice announcement every 5 minutes DUT1: CCIR code by lengthening
MSF	Rugby UK	52°22'N 01°11'W	National Physical Laboratory	0.1	60	continuous (except 10 h to 14 h on first Tuesday of each month)	Second pulses: carrier interruption for 0.1 s Minute pulses: carrier interruption for 0.5 s DUT1: CCIR code by double pulse

Station	Location	Coordinates	Organization	Frequencies (kHz)	DUT (s)	Schedule	Remarks
MSF	Rugby UK	52°22'N 01°11'W	National Physical Laboratory	2500 5000 10000	1.0	between minutes 0 and 5, 10 and 15 20 and 25, 30 and 35 40 and 45, 50 and 55	Second pulses: 5 cycles of 1 kHz tone; Minute pulses: prolonged; DUT1: CCIR code by double pulse
RWM	Moscow, USSR	55°19'N 38°41'E	Conseil des Ministres de l'URSS	10 000 15 000	0.5	10000 kHz from 1 h 30 to 3 h and 17 h 50 to 24 h; 15000 kHz from 3 h 50 to 17 h	Second pulses: pulses between 30 m and 35 m, 41 m and 45 m, 50 m and 60 m; Minute pulses: 0.5 s long; DUT1 + dUT1: Morse code
VNG	Lyndhurst, Australia	38°03'S 145°16'E	A.P.O Research Laboratories	4500 7500 12000	1.0	4500 and 7500 kHz, 9 h 45 to 21 h 30 continuous. 12000 kHz, 21 h 45 to 9 h 30	Second pulses: 50 cycles of 1 kHz tone. Seconds 55 to 58 shortened to 5 cycles, second 59 omitted; Minute pulses: 500 cycles long. At 5th, 10th, 15th etc. minute pulses from 50th to 58th second shortened to 5 cycles; Voice announcement in 15th, 30th, 45th, 60th minute; DUT1: CCIR code by 45 cycles of 900 Hz tone
WWV	Fort-Collins, USA	40°41'N 105°02'W	National Bureau of Standards	2500 5000 10000 15000 20000 25000	0.1	continuous	Second pulses: 5 cycles of 1 kHz tone. Seconds 29 and 59 omitted; Minute pulses: 0.8 s long (1 kHz tone); Hour pulses: 0.8 s long (1.5 kHz tone); DUT1: CCIR code by double pulse
WWVB	Fort-Collins, USA	40°41'N 105°03'W	National Bureau of Standards	60	0.1	continuous	Second pulses: amplitude reduction of carrier. Coded announcement of date and time; No CCIR Code
WWVH	Kauai, USA	21°59'N 159°46'W	National Bureau of Standards	2500 5000 10000 15000 20000	0.1	continuous	Second pulses: 6 cycles of 1.2 kHz tone. Seconds 29 and 59 Omitted; Minute pulses: 0.8 s long (1.2 kHz tone); Hour pulses: 0.8 s long (1.5 kHz tone); DUT1: CCIR code by double pulse
ZUO	Olifantsfontein, South Africa.	24°58'S 28°14'E	National Physics Research Laboratory	2500 5000 (10000)	0.1	continuous from 18 h to 4 h (continuous)	Second pulses: 5 cycles of 1 kHz tone; Minute pulses: second 0 prolonged; DUT1: CCIR code by lengthening

curvature, in the space between the ground and the ionosphere, and are much less influenced by ionospheric variations. However, due to the lower available bandwidth, much less information can be transmitted at low frequencies. It is also necessary to have fast risetime pulses, to give precise time synchronization, and at low frequencies the pulse risetime is degraded by time constants of the antenna system and the receiving equipment.

2.4 Electrical standards

2.4.1 *Standards for current*

Early standards for the ampere were based on the electrical deposition of silver, from a silver nitrate solution. The International Ampere was defined as the current which deposits silver at the rate of 1.118 mg/s, from a standard silver nitrate solution.

The 11th General Conference on Weights and Measures, meeting in 1960, redefined the ampere in SI units as the constant current which, when maintained in two straight parallel conductors of infinite length, and of negligible cross section, placed 1 metre apart in vacuum, will produce a force between them of 2×10^{-7} N per metre length. This is also termed the Absolute Ampere. It may be measured in absolute terms using a current balance, to determine the force exerted between two current-carrying coils. Generally, however, standards for the ampere are maintained by using a standard voltage and resistance, and applying Ohm's Law. The current is passed through the current terminals of a precision four terminal standard resistor, and the voltage drop across the potential terminals of the resistor is measured by a precision potentiometer operating from a standard voltage source.

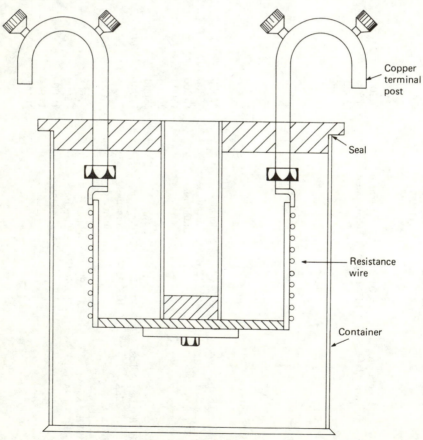

Fig. 2.8. Standard resistor.

2.4.2 Resistance standards

The absolute value of the ohm, in the SI system of units, is defined in terms of the fundamental units of length and mass. Absolute measurement of the ohm is done against a group of resistance standards, which are maintained at the International Bureau of Weights and Measures at Sevres. This group is periodically checked against each other, and with similar groups of standards kept by other standards authorities around the world.

Fig. 2.8 shows the construction of a typical standard resistor. The resistance wire is usually made from an alloy, like manganin (Ni 4%, Cu 84%, Mn 12%), which has a resistivity of $48\,\mu\Omega\cdot$cm, a low temperature coefficient of resistivity of $\pm 15\,$ppm/$^\circ$C, and a thermal emf with copper of below $3\,\mu$V/$^\circ$C. For high resistance values Evanohm (Ni 74.5%, Cr 20% and Al, Fe, or Cu for the remainder) is sometimes used. It has a resistivity of $133\,\mu\Omega\cdot$cm, a temperature coefficient of resistivity of $\pm 20\,$ppm/$^\circ$C, and a thermal emf with copper of about 2μV/$^\circ$C.

The resistance coil is wound to reduce stress, and may be immersed in oil for temperature stability. It is maintained in a sealed moisture free container. Calibration reports on standard resistors, from national standard laboratories, include values for the thermal coefficients α and β. Usually these values are less than 10^{-5} and -5×10^{-7} respectively, so that resistance change over a temperature band of $10\,^\circ$C is less than $50\,$ppm. If desired, correction for temperature can be made by using the formula:

$$R_t = R_{25} + \alpha(t - 25) + \beta(t - 25)^2 \qquad (2,4)$$

where R_{25} is the nominal value of resistance, which is usually specified at $25\,^\circ$C, and R_t is the resistance at $t\,^\circ$C.

Errors occur in standard resistors due to skin effect, when they are used on high frequency a.c. These are minimised by limiting the wire diameter size, for any specified frequency. Errors due to stray inductance and capacitance are reduced by arranging the windings so that adjacent turns carry current in opposite directions, and the adjacent turns are widely spaced and have minimal potential difference between them. The Ayrton–Perry method of winding the resistance achieves these objectives, as shown in Fig. 2.9, and is used for resistors below about 100 ohms.

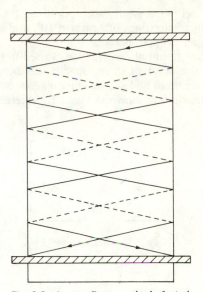

Fig. 2.9. Ayrton–Perry method of winding resistors to minimise inductance and shunt capacitance.

Errors also arise in low valued resistors, due to contact resistance of the terminals. This is reduced by using a four terminal construction, in which current is fed between two outer terminals and the voltage reading, for resistance value, is obtained between two inner terminals. The current drawn from the inner terminals must be kept to a minimum, for accurate readings. (See Section 13.3 and Fig. 13.5.)

2.4.3 Capacitance standards

The unit of capacitance in the SI system is the farad. It is defined as the capacitance, between the plates of which there is a potential difference of one volt, when it is charged by a quantity of electricity equal to one coulomb. The coulomb is the quantity of charge, transported in one second, by a current of one ampere.

Standard capacitors are constructed from interleaved multiplates, suspended in a gaseous dielectric. The plates are made from a low temperature coefficient of expansion material, such as invar, and they are annealed and mounted strain-free. The plates are sealed in a dry air or nitrogen environment, which acts as the dielectric. The area of the plates and the distance between them, is precisely defined, so that the capacitance value can be calculated.

This technique yields capacitors up to 1000 pF, having an accuracy of 2 parts in 10^8, and a drift of under 20 ppm per year.

Three terminal construction is used to minimise the effect of stray capacitances between each terminal and surrounding objects. Such a construction is shown in Fig. 2.10 where C_3 is

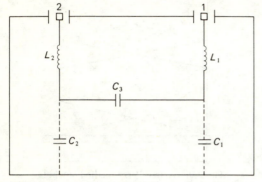

Fig. 2.10. Equivalent circuit of a standard capacitor.

the desired capacitor, C_1 and C_2 are the terminal capacitances, and L_1 and L_2 are residual inductances.

Working standard capacitors are available in a variety of sizes. The smaller values use an air dielectric, whereas larger capacitors use a solid dielectric. Silver–mica capacitors make good secondary standards. They have a relatively large capacitance value in a small size, are stable, have a low dissipation factor, little ageing and a small temperature coefficient.

2.4.4 *Voltage standards*

Voltage standards are defined in SI units as volts. A volt is the electrical potential between two points of a wire carrying a current of one ampere, when the power dissipated between the two points is equal to one watt. A watt is the power which gives rise to energy at the rate of one joule/second, and a joule is the work done when a force of one newton is displaced by one metre.

2.4.4.1 Weston cadmium cell

The three main types of voltage standards in use are the Weston cadmium cell, the Josephson-effect standard and zener diode standards. The Weston cadmium cell is the conventional standard used in most large standards laboratories. There are two types of Weston cells, the saturated and the unsaturated cell.

Fig. 2.11 shows the construction of a Weston cadmium saturated cell. The positive electrode consists of metallic mercury, and the negative electrode of a cadmium–mercury amalgam hav-

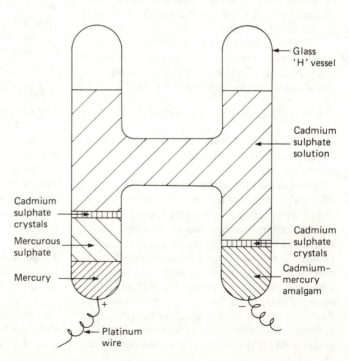

Fig. 2.11. Construction of a saturated Weston cadmium cell.

ing 10% cadmium. Electrical connections are made by coils of platinum wire, sealed into the glass envelope. A layer of mercurous sulphate, above the positive mercury electrode, acts as a depolarizer. The cell electrolyte consists of a solution of cadmium sulphate, with an excess of cadmium sulphate crystals to maintain the solution in a saturated state. A little sulphuric acid is usually added to acidify this solution. This reduces the emf of the cell by an amount given by:

$$E_{(20)} = 1.018\,36 - 6.0 \times 10^{-4} N - 5.0$$
$$\times 10^{-5} N^2 \tag{2.5}$$

where $E_{(20)}$ is the emf of a normal cell at 20 °C, and N is the normality of the acidified solution. Usually N is between 0.04 and 0.08.

The saturated Weston cell has a cell voltage which is reproducible, to within a few microvolts of the value given by (2.5). It has a relatively large temperature coefficient, of about $-39.4\,\mu V$ per °C at 20 °C, and is therefore usually operated in an air or oil bath, which maintains its temperature at a known value, to within 0.001 °C. The value of the cell voltage at any temperature t can be related to that at 20 °C by (2.6)

$$E_{(t)} = E_{(20)} - 4.6 \times 10^{-5}(t - 20)$$
$$- 9.5 \times 10^{-7}(t - 20)^2$$
$$+ 1.0 \times 10^{-8}(t - 20)^3 \tag{2.6}$$

The voltage drift of the saturated Weston cell is less than one microvolt per year.

The unsaturated Weston cadmium cell is similar in construction to the saturated cell, shown in Fig. 2.11, except that the concentration of electrolyte used in such that it saturates at 4 °C. Therefore at normal room temperature the cell remains unsaturated. The unsaturated cell has a lower accuracy than a saturated cell (about 0.005%), but also has a much lower temperature coefficient (below $-10\,\mu V/°C$). It can therefore be used in a simpler construction, and is more rugged and portable, making it an ideal secondary or working standard for industrial laboratories. The range of the unsaturated cell varies from 1.0180 V to 1.0194 V, and the exact value, at a given temperature, is stamped on the cell during calibration. The change in cell voltage is about 0.01% per year.

The internal resistance of a Weston cadmium cell is between 500 ohms and 800 ohms. It will therefore give an incorrect reading if too much current is drawn from it, and it can be damaged by excessive currents or wide variations in operating temperature. The current drawn from the cell should not exceed $100\,\mu A$ and the temperature should be maintained between 4 °C and 40 °C.

2.4.4.2 Josephson-effect standard

A current flows between two pieces of superconducting material which are separated by a thin layer of insulator. This is known as the Josephson effect, after the British physicist Brian D. Josephson who predicted the current flow in 1962.

Voltage standards, based on the Josephson effect, are relatively new, and have only recently begun to be used in some standards laboratories. They consist of a thin film of lead separated by a lead oxide insulating barrier, and mounted on a glass substrate. This probe is then placed in a superinsulated helium Dewar system. The tunnelling junction is excited by a microwave source of approximately 9 GHz, and it is biased by a supply current source. Under these conditions the Josephson junction produces a voltage, given by (2.7).

$$E = Nhf/2e \tag{2.7}$$

where h is Planck's constant, f is the frequency of the microwave source, e is the electron charge and N is an integer. The value of N can be varied, by control of the bias current, to give an output which changes in discrete steps.

Josephson-effect voltage standards have an accuracy of about 0.05 ppm, and have been used to calibrate other voltage standard sources, using potentiometer bridges.

2.4.4.3 Zener diode voltage standards

Zener diodes are reverse biased pn junctions, which have a sharp, and well defined, reverse breakdown voltage. For voltage standard applications the diodes should also have a low dynamic impedance, and a low temperature coefficient of impedance. Both alloy junction and diffused junction zener diodes are used, the former in the range 2.4–12 volts, and the latter from 6.8 to 200 volts. Diffused junction diodes have a lower dynamic impedance in the region of the operating point.

Zener diodes are available as laboratory working standards, called transfer standards. The zener diode is usually connected in series with one or more forward biased diodes, such that the negative temperature coefficient of the diode compensates for the positive coefficient of the zener diode. The assembly is placed in a temperature controlled oven, held to within about $\pm 0.01\,°C$. Generally the transfer standards provide multiple output of different voltages, having a stability of about 1 ppm per month. They are portable, but need a warm up time of about $\frac{1}{2}$ hour if disconnected from the power source.

2.5 Magnetic standards

2.5.1 *Inductor standards*

The SI unit for inductance is the henry, defined as the inductance of a closed circuit which produces an induced voltage of one volt, for a current change in the circuit of one ampere per second.

Primary standards for inductance consist of single layer coils, wound on a stable fused silica former. Uniform winding pitch is ensured by using a lapped groove, and the solenoids are made physically large, since their inductance value is determined by calculation from their dimensions.

Working inductance standards consist of multi turn coils, wound on ceramic or bakelite formers or on toroidal cores. The inductances are available in sizes from about $10\,\mu H$ to $10\,H$, and are stable to within 100 ppm per year. The inductors are frequency sensitive, due to stray capacitance associated with the windings. They are therefore usually calibrated at a series of step frequencies, from 100 Hz to 1 kHz.

2.5.2 *Flux standards*

The weber is the SI unit of magnetic flux, and is defined as the flux which, linking a coil of one turn, produces in it an emf of one volt, when the flux decays to zero in one second at a uniform rate.

Primary flux standards can be obtained by the decay of current through standard inductors. A ballistic galvanometer is used to measure the effect of current reversal in a standard mutual inductor. Fig. 2.12 shows a working flux standard, called the Hibbert standard, which does not rely on external current sources. The permanent

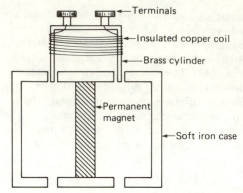

Fig. 2.12. A Hibbert flux working standard.

magnet is enclosed in a soft iron case, having a narrow cut-out on the top. The brass cylinder can move freely in the cut-out. When the cylinder is released it moves downwards, towards the magnet. The rate of change of the flux cutting the coil is uniform since the cylinder moves under gravity. Therefore the current induced in the coil is proportional to the flux in the air gap.

2.6 Thermal standards

The SI unit for thermodynamic temperature is the kelvin. This is defined as $1/273.16$ of the value of the thermodynamic temperature of the triple point of water. The triple point of water is the temperature at which equilibrium exists between ice, liquid water and water vapour.

Because of the difficulty of making temperature measurements on the thermodynamic scale, the 7th General Conference of Weights and Measures, meeting in 1927, defined an alternative temperature scale, called the practical scale, which is based on the degree celsius (°C). This scale has two fundamental fixed points, which are the boiling point of water at atmospheric pressure, equal to 100 °C, and the triple point of water at atmospheric pressure, equal to 0.01 °C. Several other primary fixed temperature points have also been established, as shown in Table 2.3. The conversion between the kelvin and celsius scales is given by

$$\begin{bmatrix} \text{Temperature in} \\ \text{degrees centigrade} \end{bmatrix} = \begin{bmatrix} \text{Temperature} \\ \text{in kelvin} \end{bmatrix} - 273.15 \qquad (2.8)$$

The values of the above temperature points are

Table 2.3. *Fixed points on the practical scale of temperature*

Fixed point	kelvin	°C
Triple point of equilibrium hydrogen	13.81	− 259.34
Boiling point of equilibrium hydrogen	20.28	− 252.87
Boiling point of neon	27.102	− 246.048
Triple point of oxygen	54.361	− 218.789
Boiling point of oxygen	90.188	− 182.962
Triple point of water	273.16	0.01
Boiling point of water	373.15	100.0
Freezing point of zinc	692.73	419.58
Freezing point of silver	1235.08	961.93
Freezing point of gold	1337.58	1064.43

accurately defined and reproducible, and can act as standard measurement temperatures. The primary standard for temperature is the platinum resistance thermometer. It is constructed from strain-free platinum, and the values between the primary and fundamental fixed temperature points can be calculated accurately, knowing the resistance properties of platinum wire. The platinum resistance thermometer is discussed in more detail in Chapter 4.

2.7 Optical standards

The SI unit of luminous intensity is the candela. It is defined as the luminous intensity, in the perpendicular, of a surface of 1/600 000 of a square metre of a blackbody, at a temperature at which platinum solidifies (1773 °C approximately), under a pressure of 101 325 N/m² (standard atmospheric pressure). A blackbody is a theoretically perfect absorber and emitter of radiation.

The primary standard of luminous intensity is a full radiator (blackbody or Planckian radiator), maintained at the above temperature. Secondary standards of luminous intensity, which require frequent calibration against the primary standard, consist of special tungsten filament lamps. The lamp power is adjusted to give an operating temperature at which the spectral power distribution of the lamp matches that of the basic standard.

3. Measurement errors

3.1 Introduction

All measurements are subject to errors, due to a variety of reasons such as inherent inaccuracies of the instrument, human error in taking readings and using the instrument in a way for which it was not designed. Many of the errors which arise are related to the type of instrument used, and these will be described in later chapters. The present chapter looks at the aspects of measurement errors which are common to all instruments, and how they can be identified and minimised.

3.2 The statistics of errors

Statistical analysis is frequently used in measurements. This section introduces four concepts which are commonly applied; averages, dispersion from the average, probability distribution of errors and sampling.

3.2.1 Averages

The most frequently used averaging technique is the arithmetic mean. If n readings are taken with an instrument, and the values obtained are $x_1, x_2, x_3, \ldots, x_n$, then the arithmetic mean is given by

$$\bar{x} = \frac{x_1 + x_2 + x_3 + \cdots + x_n}{n} \qquad (3.1)$$

or

$$\bar{x} = \frac{\sum\limits_{r=1}^{n} x_r}{n} \qquad (3.2)$$

Although the arithmetic mean is easy to calculate it is influenced unduly by extreme values, which could be false. An alternative averaging technique, called the geometric mean, is not over affected by extreme values. It is often used to find the average of quantities which follow a geometric progression, or an exponential law. The geometric mean is given by

$$x_g = \sqrt[n]{(x_1 \times x_2 \times x_3 \times \cdots \times x_n)} \qquad (3.3)$$

The harmonic mean is used when considering rates of change, such as speeds or prices. It is given by

$$x_h = \frac{n}{\sum\limits_{r=1}^{n} (1/x_r)} \qquad (3.4)$$

As a rule, when dealing with items such as P per Q, if the figures are for equal Ps then the harmonic mean is used; but if it is for equal Qs then the arithmetic mean is used. For example suppose a plane flies over three equal distances at speeds $5\,\text{m/s}$, $10\,\text{m/s}$ and $15\,\text{m/s}$. Then the mean speed is given by the harmonic mean as $3/(\frac{1}{5} + \frac{1}{10} + \frac{1}{15}) = 8.18\,\text{m/s}$. If the plane were to fly for three equal times at speeds of $5\,\text{m/s}$, $10\,\text{m/s}$ and $15\,\text{m/s}$, then the mean speed would be given by the arithmetic mean as $(5 + 10 + 15)/3 = 10\,\text{m/s}$.

3.2.2 Dispersion from the average

The average represents the mean figure of a series of numbers. It does not give any indication of the spread of these numbers. For example suppose nine different voltmeter readings are taken of a fixed voltage, and the figures read are 96, 98, 98, 100, 100, 100, 102, 102 and 104 volts. The mean voltage is therefore equal to 100 volts. If the readings are changed to 80, 90, 90, 100, 100, 100, 110, 110 and 120 the mean will still be 100 volts, although now they are more widely separated from the mean. The three techniques most frequently used to measure dispersion from the mean are the range, mean deviation and the standard deviation.

The mean deviation is found by taking the mean of the differences between each individual number in the series and the arithmetic mean, and ignoring negative signs. Therefore for a series of n numbers $x_1, x_2, x_3, \ldots, x_n$ having an arithmetic mean of $\bar{x}$ the mean deviation is given by

$$M = \frac{\sum\limits_{r=1}^{n} |x_r - \bar{x}|}{n} \tag{3.5}$$

Column 1 of Table 3.1. shows the nine volt-

Table 3.1. *Illustration of the mean and dispersion*

Voltmeter reading (volts)	Deviation from the mean (d)	d^2
96	− 4	16
98	− 2	4
98	− 2	4
100	0	0
100	0	0
100	0	0
102	+ 2	4
102	+ 2	4
104	+ 4	16
900	0	48

meter readings considered earlier. The mean of these is $900/9 = 100$ volts. The deviation from this mean is shown in the second column, where individual values which are less than the mean are given as negative. Note that the sum of the deviations from the mean is always zero. If the sign of the deviation is ignored then the mean deviation is given by $16/9 = 1.78$ volts.

Table 3.2 gives the alternative set of voltmeter readings, which also have a mean of 100 volts; but the mean deviation is now $80/9 = 8.89$ volts. Therefore, by comparing the mean deviation of the two sets of readings, one can deduce that the first set is more closely clustered around the mean, and therefore represents more consistent values. The reasons for this are explained in the next section.

The range is also used as a factor to measure dispersion. It is the difference between the largest

Table 3.2. *Second illustration of the mean and dispersion*

Voltmeter reading (volts)	Deviation from the mean (d)	d^2
80	− 20	400
90	− 10	100
90	− 10	100
100	0	0
100	0	0
100	0	0
110	10	100
110	10	100
120	20	400
900	0	1200

and smallest values. Therefore for the readings in Table 3.1 the range is $104 - 96 = 8$ volts, and for that in Table 3.2 it is $120 - 80 = 40$ volts. Neither the mean deviation nor the range are suitable for use in statistical calculation. The standard deviation is the measure of dispersion which is most commonly used for this.

The standard deviation of a series of n numbers $x_1, x_2, x_3, \ldots, x_n$, having a mean of $\bar{x}$, is given by

$$\sigma = \left(\frac{\sum\limits_{r=1}^{n} (x_r - \bar{x})^2}{n} \right)^{1/2} \tag{3.6}$$

Because the deviation from the mean is squared before summing, the signs are taken into account, so that the calculation is mathematically correct. For the example given in Table 3.1 the standard deviation is $(\frac{48}{9})^{1/2} = 2.31$ volts, and for that in Table 3.2 it is $(\frac{1200}{9})^{1/2} = 11.55$ volts.

3.2.3 *Probability distribution of errors*

If an event A, for example an error, occurs n times out of a total of m cases, then the probability of occurrence of the error is stated to be

$$p(A) = n/m \tag{3.7}$$

Probabilities vary between 0 and 1. If $p(A)$ is the probability of an event occurring then $1 - p(A)$, which is written as $p(\bar{A})$, is the probability that the event will not occur.

There are several mathematical distributions

which are used to define the spread in probabilities. The binomial, Poisson, normal, exponential and Weibull distributions will be introduced here.

In the binomial distribution the probability of an event occurring m successive times out of n selections is given by

$$p(m) = {}^{n}C_{m}p^{m}q^{n-m} \tag{3.8}$$

where p is the probability of the event occurring and q is the probability of the event not occurring. The mean of the distribution M_b and the standard deviation S_b are given by

$$M_b = np \tag{3.9}$$

$$S_b = (npq)^{1/2} \tag{3.10}$$

The Poisson distribution is used in cases where p and q cannot both be defined. For example one can state the number of goals which are scored in a football match, but not the number of goals which are not scored. In a Poisson distribution the probability of an event occurring m succes-

sive times out of n selections is given by

$$p(m) = (np)^{m}\frac{e^{-np}}{m!} \tag{3.11}$$

The mean M_p and standard deviation S_p of the Poisson distribution are given by

$$M_p = np \tag{3.12}$$

$$S_p = (np)^{1/2} \tag{3.13}$$

The normal distribution curve is bell-shaped, and is shown plotted in Fig. 3.1. The x axis gives the event and the y axis the probability of the event occurring. If AB represents the line of the mean value $\bar{x}$ then the equation for the normal curve is given by

$$y = \frac{1}{(2\pi)^{1/2}}e^{-\omega^2/2} \tag{3.14}$$

where

$$\omega = \frac{x - \bar{x}}{\sigma} \tag{3.15}$$

Table 3.3. *Area under the normal curve from* $-\infty$ *to* ω

ω	0.00	0.02	0.04	0.06	0.08
0.0	0.500	0.508	0.516	0.524	0.532
0.1	0.540	0.548	0.556	0.564	0.571
0.2	0.579	0.587	0.595	0.603	0.610
0.3	0.618	0.626	0.633	0.640	0.648
0.4	0.655	0.663	0.670	0.677	0.684
0.5	0.692	0.700	0.705	0.712	0.719
0.6	0.726	0.732	0.739	0.745	0.752
0.7	0.758	0.764	0.770	0.776	0.782
0.8	0.788	0.794	0.800	0.805	0.811
0.9	0.816	0.821	0.826	0.832	0.837
1.0	0.841	0.846	0.851	0.855	0.860
1.1	0.864	0.869	0.873	0.877	0.881
1.2	0.885	0.889	0.893	0.896	0.900
1.3	0.903	0.907	0.910	0.913	0.916
1.4	0.919	0.922	0.925	0.928	0.931
1.5	0.933	0.936	0.938	0.941	0.943
1.6	0.945	0.947	0.950	0.952	0.954
1.7	0.955	0.957	0.959	0.961	0.963
1.8	0.964	0.966	0.967	0.969	0.970
1.9	0.971	0.973	0.974	0.975	0.976
2.0	0.977	0.978	0.979	0.980	0.981
2.2	0.986	0.987	0.988	0.988	0.989
2.4	0.992	0.992	0.993	0.993	0.993
2.6	0.995	0.996	0.996	0.996	0.996
2.8	0.997	0.998	0.998	0.998	0.998
3.0	0.999	0.999	0.999	0.999	0.999

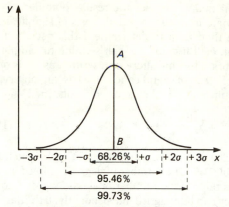

Fig. 3.1. The normal curve.

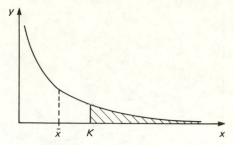

Fig. 3.2. The exponential curve.

The total area under the normal curve is unity and the area between any two values of ω is the probability of an item from the distribution falling between these values. The normal curve extends from $\pm \infty$ but 68.26% of its values fall between $\pm \sigma$, 95.46% between $\pm 2\sigma$, 99.73% between $\pm 3\sigma$ and 99.9994% between $\pm 4\sigma$.

Table 3.3 gives the area under the normal curve for varying ω. Since the curve is symmetrical the area from $+\omega$ to $+\infty$ is the same as from $-\omega$ to $-\infty$. As an example, in using Table 3.3, suppose that 5000 instruments have been put on soak test. The instruments are expected to have a mean life of 1000 hours with a standard deviation of 100 hours. It is required to predict how many instruments will fail in the first 800 hours.

From (3.15) $\omega = (800 - 1000)/100 = -2$. Ignoring the negative sign, Table 3.3 gives the probability of an instrument not failing as 0.977; so the probability of failure is $1 - 0.977 =$

0.023. Therefore 5000×0.023 or 115 instruments are expected to fail after 800 hours.

The exponential probability distribution is shown in Fig. 3.2 and has the equation

$$y = \frac{1}{x}e^{-x/\bar{x}} \qquad (3.16)$$

where $\bar{x}$ is the mean of the distribution. Table 3.4 shows the area under the exponential curve for different values of the ratio $K = x/\bar{x}$, this area being shown shaded in Fig. 3.2.

As an example suppose that the time between failures of an instrument is found to vary exponentially. If the expected mean time between failures is 1000 hours, then what is the probability that the instrument will work for 700 hours or more without a failure? Calculating K as $700/1000 = 0.7$, from Table 3.4 the area beyond 0.7 is 0.497, which is the probability that the instrument will still be working after 700 hours.

The Weibull distribution is given by

$$y = \alpha\beta(x - \gamma)^{\beta-1}e^{-\alpha(x-\gamma)^{\beta}} \qquad (3.17)$$

where α is called the scale factor, β the shape factor and γ the loading factor.

Table 3.4. *Area under the exponential curve from K to $+\infty$*

K	0.00	0.02	0.04	0.06	0.08
0.0	1.000	0.980	0.961	0.942	0.923
0.1	0.905	0.886	0.869	0.852	0.835
0.2	0.819	0.803	0.787	0.771	0.776
0.3	0.741	0.726	0.712	0.698	0.684
0.4	0.670	0.657	0.644	0.631	0.619
0.5	0.607	0.595	0.583	0.571	0.560
0.6	0.549	0.538	0.527	0.517	0.507
0.7	0.497	0.487	0.477	0.468	0.458
0.8	0.449	0.440	0.432	0.423	0.415
0.9	0.407	0.399	0.391	0.383	0.375

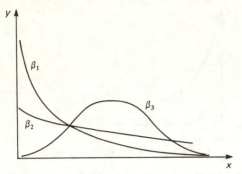

Fig. 3.3. Weibull curves ($\alpha = 1$).

The shape of the Weibull curve varies depending on the value of its factors. β is the most important, as shown in Fig. 3.3, and the Weibull curve varies, from an exponential for $\beta = 1.0$, to a normal distribution for $\beta = 3.5$. Analytical calculations using the Weibull distribution are cumbersome, and predictions are usually made using Weibull probability paper.

3.2.4 *Sampling*

Sampling techniques are often used in measurement systems. A small number of devices, from a larger population, are tested to give information on the population. For example a sample of 10 resistors from a batch of 1000 may be tested, and if these are all satisfactory it may be assumed that the whole batch is acceptable. Errors arise in sampling, which are usually evaluated on the assumption that sampling errors follow a normal distribution. Suppose a batch has n_b items, with a mean of $\bar{x}_b$. If a sample of n_s items is taken from this batch, and found to have a mean of $\bar{x}_s$ and a standard deviation of σ_s, then

$$\bar{x}_b = \bar{x}_s \pm \frac{\gamma \sigma_s}{n_s^{1/2}} \tag{3.18}$$

The value of γ is found from the normal curve depending on the level of confidence needed in specifying $\bar{x}_b$. For $\gamma = 1$ this level is 68.26%, for $\gamma = 2$ it is 95.46% and for $\gamma = 3$ it is 99.73%.

As an example suppose that a sample of 100 resistors, selected at random from a much larger population, is measured and has a mean resistance of $20\,k\Omega$ and a standard deviation of $100\,\Omega$. One can then say with 99.73% confidence that the mean value of the population lies between $20 \pm 3 \times 0.01$ i.e. between $20.03\,k\Omega$ and $19.97\,k\Omega$.

In taking samples the results obtained often deviate from the expected. Tests of significance are then used to determine if this deviation is real, or if it could have arisen due to sampling errors. The chi-square test, written as χ^2, is one such test of significance. If O is an observed result, and E is the expected result, then

$$\chi^2 = \sum \frac{(O - E)^2}{E} \tag{3.19}$$

The χ^2 distribution is given in tables, such as Table 3.5. The number of degrees of freedom in this table is the number of classes whose frequency can be assigned independently. If the data is given in the form of a table having V vertical and H horizontal columns, then the degrees of freedom is equal to $(V-1)(H-1)$.

As an example suppose that over a 24 hour period the number of instruments on test which have failed are as shown in Table 3.6. Does this indicate that most of the failures occur during the late night and early morning period, probably when the factory loading on the main supply is least? The expected value of failures, if there was no difference between time periods, would be the mean value of 5. Therefore from (3.19)

$$\chi^2 = \frac{(9 - 5)^2}{5} + \frac{(3 - 5)^2}{5} + \frac{(2 - 5)^2}{5} + \frac{(6 - 5)^2}{5}$$

$$= 6$$

There are three degrees of freedom, therefore from Table 3.5 the probability of occurrence of the result is seen to be greater than 10%. Therefore one can conclude that there is no significance in the increased number of failures in the periods 0–6 hours and 18–24 hours. If however the number of failures were each three times as large, i.e. 27, 9, 6, 18 respectively, then chi-square would be calculated as 20.67, and from Table 3.5 it is seen that the distribution of the failures is very significant, since there is a probability of less than 0.5% that it can arise by chance.

3.3 Factors influencing measurement errors

Errors arise in measurement systems due to several causes, such as human errors, or errors in using an instrument in an application for which it has not been designed. The different

Table 3.5. *The chi-square distribution*

Degrees of freedom	Probability level				
	0.100	0.050	0.025	0.010	0.005
1	2.71	3.84	5.02	6.63	7.88
2	4.61	5.99	7.38	9.21	10.60
3	6.25	7.81	9.35	11.34	12.84
4	7.78	9.49	11.14	13.28	14.86
5	9.24	11.07	12.83	15.09	16.75
6	10.64	12.59	14.45	16.81	18.55
7	12.02	14.07	16.01	18.48	20.28
8	13.36	15.51	17.53	20.09	21.96
9	14.68	16.92	19.02	21.67	23.59
10	15.99	18.31	20.48	23.21	25.19
12	18.55	21.03	23.34	26.22	28.30
14	21.06	23.68	26.12	29.14	31.32
16	23.54	26.30	28.85	32.00	34.27
18	25.99	28.87	31.53	34.81	37.16
20	28.41	31.41	34.17	37.57	40.00
30	40.26	43.77	46.98	50.89	53.67
40	51.81	55.76	59.34	63.69	66.77

Table 3.6. *Frequency distribution of instrument failures*

Time (24 hour clock)	Number of failures
0–6	9
6–12	3
12–18	2
18–24	6

types of errors are described in the next section. In the present section several definitions are introduced, which define the factors which influence measurement errors.

Accuracy and precision.   Accuracy refers to how closely the measured value agrees with the true value of the parameter being measured. For electrical meters the accuracy is usually defined as a percentage of full scale deflection. Therefore a meter having a full scale deflection of 10 volts may be stated to have an accuracy of $\pm 1\%$. This means that at any reading the pointer can be off the true value by as much as ± 0.1 volt. If the meter is reading 9 volts the true reading may be 9.1 or 8.9, i.e. an error of 0.2/9 or 2.2%. If the meter is reading 2 volts the error will increase to 0.2/2 or 10%; therefore for greater accuracy the proper range should be selected when reading instruments.

Precision means how exactly or sharply an instrument can be read. It is also defined as how closely identically performed measurements agree with each other. As an example suppose that a resistor, which has a true resistance of 32 981 ohms, is measured by two different meters. The first meter has a scale which is graduated in kohms, so that the closest one can get to the reading of resistance is 33 kohms or 33 000. The instrument is fairly accurate but it is very imprecise. The second instrument has a digital read out which gives values of resistance to the nearest ohm. On this instrument the same resistor measures 38 122 ohms. Clearly this instrument has high precision but low accuracy.

Resolution.   The resolution of an instrument is the smallest change in the measured value to which the instrument will respond. For a moving pointer instrument the resolution depends on the deflection per unit input. Resolution can be increased by using a square-law scale at full scale deflection, instead of a linear scale, and by amplification of the signal. Ultimately, however, the resolution is limited by the signal size which can be differentiated from the background noise.

Range and bandwidth. The range of an instrument refers to the minimum and maximum values of the input variable for which it has been designed. The range chosen should be such that the reading is large enough to give close to the required precision. For example with a linear scale an instrument which has 1% precision at full scale will have 4% precision at quarter scale.

The bandwidth of an instrument is the difference between the minimum and maximum frequencies for which it has been designed. If the signal is outside the bandwidth of the instrument then errors will result, since the instrument will not be able to follow changes in the quantity being measured. A wider bandwidth usually improves the response time of an instrument, but it also makes the system more prone to noise interference.

Sensitivity. The sensitivity of an instrument is defined as the ratio of the output signal, or response of the instrument, to the input signal or measured variable. Instruments which measure current are usually specified as the full scale deflection for a particular current. A meter having a range of 0–100 amperes is clearly less sensitive than one having a range of 0–1 ampere. Voltmeter sensitivity is specified in ohms per volt, and the greater the sensitivity the higher is this value.

Noise. Any signal which does not carry useful information may be called noise, and it is the cause of error. Noise may be mechanical, electrical or magnetic in form, and it can be reduced by shielding the instrument from vibration, electrostatic fields and magnetic fields. Noise is also generated internally in the instrument, and this must be reduced by careful design.

Significant figures. Significant figures usually indicate the precision of a measurement. For example a reading of 21 volts is less precise than one of 21.102 volts. However the total number of digits may not necessarily indicate the precision of a reading. For example a population may be given as 52 000, which does not mean that this is the exact value. What is meant is that the population is closer to 52 000 than to 51 000 or 53 000. A more correct method of specification in cases such as this is by the exponent, i.e. 52×10^3.

In mathematical calculations the precision of the result is determined by the lowest precision of the individual readings. For example adding resistance 21.2 ohms to 34.1356 ohms will give an answer of 55.3 ohms since the last three digits of the second reading are too precise for the result. Similarly in multiplication, a current of 2.1 amperes and a voltage of 4.2136 volts gives a wattage of $2.1 \times 4.2136 = 8.8$ watts, i.e. to two significant figures.

A method of specifying the variation in error of a value is by the plus/minus notation. A voltage of $V_1 = 31 \pm 5$ volts added to $V_2 = 29 \pm 3$ volts, gives $V_1 + V_2 = 60 \pm 8$ volts. Similarly in subtraction, $V_1 - V_2 = 2 \pm 8$ volts. Clearly one should avoid measurements which require subtraction of numbers since the result can be very inaccurate, in this case 8 volts in 2!

3.4 Types of errors

Errors always occur in any measurement, but it is important to isolate the cause of the error so that it can be minimised. Measurement errors are usually of four types, human, systematic, random and applicational.

Human error most frequently arises due to carelessness in taking or recording instrument readings. For example the wrong instrument scale is read, or the reading is taken correctly but falsely written down. Errors due to human mistakes can be minimised by adopting proper practices, and by taking several readings, preferably by different operators.

Systematic errors are faults in the measuring instrument, caused by several factors which give a systematic error in the results. For example the instrument may have an electrical or a mechanical fault, such as worn bearings or an irregular spring tension. Calibration errors, which cause the instrument to read high or low over the whole range, also fall into this class of error. Another cause of systematic error is environmental, such as the effects of changes in temperature, humidity and pressure, or of electric or magnetic fields, on the performance of the instrument. These errors can be reduced by operating sensitive instruments in air conditioned or screened enclosures.

Limiting errors are a form of systematic error, caused by the limits on the accuracy of the components which make up the instrument. For example a decade resistance box, made from resistors having 0.1% accuracy, cannot have an overall accuracy which is greater than this value. Limiting errors are usually quoted for each type of instrument and must be allowed for in the final reading.

Random errors are unpredictable and occur even when all the known systematic errors have been accounted for. These errors are usually caused by noise and environmental factors. They tend to follow the laws of chance, and although their effect cannot be eliminated, they can be minimised by taking many readings, and then finding their arithmetic mean and deviation, or by applying other statistical techniques.

Applicational errors are caused by using the instrument for measurements for which it has not been designed. For example the instrument may be used to measure a signal which is outside its bandwidth, so that it would be too slow to respond and would give a false reading. Another common cause of faulty application is to use an instrument with an internal resistance which is comparable in value to that of the circuit being measured. The instrument would then load the circuit and give a false reading. Applicational errors are usually made by novices, who are not familiar with electronic measuring instruments. They can be avoided by being fully aware of the characteristics of the instrument being used.

4. Transducers

4.1 Introduction

A transducer is an element which converts one form of energy into another. There are two basic types of transducers, active and passive. The active, or self generating, transducer converts energy directly from one state to another, without the need for an external power source, or excitation. An example of this is a thermocouple which gives an electrical signal output when one of its ends is heated. A passive transducer, on the other hand, does not convert energy directly, but controls the energy, or excitation, which comes from another source. Fig. 4.1 illustrates a simple

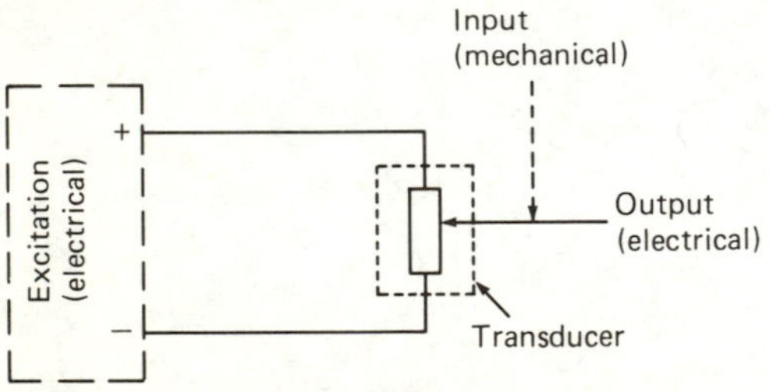

Fig. 4.1. Illustration of a passive transducer.

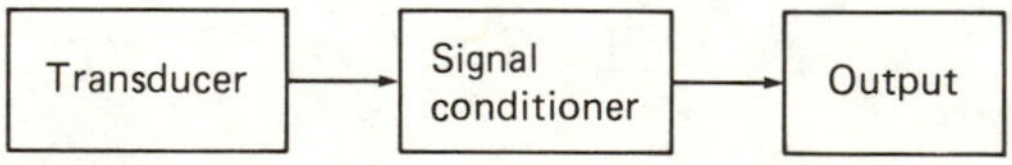

potentiometer transducer in which the mechanical input causes the transducer to give an output which is in proportion to the input excitation.

If one classifies energy into six different types then Fig. 4.2 gives the combinations of input or measurand, output, and excitation, for any transducer. This indicates the very large number of different transducer types which are available. In this chapter only those transducers most commonly used in electronic measuring systems will be discussed. In these the output and excitation are both electrical.

A transducer is usually the first element in a measurement system, as shown in Fig. 4.3. It may be located within the body of the instrument, or remote from the instrument, for example in harsh environments. The signal conditioner modifies the signal from the trans-

Fig. 4.3. Simplified measurement system using a transducer.

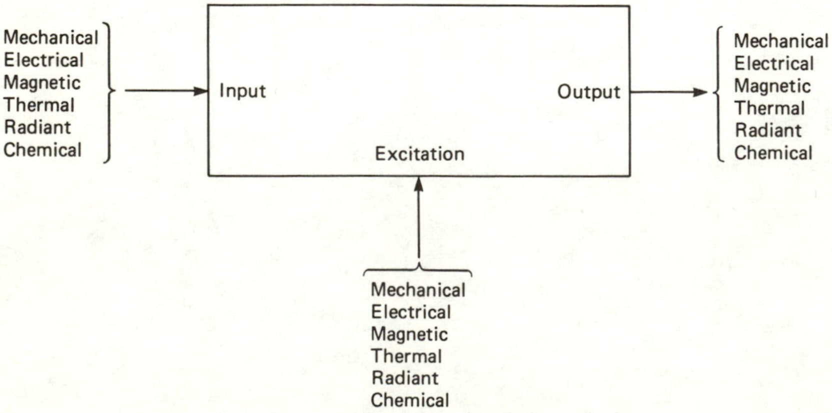

Fig. 4.2. Transducer combinations.

ducer, such as by amplification or wave shaping, to suit the requirements of the output device. This device may be an indicator or a storage medium.

4.2 Transducer selection

Generally there are three considerations in the selection of transducers; the required characteristics, the type of transducer technology and the force-summing element used. These three factors are described in this section.

4.2.1 Transducer characteristics

There are several parameters which need to be considered when selecting a transducer for an application. The *sensitivity* of the transducer is the output, usually in volts, which it produces for a given input signal and excitation level. The *resolution* is the smallest quantity which can be measured, and the *repeatability* indicates how closely two measurements of the same value correspond to each other.

The *accuracy* of the transducer is usually specified in terms of its operating conditions. *Environmental* effects, such as the influence of temperature, acceleration, shock, vibration and corrosion, on the performance of the transducer, must be considered.

The *frequency response* of a transducer is the change of output with input frequency, and this should be flat over the measurement range. The transducer's *dynamic characteristic* determines its response to step input signals. The *time constant* is the time taken by the transducer to reach 63% of its final value, and the time to reach about 90% of the final value is called the *response time* of the transducer.

The transducer should have good *linearity* over the output range, and its *hysteresis* must be considered. Other factors to be taken into account are the input *excitation requirements*, and the type of *output signal*, since this should match the characteristics of the measuring instrument. The *resonant* frequency of the transducer should be known, and this should not be approached by the range of measurements. The transducer should also be able to resist *noise*, which is usually all signals apart from the measurand.

4.2.2 Transducer classification

Transducers are usually classified by the application in which they are used, or by the type of technology in which they operate. In this chapter transducers will be mainly grouped by technology, and their principle will be illustrated by reference to typical applications. Where trans-

Table 4.1. *Transducer type–application matrix*

Type	Pressure (force)	Displacement	Position	Velocity	Acceleration	Vibration	Temperature	Magnetic flux	Optical
Strain gauge	●	●	●	●	●	●			
Potentiometric	●	●	●	●	●				
LVDT	●	●	●	●	●				
Variable inductance		●	●	●	●	●			
Hall effect		●	●					●	
Eddy current		●	●	●					
Magneto resistive		●	●					●	
Capacitive	●	●	●		●	●			
Piezo electric*	●	●		●	●	●			
RTD							●		
Thermistor							●		
Thermocouple*							●		
Photoemissive cell									●
Photoconductive cell									●
Photovoltaic cell*									●

*Self-generating or active devices

Table 4.2. *Comparison of techniques most frequently used in pressure measurement*

Transducer type	Pressure range (PSI)	Frequency response	Temperature range (°C)	Temperature coefficient (%/°C)	Accuracy (%)	Stability (%/Yr)	Resistance to shock and vibration*	Excitation	Output level	Price*
Unbonded strain gauge	$0.5–10^4$	0–10 kHz	−200 to +400	0.005	0.25	0.5	3	Regulated 10V a.c./d.c.	5 mV/V	4
Bonded foil strain gauge	$5–10^4$	0–10 kHz	−40 to +150	0.005	0.25	0.5	2	10V a.c./d.c.	3 mV/V	3
Thin film strain gauge	$15–10^4$	0–10 kHz	−200 to +400	0.005	0.25	0.25	2	10V a.c./d.c.	3 mV/V	4
Diffused semiconductor strain gauge	$15–5 \times 10^3$	0–10 kHz	−40 to +150	0.005	0.25	0.25	2	10 to 30V d.c.	20 mV/V	2
Bonded semiconductor strain gauge	$5–10^4$	0–10 kHz	−40 to +150	0.01	0.25	0.5	2	10V a.c./d.c.	20 mV/V	1
Potentiometric	$5–10^4$	0–100 Hz	−40 to +200	0.01	1.0	0.5	5	Regulated a.c./d.c.	—	3
Capacitive	0.01–20	0–50 Hz	0 to 1000	—	0.5	.05	4	a.c./d.c.	—	5
Differential transformer (LVDT)	$30–10^4$	100 Hz	0 to 100	—	0.5	0.25	5	a.c.	—	2
Piezoelectric	$0.1–10^4$	1–100 kHz	−300 to +300	0.01	1.0	1.0	1	—	—	4

*1 = Best or lowest

ducers are exclusively used for a particular application, such as thermal measurements, they are described under that section. Transducers used for special applications, such as chemical, biological or nuclear will not be considered here. Table 4.1. gives a type–application matrix for the selection of transducers. Two of the most frequent uses of transducers are in pressure or temperature measurement systems; Table 4.2 and Table 4.3 give the characteristics of selected transducers for these applications.

4.2.3 Force-summing elements

Mechanical measurement transducers require the transmission of displacement or stress, in order to generate an electrical signal, and this is done by a force-summing element. Fig. 4.4 shows a selection of such elements, the choice depending on the type of force being measured and on the characteristics of the transducer technology. For example the flat diaphragm, capsule, or corrugated diaphragm force-summing elements can all be used to operate a capacitive or a piezoelectric transducer. If a large force must be transmitted, or more displacement is needed, then the element shown in Fig. 4.4(b) or (c) should be used in preference to that in (a). The natural frequency of resonance of the diaphragm f_r is given by (4.1)

$$f_r = \frac{1}{2\pi}\left[\frac{C}{m}\right]^{1/2} \tag{4.1}$$

where C is the spring constant of the diaphragm, and m is its mass. Therefore if a high resonance frequency is needed, to make the transducer resistant to the effects of shock and vibration,

then a high value of C and a low value of m is required, so that the element shown in Fig. 4.4(a) is preferred.

The seismic transducer construction is also used in many applications, as shown in Fig. 4.5, where no fixed point of reference is available. If the unit is accelerated upwards, the inertia of the mass will cause it to resist motion, until a point when the force on the spring equals the product of the mass and the acceleration. Therefore the mass will be displaced by an amount which is proportional to the acceleration, and it can operate a transducer element to give an output signal. If this signal is integrated with respect to time, then the output is proportional to velocity. The resonant frequency of the system is given by (4.1), and for accurate readings the acceleration frequency must at least be less than one third of this resonant value.

4.3 Resistive transducers

This section describes those transducers in which the measurand, either directly or indirectly, usually through a force-summing device, causes an ohmic change in a resistive element. The types of transducers most commonly used are potentiometric, the strain gauge, magnetoresistive, photoconductive, resistance temperature detector, and the thermistor. Magnetoresistive transducers are described in Section 4.6, photoconductive in Section 4.8, and the resistance temperature detector and the thermistor in Section 4.9.

4.3.1 Potentiometric transducer

Fig. 4.6 shows a simplified representation of a

Table 4.3. *Temperature measurement transducer characteristics*

Parameter	Resistance temperature detector (RTD)	Thermocouple	Thermistor
Sensitivity	$0.1-10\,\Omega/°C$	$10-50\,\mu V/°C$	$0.1-1.0\,k\Omega/°C$
Stability (drift per year)	.01%	0.5 °C	1%
Repeatability	.05 °C	5 °C	0.5 °C
Temperature range	$+150\,°C$ to $+850\,°C$	$-200\,°C$ to $+1600\,°C$	$-100\,°C$ to $+350°C$
Linearity*	1	2	3
Minimum size	5 mm dia. × 5 mm length	0.4 mm dia.	0.4 mm dia.
Accuracy*	1	2	3
Cost	3	1	2

*1 = Best or lowest

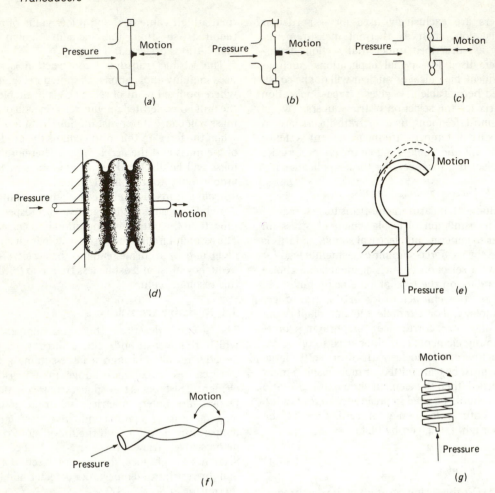

Fig. 4.4. Force-summing elements; (*a*) flat diaphragm, (*b*) corrugated diaphragm, (*c*) capsule, (*d*) bellows, (*e*) circular Bourdon tube, (*f*) twisted Bourdon tube, (*g*) helical Bourdon tube.

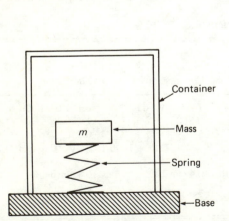

Fig. 4.5. Seismic transducer construction.

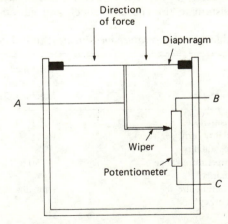

Fig. 4.6. Simplified representation of a potentiometric transducer used to measure force.

potentiometric transducer, in which the diaphragm acts as a force-summing element to move a wiper across the body of the potentiometer. A variable resistance, proportional to the amount of slider movement, can be measured in a bridge circuit at terminals A and B, or A and C. Alternatively an excitation voltage can be applied across B and C, and a voltage read at A.

Either a.c. or d.c. may be used for measurements in a potentiometric transducer. A.C. is preferred since it is easier to amplify later, but at high frequencies the a.c. resistance of the potentiometer changes with frequency, due to skin effect, and this needs to be compensated for. The inductance and stray capacitance effects of the potentiometer winding must also be considered at high frequencies.

Potentiometric transducers are popular since they give an output which can be used without further amplification. They have a limited life, in the region of 10^6 cycles, and are prone to noise due to the mechanical movement of the wiper. Potentiometric transducers also need a relatively large force, and amount of diaphragm movement, for the wiper to traverse the length of the potentiometer. Both wirewound and conductive plastic potentiometer constructions are used. The wirewound structure is more robust and reliable, but the resolution is limited to a value equal to the resistance divided by the number of turns. Conductive plastic devices have a resolution of about 0.002%.

To avoid loading effects the internal resistance of any instrument, connected to a potentiometric transducer, should be several times that of the potentiometer. Fig. 4.7 shows a transducer of resistance R_1 connected to an instrument of resistance R_2. If the potentiometer is set at a fraction α of its full scale reading, the error in the

reading is given by (4.2), where $\beta = R_1/R_2$.

$$\text{Error} = [1 - [1 + \beta\alpha(1 - \alpha)]^{-1}] \times 100\% \quad (4.2)$$

The error is zero at the two ends of the potentiometer, and is a maximum at its mid setting. For a linear response R_2 should be several orders of magnitude larger than R_1. In the limit, when R_2 is infinite, β equals zero, and there is no error.

4.3.2 *Strain gauges*

Strain is the change in length per unit length of an element, and strain gauges measure the result of a force, such as stress and displacement, by converting the strain caused by this mechanical force into a change in resistance. Strain gauges are available in many sizes, down to about 0.025 cm in length, and can measure strains as low as 10^{-6}. They may be surface mounted or imbedded in the material whose strain is being measured.

There are three principal types of strain gauges, wire, foil and semiconductor. Wire gauges can also be of the bonded or unbonded type, and semiconductor gauges may be bonded or diffused.

4.3.2.1 Operating principle

Wire and foil gauges work on the principle that when a metal is subject to stress it changes its length and cross sectional area, giving a change in resistance measured as in (4.3)

$$R = \rho\frac{l}{a} \quad (4.3)$$

where ρ is the resistivity of the material, l is its length and a its cross sectional area. The sensitivity of a strain gauge is measured as the ratio of the change in resistance to the change in length. This is given by the gauge factor (K), as in (4.4)

$$K = \frac{\Delta R/R}{\Delta l/l} \quad (4.4)$$

It can also be shown that the gauge factor is related to the Poisson's ratio for the metal (μ) by (4.5)

$$K = 1 + 2\mu \quad (4.5)$$

Most metals have a μ between 0.25 and 0.35, giving a K of 1.5 to 1.7. However the alloys used for strain gauges have gauge factors between 2 and 5.

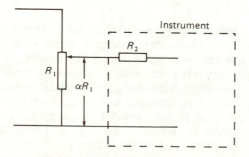

Fig. 4.7. The effect of an instrument's internal resistance on the accuracy of a potentiometric transducer.

In a semiconductor strain gauge the stress produces a change in the crystal structure, called the piezo electric effect, giving a greater change in resistance than metal gauges, and a gauge factor of between 50 and 200.

4.3.2.2 Construction

Wire gauge. The unbonded wire gauge usually consists of a wire, in four sections, wound on posts. The posts are pivoted such that pressure normal to their plane will cause them to tilt, and increase the stress on two of the wires, whilst decreasing it on the other two. The wires may be connected to form the four arms of a bridge, giving greater sensitivity, and a degree of temperature compensation. A stress of $1000 \, kg/cm^2$ would give a strain of about 0.1%, so that the bridge must be carefully designed to detect this.

Unbonded strain gauges are less sensitive than bonded types, and are also bulkier, so they are usually only used where the gauge forms part of another unit, such as a load cell or an accelerometer.

Bonded wire gauges are made by drawing wire, of diameter 0.0025 cm or smaller, and forming this into a zig-zag pattern, as in Fig. 4.8. The wire is fixed to a mount which provides mechanical support and dimensional stability.

The mount may form a permanent part of the gauge, as in Fig. 4.8, in which case it must be less than 0.0025 cm thick, and flexible, to enable it to make mechanical contact with the surface on which it is bonded. Alternatively the mount may be a temporary carrier, which is peeled off and discarded when the gauge is used. The advantage of this is that better mechanical and thermal contact is now made between the gauge and the specimen to which it is attached.

Although gauges respond mainly to strain along their length, they also have transverse sensitivity. The long narrow shape measures strain mainly in one direction, and the loops, back and forth, increase the length within a small overall dimension. However the ends of the loops are sensitive to transverse strain, so these should be minimised. Strain can be measured in several directions by a rosette construction. These may be stacked, with individual elements on top of each other, or planar, where elements do not overlap. Fig. 4.9 shows two examples of stacked rosettes.

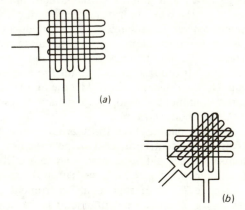

Fig. 4.9. Bonded wire gauge stacked rosettes; (*a*) 90°, (*b*) 60°.

Bonded wire gauges have a small surface area, which reduces the leakage current at high temperature or high potential. Except when this property is required, modern gauges use the foil construction, which has a greater surface area to cross section ratio, and is therefore more stable over long periods. Bonded strain gauges are available in various values, such as 120, 350, 500, 1000 and 5000 ohms, to match the standard types of bridges, excitations and readouts.

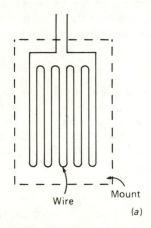

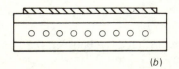

Fig. 4.8. A wire strain gauge; (*a*) plan view, (*b*) cross section.

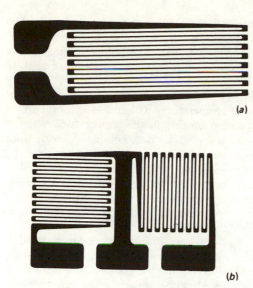

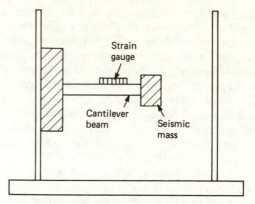

Fig. 4.11. Strain gauge used in an accelerometer.

Fig. 4.10. Foil gauges; (a) unidirectional, (b) 90°.

Foil gauge. Metallic foil gauges are made by photo etching thin sheets, 0.0005 cm or less in thickness, of heat treated metal. Fig. 4.10 shows two examples of foil gauge patterns. The ends of the loops have their cross sections increased so as to reduce their resistance, and so minimise the effect of unwanted transverse strain. The metal foil may now be stuck onto a permanent or temporary mount, or it can be formed by metallic deposition and photo etching, directly onto a permanent mount.

The foil gauge is rugged, and has excellent linearity and low hysteresis. It is less sensitive (lower K) than a silicon gauge, but can better withstand the effects of shock, vibration and temperature variation. To minimise temperature effects two identical gauges are usually used in the adjoining arms of a measurement bridge. Both gauges are kept at the same temperature, but only one is under stress. The elongation of the gauge varies from about 0.5%, when cermet backing is used for high temperature operation, to 5% when the backing is flexible polyimide or paper.

Fig. 4.11 shows a strain gauge used in an accelerometer. The deflection of the seismic mass produces strain in the cantilever beam, which is measured by the gauge. Two gauges may be used for greater sensitivity, one mounted above the beam and one below it, the gauges being connected in opposite arms of a measurement bridge.

Semiconductor gauge. Semiconductor strain gauges are made from silicon crystals cut into filaments. Both bonded and diffused structures may be used. The bonded semiconductor gauge is similar in construction to a bonded metallic gauge, and gives an output of about 15 mV/V of excitation. Diffused semiconductor gauges are made by diffusing the gauge element into the surface of a diaphragm, which is made from a slice of mono-crystalline silicon. This gives better linearity (0.05%) and less hysteresis (0.01%) than the bonded type, but the output level is also lower (10mV/V). Fig. 4.12 shows one structure used for a bonded silicon gauge.

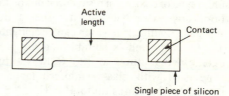

Fig. 4.12. Bonded silicon gauge.

Semiconductor strain gauges are much more sensitive than metallic gauges, but are also less linear, and have a higher temperature sensitivity. They are therefore usually used with compensation circuits, which are either diffused into the same silicon die or mounted in the same hybrid package.

4.3.2.3 Materials

The three materials of importance in gauges are the gauge metal, the backing for the metal, and the adhesive used to attach the gauge to the specimen under stress.

Gauge metals. The metals used for the gauge need a high stability, that is, a low temperature coefficient of resistance. This is specially important for static strain measurements. The four most commonly used gauge alloys are as follows:

(i) Copper–nickel alloy, such as constantan. This has a low and controllable temperature coefficient of resistance, and is used for static strain measurements, where the strain level is below $\pm 1500\,\mu$cm/cm and the temperature limits are $-70\,°$C and $+230\,°$C.

(ii) Nickel–chrome alloy, such as stabiloy. This has good stability from cryogenic temperatures up to $350\,°$C, and good fatigue life.

(iii) Nickel–iron alloy, such as dynaloy. The material has a high gauge factor, but poor temperature stability, and is mainly used to measure dynamic strain.

(iv) Platinum–tungsten alloys. These have excellent stability and good fatigue life, and are used for static measurements up to $650\,°$C and dynamic measurements up to $820\,°$C. The material has a large temperature coefficient, and this must be compensated for.

Backing materials. Strippable vinyl is usually used as the backing material for temporary carriers. Four materials are commonly used for permanent backings, as follows:

(i) Polyimide resins. These can be used as a cast film type of coating of resin cured on the film without any extra backing. It forms a thin flexible backing material which can be elongated by up to 20%, and is suitable for use from cryogenic temperatures to $200\,°$C. Resins may also be used with glass reinforced laminate carrier for temperatures up to $400\,°$C.

(ii) Paper impregnated with phenolic resins is used at up to $260\,°$C, and it is bonded to the foil or wire under heat and pressure. Glass fibre reinforcement may be added to the backing, which gives dimensional stability but reduces the strain range.

(iii) The strain gauge may be attached to a metal shim, about 0.005 cm thick, by an insulating adhesive. The shim is fixed to the specimen by spot welding, and it should have a temperature coefficient which matches that of the specimen.

Adhesives. The adhesives used to attach the strain gauge to the specimen must have high shear strength over the temperature range, to ensure that the strain is transmitted accurately from the specimen to the sensing element, and have a high insulation resistance, in the region of 10^{10} ohms. Four types of adhesives are used, as follows:

(i) Nitrocellulose adhesive. This is cured at room temperature, within 2–48 hours, and it can then operate in the temperature range of -70 to $+80\,°$C. It is compatible with most specimen materials, except those affected by ketone solvents; the gauge can be removed, if necessary, with a ketone. Nitrocellulose is hygroscopic, so the assembly must be covered with a moisture resistant coating to give long term stability.

(ii) Cyano-acrylate adhesives are compatible with most specimens and carriers, and as they are contact setting the gauge can be installed in minutes. They operate over a temperature range of -70 to $+620\,°$C, but are sensitive to moisture, so they must be protected.

(iii) Epoxy resins. Some types of resins set at room temperature within 2–10 hours, whilst others need a curing temperature of 100–$200\,°$C for 1–5 hours. The adhesive has a working temperature range of -250 to $+300\,°$C, and is unaffected by most chemicals and moisture.

(iv) Ceramic cement is used as an adhesive for gauges having a temporary carrier, and working at temperatures above $400\,°$C. The cement, which is an insulator, is applied to both the gauge and the specimen, and is then baked at $320\,°$C. This forms a hard porous covering, which can work over the range -450 to $+600\,°$C.

4.4 Capacitive transducers

The capacitance of a parallel plate capacitor is given by

$$C = \varepsilon_0 \varepsilon_r \frac{(n-1)A}{d}$$

where n is the number of plates, A the area of one side of a plate, d is the thickness of dielectric, ε_r is the relative permittivity of the dielectric, and ε_0 is the permittivity of free space, equal to 8.85×10^{-11} F/m. In a capacitive transducer either the plate separation, effective plate area (or overlap), or the constant of the dielectric is varied by the measurand to give a changing capacitance, which can be sensed.

Fig. 4.13 shows some capacitance transducer

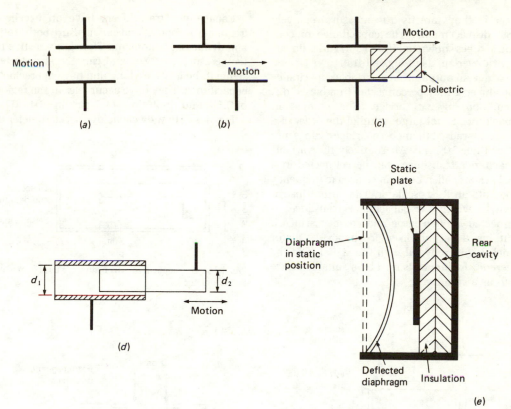

Fig. 4.13. Capacitive transducer arrangements; (a) parallel plate with variable separation, (b) sliding plate with variable overlap, (c) variable dielectric, (d) concentric tube, (e) capacitive microphone type.

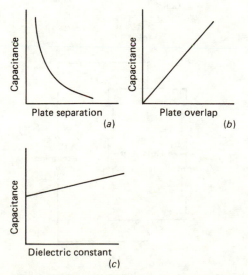

Fig. 4.14. Variation of capacitance; (a) with plate separation, (b) with plate overlap, (c) with dielectric constant.

arrangements. In Fig. 4.13(a) the applied force varies the plate separation, and Fig. 4.14(a) shows the resulting capacitance curve. Highest resolution is obtained when the plates are close together, so large plates with small spacing is desirable. The gap is usually greater than $100\,\mu m$ since the plates may not be flat and parallel, and are therefore in danger of touching. The capacitance curve can be linearised by means of electronic circuits. Flat plate transducers are used for small but accurate measurements, below about one millimetre.

Fig. 4.13(b) shows a transducer in which the effective plate area is changed by varying their overlap, and Fig. 4.14(b) indicates that the capacitance variation is linear. Similarly varying the dielectric constant, by changing the amount of dielectric between the plates, will give a linear capacitance change as indicated in Figures 4.13(c) and 4.14(c). Both these last two transducer methods are used to measure relatively large displacements. The capacitance of concentric tubes shown in Fig. 4.13(*d*) varies linearly with overlap, as in Fig. 4.14(*b*).

To give an output from a variable capacitance transducer it may be connected in a biased d.c. circuit or an a.c. bridge. The change in capacitance is not large, being of the order of a few picofarads, but modern bridges can sense better than 10^{-5} pF. Alternatively the variable capacitance transducer can be connected in a feedback amplifier, or it can be used to frequency modulate an R.F. oscillator. The output instrument to which the capacitance transducer is connected is often large and complex, since it contains a second fixed frequency oscillator which is heterodyned with the signal, and the difference frequency is read by a suitable device such as a counter.

Capacitance transducers have an excellent frequency response, and can measure both static and dynamic phenomena. They are sensitive to temperature changes, and can give a distorted signal if the leads are long and have appreciable capacitance. They have accuracies in the region of 0.5% and resolutions of 1 part in 20 000.

Fig. 4.15 shows a capacitive accelerometer in

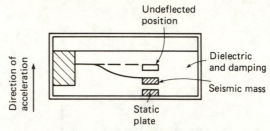

Fig. 4.15. A capacitive transducer used in an accelerometer.

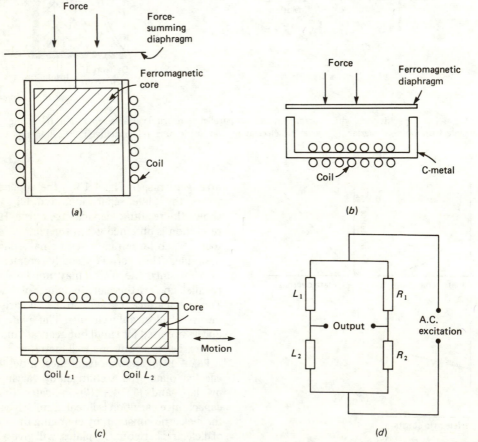

Fig. 4.16. Variable inductance transducer; (*a*) single coil arrangement, (*b*) diaphragm type, (*c*) two coil arrangement, (*d*) bridge for two coil arrangement.

which the seismic mass moves relative to the static plate, giving a change in spacing between the two capacitance plates.

4.5 Inductive transducers

Inductive transducers work on the principle of changing the self or mutual inductance of a coil. Fig. 4.16(a) shows a simple arrangement in which the ferromagnetic core is attached to a force-summing diaphragm, and moves relative to the coil. The inductance of the coil changes with the position of the core, and this change can be measured in an a.c. inductance bridge, or as a changing frequency if the inductance forms part of an oscillator circuit.

Fig. 4.16(b) shows an alternative arrangement in which the ferromagnetic diaphragm moves with the force, and causes a change in inductance of the coil due to a variation of the air gap. The ferromagnetic diaphragm can be part of the force-summing member, but the performance of the assembly is degraded because of the trade off needed between the mechanical and magnetic properties of the diaphragm. Two coils may be used, as in Fig. 4.16(c), in which the inductance of one coil increases whilst that of the other decreases. The two coils may form one arm of a bridge, as in Fig. 4.16(d), the bridge being in balance when the core is in the centre, between the coils.

An inductive transducer which is extensively used, especially for displacement measurements, is the linear variable differential transformer (LVDT). This is shown in Fig. 4.17, and uses the change in mutual inductance between magnetically coupled coils, rather than the change in self inductance, as the parameter being measured. Three coils are wound on a former, the a.c. supply applied to the primary inducing a voltage in the two secondaries. The secondaries are connected in series opposition, so that the output voltage is the difference of the voltages induced in the two coils. When the core is in the centre of the coils the voltages induced in the two secondaries cancel each other, so there is no output. This is the balance or null position. When the core moves the voltage induced in the secondary nearest the core predominates, and gives a finite output. Fig. 4.17(c) shows a plot of output voltage against position of the core. For small displacements the voltage varies linearly with motion, and undergoes a 180° phase change as the core moves through the central position, giving an indication of direction of motion.

Inductive transducers have an advantage over potentiometric transducers since there are no moving parts to rub and wear down, and to provide extra friction to the system. The inductive transducer responds to static and dynamic measurands. Its frequency response is limited by the construction of the force-summing member, and it can be affected by external magnetic fields, requiring shielding in certain applications. The transducer is also sensitive to shock and vibration, and has poor thermal stability and a limited temperature range, so that it is mainly used in low accuracy systems.

The advantage of a differential output, as in an LVDT, is that it gives a greater output for equal displacements, and lower variation of output with changes in temperature, magnetic fields, supply voltage or supply frequency. The usual excitation frequency band of an LVDT is from 40 Hz to 200 kHz, although it may go up to 1 MHz. Generally the frequency should be at least ten times greater than the maximum frequency present in the mechanical motion being measured. The sensitivity and efficiency of the transducer both increase with frequency, being optimum at between 1 kHz and 5 kHz.

The LVDT can be designed to operate over distances of ± 0.01 cm to ± 30 cm. The excitation voltage is in the region of 5–100 volts rms, and the output voltage varies from 50 –500 mV per millimetre of movement, with a linearity of about 0.025%.

Commercially available LVDT transducers are available with an internal oscillator, operating from a d.c. input supply, to provide its a.c. excitation. The output from the secondaries is demodulated within the transducer to give a d.c. voltage, which is proportional to the transducer's displacement. Since the output of an LVDT is proportional to the primary voltage, this voltage must be regulated for good accuracy. The frequency source must also have a low output impedance, to minimise the output voltage changes resulting from variations in the LVDT impedance with different core positions, and with temperature.

4.6 Magnetic effect transducers

Two types of transducers which are often used to measure the strength of a magnetic field are

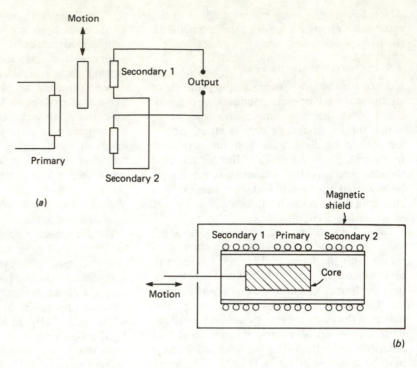

Fig. 4.17. Linear variable differential transformer; (*a*) circuit connection, (*b*) construction, (*c*) characteristic.

described in this section. These are Hall effect devices, and components which depend on the magnetoresistive effect for their operation.

4.6.1 *Hall effect transducers*

Although several materials exhibit the Hall effect, this effect is not very pronounced in conductors, so that all commercial devices are made from semiconductors. Fig. 4.18 shows its principle of operation. When a thin piece of semiconductor (thickness *t*) is placed in a magnetic field (*B*), and a current (*I*) is passed through it such that the electric field is at right angles to the

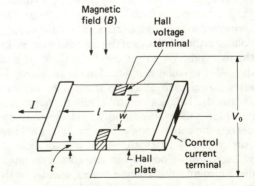

Fig. 4.18. Operating principle of a Hall-effect transducer.

magnetic field, then a voltage will appear across the remaining transverse faces of the material. The value of this open circuit Hall voltage is given by (4.7)

$$V_0 = R_H IB/t \tag{4.7}$$

R_H is called the Hall constant of the material and is given by (4.8)

$$R_H = [ne]^{-1} \tag{4.8}$$

n is the concentration of electrons in the material, and e is the electron charge. Semiconductors have a value of $n = 10^{22} \, \text{m}^{-3}$, which is almost 10^7 times less than that of conductors, so that from (4.8) and (4.7) it is seen that semiconductors will show a corresponding larger Hall voltage. The effective length (l) and width (w) of the Hall device should be such that $l/w = 2$–3, since for smaller values of this ratio the Hall voltage is less than that given by (4.7). The device should also be as thin as possible (small t).

Materials used for Hall transducers need a high electron mobility and energy gap. Two materials commonly used are indium arsenide and indium arsenide phosphide. Doped indium arsenide has an electron concentration of 5–$7 \times 10^6 \, \text{cm}^{-3}$, room temperature mobility of $20\,000 \, \text{cm}^2/\text{Vs}$, and a Hall constant of $100 \, \text{cm}^3/°\text{C}$. Indium arsenide phosphide has about half this mobility but a higher energy gap, and it can run at higher temperatures. Indium antimonide is also sometimes used for Hall transducers, but it has a high temperature coefficient of R_H, of about $2\%/°\text{C}$.

The Hall plate semiconductor is made by one of three techniques. In the first the semiconductor is cut from bars of the material, and ground and etched to a wafer 5–$100 \, \mu\text{m}$ thick. This is then stuck onto a substrate by a 1–$2 \, \mu\text{m}$ layer of epoxy resin, which fills the cracks in the material and gives good heat transfer. In the second method semiconductor material is vapour deposited onto a substrate, to a thickness of 2–$3 \, \mu\text{m}$. This device is suitable for use at very low and very high temperatures. In the third method a layer of about $10 \, \mu\text{m}$ of gallium arsenide is grown epitaxially from the gas phase, onto semi-insulating gallium arsenide. This gives a device with high stability and low temperature coefficient, and which is suitable for precision measurements.

The substrate used for the Hall plate must be rigid in thin sections, must have a high resistivity, and it should have a high thermal conductivity to remove the heat generated in the Hall plate. The temperature coefficient of expansion of the substrate must also match that of the plate. Beryllia and alumina are most frequently used in Hall effect devices.

Electrical contacts may be soldered to the Hall plate although usually the metal is evaporated onto the semiconductor, and then alloyed by heating. The electrical contacts should extend across the width of the Hall plate, so that the lines of current flow are straight and parallel. The material used in the Hall contact has a higher conductivity than the semiconductor, which causes the lines of current flow to become distorted near the Hall electrodes, giving a non-linear relationship between B and V_0 in (4.7). The non-linearity can be minimised by making the electrode length less than 10% that of the Hall plate, or by loading the Hall generator with a linearising resistor R_{LL}. This, however, reduces the sensitivity of the device, where sensitivity S is given by (4.9)

$$S = V_0 [IB]^{-1} \tag{4.9}$$

The Hall constant R_H varies with temperature, and with the strength of the magnetic field, although it does not show a significant change for fields up to about $15 \, \text{T}$ ($5 \, \text{T}$ for indium antimonide). The Hall device cannot be destroyed by large magnetic fields, but the field will increase the internal resistance of the device, so that its control current must be kept below its rated value to limit device dissipation.

The Hall voltage decreases with load, and optimum linearity is obtained at a value of resistive termination R_{LL} which is quoted in the data sheets for the device. The Hall generator can be considered to have an internal resistance R_1 in the electrical control side, and R_2 in the Hall voltage side. Resistance R_1 is measured between the control leads when the Hall circuit is open. The data sheets give the resistance value when $B = 0 \, (R_{10})$, and a curve showing how R_1 varies with the strength of the magnetic field. Resistance R_2 is defined as the resistance measured between the Hall leads, with the control circuit open. This is also a function of B and is specified at $B = 0 \, (R_{20})$.

As a result of the process used to make Hall

devices, a small resistive component of voltage is superimposed onto the Hall voltage. With zero field the voltage is given by $R_0 I$, where I is the control current and R_0 is called the resistive zero component of the Hall device. The voltage due to R_0 can be compensated for in a Hall effect device application by external circuitry.

The lead wires of the Hall electrodes form an inductive loop of area A_0 which cannot be completely eliminated. If the magnetic flux is changing at the rate dB/dt then even with zero control current I a voltage E given by (4.10) will be induced in the loop.

$$E = A_0 \, dB/dt \qquad (4.10)$$

A_0 is called the inductive zero component of the Hall device, and E typically has a value of $500 \, \mu V$ for a field of 1 Tesla and a frequency of 50 Hz.

4.6.2 *Magnetoresistors*

Magnetoresistive elements operate on the law of electrodynamics which says that Lorentz forces act on mobile charge carriers in a magnetic field, causing the electrons to move in an indirect route, so lengthening the current path and the resistance of the material. The amount of deflection of the electrons depends on electron mobility; it is highest in a semiconductor such as indium antimonide, which has a mobility of 78 000 cm²/Vs compared to 50 cm²/Vs for a metal. Magnetoresistors are made from indium antimonide or indium arsenide, which has a mobility of 24 000 cm²/Vs.

Magnetoresistors are sensitive to the total magnetic field, and not to its rate of change. This is important in hand-held sensing systems, which may move at an undefined speed. Fig. 4.19 shows the magnetoresistive characteristics. H_0 is the effective anisotropy field in the material, and is the sum of the demagnetisation

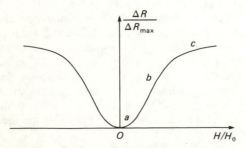

Fig. 4.19. Basic magnetoresistive characteristic.

anisotropy field and the anisotropy field induced during elemental deposition. With no field applied the domain magnetisation is along the elemental length at o. As the field increases the resistance increases, until at b the elements are rotated by about $45°$ to the elemental length. Further increase in field leads to saturation at c. The magnetoresistor may be operated at O, or at b by means of an external bias magnet. Operation at b gives a linear characteristic.

The magnetoresistive effect is increased as the ratio of length to width (l/w) of the material decreases. This is shown in Fig. 4.20. In order to

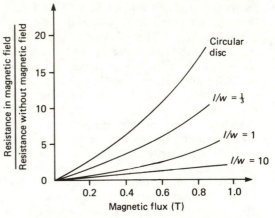

Fig. 4.20. Magnetoresistive curves for different geometry devices.

get a high intrinsic resistance at zero magnetic field, and a large change in resistance with increasing magnetic field, several devices with low ratios can be connected in series. This may be done by depositing metallic film onto a long piece of semiconductor, and etching out areas to leave a rastered pattern of short circuiting conductors, as in Fig. 4.21(*a*), When a magnetic field is applied the current path is rotated by the Hall angle θ, as in Fig. 4.21(*b*). Since the metallic conductors are equipotential surfaces, the current flows back to the other side of the material, and reappears rotated by θ in the next semiconductor section. A wide range of intrinsic resistance is possible by choice of the ratio l/w.

Instead of using external metallic short circuiting lines across the semiconductor, the material, indium antimonide, can be made with parallel oriented crystals of metallic nickel antimonide,

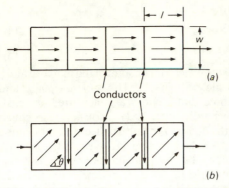

Fig. 4.21. Current path in a rastered magnetoresistive element: (a) without magnetic field, (b) with magnetic field.

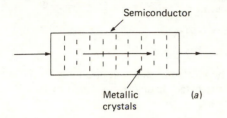

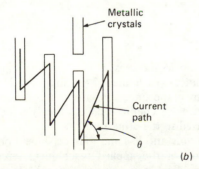

Fig. 4.22. Semiconductor with internal short circuiting metallic crystals; (a) current flow without magnetic field, (b) current flow with magnetic field.

which act as internal short circuiting lines. Without a magnetic field the current path is at right angles to the equipotential surfaces of the metallic crystals, but is rotated by the Hall angle when a field is applied, and zigzags through the material, as in Fig. 4.22.

Magnetoresistors are built by depositing a film of about $25\,\mu$m of indium antimonide/nickel antimonide onto a 0.1 mm thick substrate. The film is in the form of a meander, and the value of resistance at zero magnetic field can be varied by changing the dimensions of this meander, and

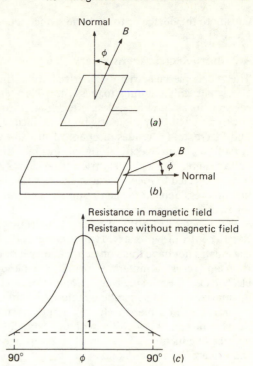

Fig. 4.23. Effect of magnetic field direction on a magnetoresistor; (a) and (b) field direction, (c) resistance curve.

the number of loops. The film is insulated from the substrate, which may be made from a magnetic base coated with a thin insulating film of a non-magnetic material, such as ceramic or plastic.

The value of resistance depends on the direction of the magnetic field, as indicated in Fig. 4.23. The maximum change in resistance occurs when the nickel antimonide crystals are parallel to each other, and the magnetic and electric fields are at right angles to each other. The temperature response of the magnetoresistor depends on the material doping. Generally resistance decreases with temperature, and the drop in resistance is greater after the application of the magnetic field, so that the ratio of resistance in a magnetic field to resistance without a field decreases with temperature.

Magnetoresistors have low noise, unless cracks appear in the material due to faulty mounting. Since the device is a bulk resistor it does not depend on surface effects, so there is very little ageing. Most of the ageing is due to deterioration of the epoxy resins used to encapsulate the device or to stick the semiconductor to the base.

4.7 Piezoelectric transducers

The word piezoelectricity is derived from the Greek 'piezein', meaning 'to press', and it is often referred to as 'pressure electricity'. It was discovered in 1880 by the Curie brothers, Jacques and Pierre, and it relates mechanical stress in a crystal with an electrical signal. An electric voltage applied to a piezoelectric crystal causes it to be subjected to mechanical stress; or a mechanical stress applied to the crystal will generate an electric voltage across it.

Piezoelectricity occurs in crystals which do not have a centre of symmetry. Twenty one classes of crystals do not have this symmetry, and over one thousand crystal materials exhibit the piezoelectric effect. Both single crystal material, such as quartz, and polycrystalline material, such as ceramics, exhibit piezoelectricity. Single crystal quartz has a higher temperature stability, greater chemical inertness and is more rugged than piezoelectric ceramic. But, to get good piezoelectric properties, single quartz crystals must be accurately oriented, and only perfect crystals must be used. The range of shapes available is also limited to simple structures like plates and discs.

Unlike natural crystals, such as quartz or tourmaline, piezoelectric ceramics have no piezoelectric properties when first made, due to the random orientation of their electric dipoles. Subjecting them to an electric field of between 10 and 30 kV/cm, at a temperature just below the Curie temperature, 'poles' the material so that it acts as a single crystal. The advantages of piezoelectric ceramics are that they can be formed into complex shapes, they are chemically inert and they can be made using standard ceramic technology.

Depolarisation or depoling can occur in ceramic materials, with loss of piezoelectric properties, if the material is exposed to a strong a.c. field, or a d.c. field opposed to the original direction of poling, or if the temperature rises above the Curie point, or if the mechanical stress on the material exceeds specified limits.

In a piezoelectric material the relationship between an electric field E and the resultant displacement D, and between a mechanical force T and resultant strain S in the material are given by (4.11) and (4.12)

$$S = dE + \gamma^E T \tag{4.11}$$

$$D = dT + \varepsilon^T E \tag{4.12}$$

ε is the dielectric constant of the material and γ its elastic compliance. The superscripts T and E denote that ε and γ are measured at constant mechanical force and electric field respectively. d is the charge per unit applied stress, at constant electric field, or alternatively it is the strain, per unit applied field, at constant stress. Therefore d is referred to as the piezoelectric charge constant of the material, and has units of metres per volt.

An alternative constant, the piezoelectric voltage constant g, is defined by (4.13) and (4.14), and is measured in units of volt metres per newton.

$$E = -gT + D\varepsilon^{-T} \tag{4.13}$$

$$S = \gamma^D T + gD \tag{4.14}$$

The electromechanical coupling coefficient k of a piezoelectric device is defined by (4.15) and (4.16)

$$k^2 = d^2 \varepsilon^{-T} \gamma^{-E} \tag{4.15}$$

or

$$k^2 (1 - k^2)^{-1} = g^2 \varepsilon^T \gamma^{-D} \tag{4.16}$$

At low frequencies, below the frequency of mechanical resonance of the material, k^2 is a measure of the amount of energy supplied in one form (i.e. mechanical or electrical) which is converted to the other form.

The constants d, g and k are dependent on the direction of the applied mechanical force or electrical field in the crystal. Fig. 4.24(a) shows the six possible directions where 4, 5 and 6 represents shear about the three axes. The constants are represented by two numbers such as g_{31} where the first subscript is the direction of the generated field, and the second is the direction of the applied mechanical stress. Fig. 4.24(b) shows an example where the electrical field and mechanical stress are about axis 3 and in Fig. 4.24(c) the mechanical stress is about the shear axis 4 while the electrical field is at axis 2. Table 4.4 gives constants for some piezoelectric materials.

Piezoelectric transducers are self generating and can only be used for dynamic measurements. Their most frequent application is in accelerometers, and for the measurement of

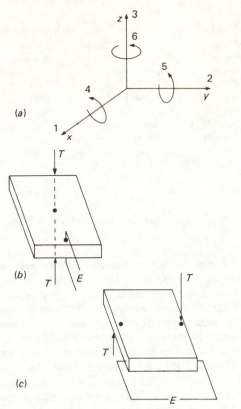

(a)

(b)

(c)

Fig. 4.24. Directional properties of piezoelectric constants; (*a*) axis rotations, (*b*) g_{33}, (*c*) g_{24}.

shock and vibration. Typical characteristics of the device are a sensitivity of $10-50\,\text{mV/g}$, frequency of $2\,\text{Hz}-100\,\text{kHz}$, range of

$1000-20\,000\,\text{g}$, and an impedance of $1\,\text{nF}-500\,\text{pF}$. The output from piezoelectric transducers has a low amplitude and high impedance, so that it needs signal conditioning before it can be used. Often this conditioning circuit is included within the body of the transducer.

4.8 Optical transducers

A large number of different optical transducers have been developed, using a variety of techniques, and this is the area of sensing in which most of the current research is taking place. Many of the transducers are based on optical fibres to receive and transmit light, and often the light is modulated rather than interrupted by the measurand. Optical transducers can use encoders, with coded discs between the emitter and detector, for position or speed sensing. They can use techniques such as Moiré fringe modulators to measure distance, or laser interferometry to measure length, position and displacement.

In this section only the basic types of optical detectors will be considered. These may be photoemissive, junctionless or bulk, or they can be based on junction devices.

4.8.1 *Photoemissive detectors*

In photoemissive devices, such as the photomultiplier tube, the anode current is proportional to the intensity of the incident light. The disadvantage of photoemissive sensors is that they are large and expensive, and need between $300\,\text{V}$ and $2500\,\text{V}$ for operation. Their advantage is

Table 4.4. *Constants for some piezoelectric materials*

Parameter	Quartz X-cut	Lithium sulphate Y-cut	Barium titanate	Lead zirconate-titanate	
				PZT–4	PZT–5
Maximum operating temperature (°C)	550	75	90	250	300
Density (kg/m³)	2.7×10^3	2.1×10^3	5.6×10^3	7.6×10^3	7.7×10^3
Dielectric constant	4.5	10.3	1700	1300	1700
Piezoelectric charge constant d_{33} (m/V)	2.3×10^{-12}	1.6×10^{-11}	1.5×10^{-10}	2.9×10^{-10}	3.7×10^{-10}
Piezoelectric voltage constant g_{33} (Vm/N)	0.058	0.18	0.014	0.026	0.025
Electromechanical coupling coefficient k_{33}	0.1	0.35	0.48	0.65	0.68
Young's modulus (N/m²)	8.0×10^{10}	1.2×10^{11}	8.2×10^{10}	6.8×10^{10}	2.9×10^{10}
Curie temperature (°C)	575	—	120	320	365

Table 4.5. *Comparison of some typical photodetectors*

Parameter	Photo-conductive	Photo-emissive	Photo-voltaic	Photo-diode	Photo-transistor	Photo-thyristor
Maximum temperature (°C)	75	80	150	125	125	100
Maximum voltage (V)	1000	2800	0.5	200–2000	100	200
Maximum current	1 A	10 mA	1 A	5 mA	50 mA	1.5 A
Power dissipation	20 W	0.01–1 W	400 mW	50 mW	400 mW	2W
Switching times	1–100 ms	0.1 μs	1–100 μs	1 ns–1 μs	2–100 μs	2 μs
Maximum frequency	1 kHz	10 MHz	50 kHz	10 MHz	100 kHz	1 kHz
Operating light level (mW/cm²)	0.001–70	10^{-9}–1.0	0.001–1000	0.001–200	0.001–20	2–200
Long term stability*	3	2	1	2	2	2
Peak spectral response (μm)	0.6–2.2	0.1–1.0	0.85	0.85	0.8	0.85
Size*	3	4	3	1	2	2

*1 = Best or smallest

that they have a high frequency response, and are very sensitive. The spectral response of photoemissive sensors can be varied from 100 nm to 1000 nm by changing the cathode material. Generally these types of transducers are not used in general purpose electronic systems, and they are not considered further, although their key parameters are summarised in Table 4.5.

4.8.2 *Junctionless detectors*

These are photoconductive devices, and are known as light dependent resistors (LDR). They operate on the principle that, when exposed to light hole-electron pairs are generated in the material, and electrons jump into the conduction band, reducing the bulk resistance.

The materials most commonly used for visible light are an aluminum oxide ceramic substrate, coated with a layer of cadmium sulphide (CdS) or cadmium selenide (CdSe). The devices are made by a variety of techniques, such as sintering, firing, or by vapour, chemical or vacuum deposition. They are made with a large sensitive area, and need a high ratio of dark to light resistance, which typically varies from 10 MΩ to 10 kΩ. The spectral response of cadmium sulphide is 0.6 μm and for cadmium selenide it is 0.72 μm. The two materials can be mixed to give a response of about 0.66 μm. For infrared sensing, lead sulphide (PbS) or lead selenide (PbSe) may be used, which have a peak spectral response of about 2.2 μm. The curve for lead selenide is relatively flat, giving a good response from 1.8 μm to 3.6 μm.

Photoconductive transducers are relatively cheap, can be used bidirectionally, and can withstand a high voltage and large dissipation. Their main disadvantage is that they are slow, the time constant for cadmium selenide being about 10 ms and for cadmium sulphide about 100 ms. Response time for lead selenide is about 10 μs and for lead sulphide about 500 μs. An LDR also exhibits hysteresis, which means that the conduction is a function of the cell's previous history of exposure to light intensity and duration. The hysteresis effect is most noticeable at low light levels. Due to their low cost and high sensitivity LDRs are used in applications such as street lighting, camera exposure control and low speed industrial counting.

4.8.3 *Junction devices*

Photodiodes.   Fig. 4.25 shows the characteristic of a single junction. With no illumination it behaves like a conventional diode, and as the illumination level increases it gives rise to a series of curves. If the photodiode is reverse biased by an external supply E, and connected to a load, it will operate in quadrant III and the load line is EA. Very little current flows through the reverse biased PN junction in the absence of light, and this is called the dark current. Incident light creates hole-electron pairs within the junction, and these are carried across the junction by the external bias. The performance of the photodiode measured in terms of electron emission for a quantum of light, called the quantum efficiency, is about 95%.

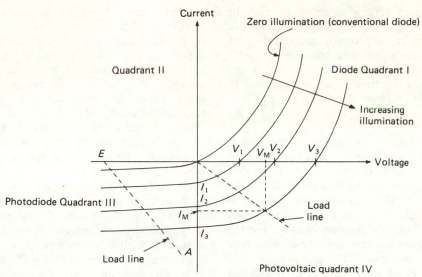

Fig. 4.25. PN junction photodetectors.

Noise generated in a photodiode is measured as its noise equivalent power (NEP), which is the amount of light needed to produce a signal equivalent to the noise level. Diode noise increases with junction area so detectivity, defined as NEP per active area, is often used as a measure of noise performance.

Photodiodes are made from silicon and can have several different structures. PIN diodes, which have an intrinsic layer between the P and N regions, have a lower junction capacitance, and can operate faster than conventional photodiodes. Avalanche photodiodes (APD), which work on the principle that the electrons generated by the incident light are rapidly multiplied due to the avalanching effect through the junction, are used for very high speeds. The disadvantage of the APD is that it is relatively noisy. For large area detectors Schottky barrier diodes are used, in which a thin layer of gold is evaporated onto the silicon to form the junction. The disadvantage of a Schottky device is that it cannot be used at high temperature or high light levels.

Phototransistors. The phototransistor can be regarded as a photodiode connected to an amplifying transistor. It is formed as part of the reverse biased collector-base junction of the transistor, and this junction is designed to have a large area to enhance its efficiency as a photodetector. The sensitivity of the phototransistor is 100–1000 times greater than that of a photodiode, but the dark current is also amplified so that the signal to noise ratio is no better. The phototransistor is also slower than a photodiode, partly due to the parasitic collector-base capacitances, and partly due to limitations in current gain–bandwidth product.

For higher sensitivities, in excess of about 10 000 times that of a photodiode, a photo Darlington may be used. In essence it consists of a phototransistor feeding a transistor, both devices being built onto the same silicon die. Field effect transistor photosensors have about ten times higher sensitivity than bipolar phototransistors, since the higher input impedance gives larger voltage swings from a lower gate photocurrent. The FET phototransistor uses the gate-to-channel junction as a photodiode. It has very low offset voltage, but is slower than a bipolar phototransistor.

Photothyristors. When a photothyristor, also called a light activated SCR or LASCR, is illuminated hole-electron pairs are generated at its blocking junction, and these can act as a trigger current, turning the device on. Usually the gate lead is brought out of the package, and a bias may be applied to it to vary the threshold light level for turn on.

The photothyristor is made from thin layers to enable greater light penetration, which unfortunately results in a lower blocking voltage. The device also has a larger junction, to increase light sensitivity, but this also makes it more sensitive to variations in temperature and voltage, and gives it a longer turn off time than a conventional thyristor. Resistors connected between gate and cathode of the thyristor reduce its susceptibility to noise and dV/dt effects, but also reduce its light sensitivity.

Photovoltaic cells.   If no external supply is connected to a photodiode then the incident light will create a potential across the junction, so that the P layer acquires a positive potential relative to the N layer. This will cause a current to be driven through an external load resistor, such that the voltage is positive but the current through the diode is reversed. The device is now called a photovoltaic diode, or solar cell, and operates in quadrant IV of Fig. 4.25. Voltages V_1, V_2, V_3 are the open circuit voltages of the photovoltaic diode and I_1, I_2, I_3 are its short circuit currents.

The maximum value of the open circuit voltage of a photovoltaic diode is close to the band gap energy of the material. If V_3 is the open circuit voltage and I_3 the short circuit current at a given light level, then the theoretical maximum power output of the cell is $V_3 I_3$. In practice, due to the curvature of the characteristic, this is equal to $V_M I_M$. The fill factor of the cell is defined as the ratio $V_M I_M / V_3 I_3$.

If I_p is the intensity of the incident light, measured in milliwatts per square centimetre, then the efficiency of the cell is given by $V_M I_M / I_p$. Photovoltaic cells have efficiencies in the region of 5–25%.

Several materials may be used to make photovoltaic cells. Silicon cells have a very long operating life provided they are carefully encapsulated. The cells can degrade in performance by between 2% and 10% in bright light and under high temperature. The band gap of silicon is 1.12 eV, and this gives a cell with a maximum open circuit voltage of 0.6 V. The efficiency of a silicon cell varies between 12% and 18%, with a theoretical maximum of 22%. There are several variations to the basic silicon cell, and these are compared in Table 4.6.

Gallium arsenide absorbs light near its surface, whereas in silicon this absorption occurs much further down in the material. Therefore in silicon the photogenerated carriers need to travel longer distances without recombination, and the material needs to be purer than gallium arsenide to avoid defects at which recombination can occur.

The band gap of gallium arsenide is 1.35 eV, which is close to the optimum for maximum conversion efficiency in terrestrial sunlight. The cell is also very good for use in concentrating systems. This would normally raise the cell temperature and reduce its photovoltaic efficiency, but the photothermal efficiency would increase, giving a cell with a high overall efficiency, up to about 25% for 250–300 °C and a concentration factor of 1000 Suns. The disadvantages of gallium arsenide are that the material is expensive, its supply is limited, and the method currently used to make it from the liquid phase epitaxy is expensive.

The cadmium sulphide photovoltaic cell consists of a base layer of N type cadmium sulphide

Table 4.6. *Comparison of some photovoltaic cells*

Material	Structure	Open circuit voltage (V)	Short circuit current density (mA/cm²)	Typical efficiency (%)
Silicon	non-reflecting	0.6	40	18
	heterojunction	0.5	25	10
	Schottky barrier	0.6	25	10
Cadmium sulphide	conventional	0.5	25	6
Gallium arsenide	heterojunction	0.9	40	25
	Schottky barrier	0.9	20	15

(CdS) and a thin surface layer of P type copper sulphide (Cu_2S). Both materials are strongly absorbent to light, and so a much thinner cell than that using silicon can be made.

The chief merit of the cadmium sulphide cell is that it can be manufactured relatively cheaply. The cell has a theoretical maximum efficiency of 18–25% but in practice the efficiency is close to 5%. The cell is also unstable, and high temperature and illumination can cause a degradation of the material, and a fall in its power output. The cause of degradation is the slow conversion of Cu_2S to CuS and to Cu. Even under normal conditions the cell loses about 10% of its power output after one year of normal exposure, due to the effects of water vapour and oxygen.

4.9 Temperature measurement transducers

There are three principal types of temperature measurement devices; resistance temperature detectors, thermocouples and thermistors. Their properties are summarised in Table 4.3. In addition three other techniques for temperature measurement are introduced in this section, semiconductor junction, quartz crystals and radiation pyrometers.

Temperature measurement transducers may be surface mounted or immersed in the fluid being measured. Often they are protected from the environment by a case. It is important that the measurement device does not affect the temperature of the surface, such as by conducting heat away from it.

4.9.1 *Resistance temperature detectors (RTD)*

The Callendar equation for temperature measurement in a conductor was extended in 1924 by M.S. Van Dusen, as in (4.17)

$$t = \frac{1}{\alpha}\left(\frac{R_t}{R_0} - 1\right) + \delta\left(\frac{t}{100}\right)\left(\frac{t}{100} - 1\right)$$
$$+ \beta\left(\frac{t}{100} - 1\right)\left(\frac{t}{100}\right)^3 \qquad (4.17)$$

R_t is the resistance at $t\,°C$, R_0 the resistance at $0\,°C$, and α, δ and β are constants found from resistance measurements taken at the steam point of water ($100\,°C$), the freezing point of zinc ($419.58\,°C$) and the boiling point of oxygen ($-182.96\,°C$) respectively.

The RTD is usually made as a wire wound construction, although metal film foil has been used where faster response is needed. The material should have a high temperature coefficient of resistance; high resistivity, so that large values of resistance can be obtained in a small physical size; stable characteristics, showing little variation with repeated heating or cooling, and with time; a good response; and resistance to shock and vibration.

High purity annealed platinum wire is the most frequently used material for RTD applications. It shows a linear variation of resistance over a wide temperature range. Platinum is used for the international thermometer scale from the liquid oxygen point ($-182.96\,°C$) to the antimony point ($+630.74\,°C$). It is likely that this will be extended to the melting point of gold ($+1063\,°C$).

The temperature for a platinum RTD can be found from the Callendar equation (4.17). Errors from this are less than $\pm 0.1\,°C$ at temperatures below $500\,°C$, but can reach $\pm 1\,°C$ at about $850\,°C$. Platinum resistance also has long term stability, high accuracy, and repeatability. Its calibration drift is less than $0.1\,°C$ per year.

Nickel and nickel alloys are also used in RTDs. Its temperature coefficient is about twice that of platinum, so it is more sensitive. The temperature coefficient of nickel is about $0.0062\,\Omega/°C$. Copper has a temperature coefficient of resistance which is slightly higher than platinum. It can operate over a temperature range of -200 to $+150\,°C$ since copper oxidises at a higher temperature. Because of its low resistivity fine wires must be used to limit the detector's size, and improve its response time. Copper has a linear temperature coefficient of resistance over the range -50 to $+150\,°C$, and is sometimes used in preference to platinum for measurements near ambient temperature. Tungsten has occasionally been used in RTDs; it has a good temperature coefficient of resistance, but is difficult to manufacture, and is too brittle to be reliable.

RTDs have the advantages of being reproducible, stable, accurate, and having a linear output. Their disadvantages are the relatively large size and slow response time, and higher cost. An RTD has a resistance of typically 100 ohms at $0\,°C$ and changes by about $0.3\,\Omega/°C$. The leads connecting the detector to the measuring instrument can have a resistance of several tenths of an ohm, so that four terminal measurement techniques must be used to minimise lead errors.

4.9.2 *Thermocouples*

When two dissimilar metals are joined at one end (sensing junction), and the other end is kept at a fixed known temperature (reference junction), a current will flow in the circuit. The magnitude of this current is proportional to the characteristics of the materials, and the difference between the two end temperatures. The effect which causes current flow is called the thermoelectric or Seebeck effect, and the voltage generated in the circuit can be measured to give an indication of temperature. Usually the reference junction is at 0 °C, or electrical cold junction compensation is used to simulate a reference temperature. Multiple junctions of the same material can be connected in series, and kept close to each other, to give an output which is a sum of the individual outputs. This is known as a thermopile.

Several combinations of materials may be used to make thermocouples. Standards such as ANSI Standard C96.1 and BS4937:1973 give better designations for most of these, and define their output voltage. Fig. 4.26 shows the curves for some of these materials. Platinum–rhodium (Type R) has a maximum operating temperature of 1800 °C, and an output of about 10 μV/°C.

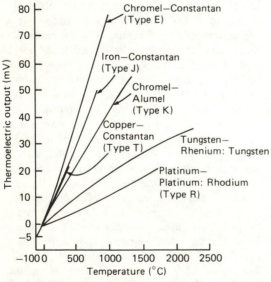

Fig. 4.26. Thermocouple output curves for a 0 °C reference temperature.

It shows good stability and corrosion resistance. Tungsten–rhenium has a maximum temperature of 2800 °C and an output of about 20 μV/°C. It must operate in a vacuum or inert atmosphere. Iron–constantan (Type J) has a maximum operating temperature of 800 °C, and an output of 65 μV/°C, and it can be used in an oxidising atmosphere. Copper–constantan (Type T) has a maximum temperature of 350 °C, since copper oxidises above 350 °C, and an output of 60 μV/°C. Chromel–constantan (Type E) has the highest output of 80 μV/°C, and a maximum temperature of 700 °C.

Thermocouples are fragile and easily corroded. They are usually protected in a metal or ceramic tube, which is chemically inert and vacuum tight. The leads connecting the thermocouple to the measuring instrument are called compensating wires, and these should be of the same material as the thermocouple for maximum accuracy. Thermocouples have the widest temperature range of the measurement transducers, and their small mass gives them a good response and sensitivity.

4.9.3 *Thermistors*

Although both positive and negative temperature coefficient thermistors are available, the latter are used almost exclusively in temperature measurement applications. The most common materials used for negative temperature coefficient (NTC) thermistors are the oxides of the iron group of elements, such as chromium, manganese, iron, cobalt and nickel. These materials normally exhibit a high resistivity, but are converted into semiconductors by the addition of small amounts of ions having a different valency, for example a few iron ions in iron oxide being replaced by titanium.

NTC thermistors are available in a variety of shapes and sizes, and have resistance values, which are usually quoted at 25 °C, of between 0.1 Ω and 100 MΩ. They can be designed to operate in the range of -100 to $+600$ °C, and a single device can operate over an approximate 200 °C range, changing its resistance value by a factor of 1000. Devices are available commercially as very small thermistor beads, below 0.2 mm in diameter, to rods or discs of over 25 mm diameter. The beads have a very fast thermal response, and are made by sintering a drop of thermistor paste onto two platinum alloy wires. The beads are then coated with glass to give a hermetic seal, or are mounted in an evacuated or gas filled tube.

The temperature–resistance characteristic of a thermistor is given by (4.18)

$$R_T = R_\infty \exp(\beta T^{-1}) \qquad (4.18)$$

R_T is the resistance at temperature T kelvin, R_∞ is the resistance at an infinitely high temperature, and β is a constant of the device. Equation (4.18) can be re-written as in (4.19)

$$R_2 = R_1 \exp[\beta(T_2^{-1} - T_1^{-1})] \qquad (4.19)$$

R_1 and R_2 are the resistances at T_1 and T_2 kelvin respectively. Constant β can vary up to about 10^4, depending on the thermistor material, and it varies with temperature so that the thermistor equations are only useful over a limited temperature range.

Thermistors have a temperature coefficient of resistance of between 3% and 6% per °C, compared to about 0.4% per °C for platinum, so they are sensitive detectors. Thermistors also have the advantage of small size, low cost, and a unit to unit variation of less than ± 0.2 °C, over the range 0–70 °C. The disadvantage of thermistors is that they have a non-linear temperature–resistance characteristic, and poor stability, which can be as much as ± 0.5 °C per year at 300 °C.

4.9.4 *Other temperature sensors*

Three other temperature transducers are briefly introduced in this section: the quartz sensor, the semiconductor sensor and the radiation pyrometer.

Quartz sensors measure the change in the resonant frequency of a quartz crystal. The device is used in an oscillator, so that a relatively large amount of electronics and readout is required, making the quartz temperature sensor

mainly a laboratory instrument. The operating range is about -80 to $+250$ °C and the device is linear from -50 °C to $+250$ °C, with an accuracy of about ± 0.04 °C.

Semiconductor temperature sensors usually detect the change in the base-emitter voltage of a transistor with temperature, which gives a temperature sensitive voltage of about 2.1 mV/°C. The base emitter voltage V_{be} varies according to (4.20).

$$V_{be} = \frac{kT}{q} \ln(I_c I_s^{-1}) \qquad (4.20)$$

k is the Boltzmann constant, T is the temperature in kelvin, q is the electron charge and I_c and I_s are the collector current and reverse saturation current of the transistor respectively.

Two transistors can be connected to give a differential output, as in (4.21).

$$\Delta V_{be} = \frac{kT}{q} \ln(I_1 I_2^{-1}) \qquad (4.21)$$

I_1 and I_2 are the currents in the two transistors. Usually I_1 is made about twice I_2 so that all the terms in (4.21) are constant, and ΔV_{be} varies linearly with temperature. Semiconductor sensors have an accuracy of about ± 2 °C over the range -40 to $+150$ °C.

The radiation pyrometer is a non-contacting instrument for reading temperature. It responds to radiated heat transfer, primarily in the infrared portion of the spectrum, with wavelengths from 0.75 μm to 1000 μm. An optical lens and mirror system, which is sensitive in the infrared region, is used to focus the radiation onto temperature sensors, usually thermoelectric or resistive. The output is calibrated to indicate the temperature of the body being measured. Radiation pyrometers are generally used for temperatures up to about 3500 °C, but may also be used for other non-contacting measurements down to about -50 °C.

5. Instrument buses

5.1 Introduction

Buses are used to communicate data at several levels within a system. Generally there are seven levels which may be identified, as follows. (i) Interconnection between components on a printed circuit board, the bus extending for about 10 cm. (ii) Connection between one board and another, over a span of about 0.1 m. (iii) Connections within an instrument cabinet, over a range of about 1 m. (iv) Connection between one or more separate instruments, covering a range of about 10 m. (v) Interconnection between a controller and its peripheral, which usually extends over 100 m. (vi) Local area networks (LANs) which interconnect whole systems together over a distance of about 5000 m. (vii) Long distance global or satellite networks.

Bus standards are designed to enable components and equipment supplied by different manufacturers to interface. These standards have largely developed as a result of co-operation between different vendors, and this has been recognised by standards organizations, such as the IEEE, who have accepted standards for competing buses. Several manufacturers now market integrated circuits which allow an interface between two different types of buses.

For a standard bus the electrical, mechanical and protocol requirements must be accurately defined. Electrical parameters must include: power supply requirements; voltage levels; method of supply, e.g. single or multi rail; the maximum speed of data transmission; the width of the address and bus lines; the line terminating impedance. The mechanical parameters must specify: the printed wiring board (if any) shape and size; the maximum length of the data bus; the number of instruments which can be connected, and the split between talkers, listeners and controllers; the connector type and size, along with its pin assignments. Protocol determines the procedure for communication over the bus, and includes: a definition of the timing and input–output philosophy; whether the bus operates in synchronous, asynchronous, or quasi-synchronous mode; whether the bus is serial or parallel; how errors are detected and handled.

Table 5.1. compares the basic parameters of a few commonly used buses. There are primarily two types of buses, serial and parallel, and these are described in further detail in subsequent sections. Buses may also operate in synchronous or asynchronous mode. Synchronous mode has a higher throughput since during a controller read operation, for example, the controller provides the peripheral with the address, and then expects to have data available within a given time. In asynchronous operation the controller provides the address, and then waits around until the peripheral provides the data in its own time. Therefore an asynchronous operating mode may be slow, although it is more flexible since it adapts to the speed of the slowest device connected to the bus.

5.2 Parallel buses

A parallel bus uses separate wires to transfer all the bits of information, so that there is one line for each data bit, address bit and control signal. Many buses multiplex the data onto a part of the address lines, and these are indicated in Table 5.1. under the 'Type' column. The data lines transfer data in or out of the system; the address lines indicate the active part of memory or input–output location; and the control lines contain the control signals, such as read–write, input–output, and interrupt.

Although several of the buses shown in Table 5.1 will be described in this section, only two are covered in relative detail, the STD bus,

Table 5.1. *Comparison of some commonly used data buses*

Bus	Level	Type	Address bits	Data bits	Maximum speed	Range (metres)	Connector type	Board size (mm × mm)	Principal vendors supporting bus
I²C (inter-IC bus)	intra-component	asynchronous serial	7	—	100k bits/s	5	—	—	Philips
D²B (digital data bus)	intra-component	asynchronous serial	12	—	100k bits/s	10	21 way (scart)	—	Philips
S-100 (IEEE 696)	intra-board intra-cabinet	synchronous non-multiplexed parallel	16–24	8 or 16	1 Mbit/s	10	100 way (edge)	—	—
STD (IEEE 961)	intra-board intra-cabinent	synchronous non-multiplexed parallel	16	8	1 Mbit/s	10	56 way (edge)	114.3 × 165.1 or Eurocard	Mostek Pro-Log
Q-bus	intra-board intra-cabinet	asynchronous multiplexed parallel	16–18	8 or 16	1 M bit/s	10	edge	214 × 263.4 214 × 147.3	DEC
Multibus II	intra-board intra-cabinet	asynchronous multiplexed parallel	24–32	8,16 or 32	100M bits/s	10	96 way (DIN 41612)	Eurocard	Intel
VME	intra-board intra-cabinet	asynchronous non-multiplexed parallel	24–32	8,16 or 32	100M bits/s	10	96 way (DIN 41612)	Eurocard	Mostek Motorola Philips
Eurobus	intra-board intra-cabinet	asynchronous multiplexed parallel	18	8,16,24 or 32	100M bits/s	10	96 way (DIN 41612)	Eurocard	MOD Ferranti
Futurebus (IEEE 896)	intra-board intra-cabinet	asynchronous multiplexed parallel	32	8,16 or 32	300M bits/s	10	96 way (DIN 41612)	Eurocard	—
IEEE 488	intra-instrument	asynchronous multiplexed parallel	32	8	10k bits/s	20	24 way (piggy-back)	—	—
RS 232	controller to terminal	asynchronous serial	—	—	10k bits/s	100	25 way (PIN)	—	—
RS 422/ RS 423	controller to terminal	asynchronous serial	—	—	1 Mbit/s	1000	37 way (PIN)	—	—
MIL STD 1553B	controller to terminal	asynchronous serial	—	—	1 M bit/s	100	—	—	—

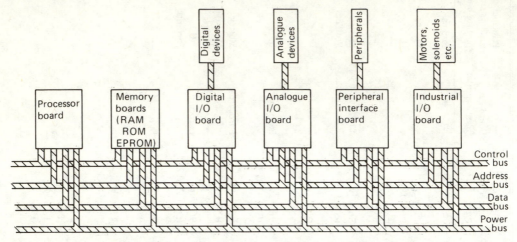

Fig. 5.1. STD bus implementation using typical printed wiring boards.

Table 5.2. *STD bus pin designations*

	Component side		Circuit side	
	Pin	Description	Pin	Description
Logic	1	logic power	2	logic power
power	3	logic ground	4	logic ground
bus	5	logic bias 1	6	logic bias 2
Data	7	data bus	8	data bus
bus	9	data bus	10	data bus
	11	data bus	12	data bus
	13	data bus	14	data bus
Address	15	address bus	16	address bus
bus	17	address bus	18	address bus
	19	address bus	20	address bus
	21	address bus	22	address bus
	23	address bus	24	address bus
	25	address bus	26	address bus
	27	address bus	28	address bus
	29	address bus	30	address bus
Control	31	write to memory or I/O	32	read memory or I/O
bus	33	I/O address select	34	memory address select
	35	I/O expansion	36	memory expansion
	37	refresh timing	38	CPU machine timing sync
	39	CPU status	40	CPU status
	41	bus acknowledge	42	bus request
	43	interrupt acknowledge	44	interrupt request
	45	wait request	46	non-maskable interrupt
	47	system reset	48	push-button reset
	49	clock from processor	50	aux. timing
	51	priority chain out	52	priority chain in
Auxiliary	53	aux. ground	54	aux. ground
power	55	aux. positive ($+12$ V D.C.)	56	aux. negative (-12 V D.C.)
bus				

which is popularly used for intra-board communication, and the IEEE 488 bus, which is used for intra-instrument communication.

5.2.1 *S-100 Bus*

This was developed in the 1970s and the draft specification for it was published in 1979. The bus defines 100 lines, not all of which are required in any one system. Of these, 16 are address lines, 8 data input lines, 8 data output lines, 8 interrupt lines, 39 control lines, 2 clock lines, 3 supply lines, and 16 lines are undefined. Up to 22 boards can be attached to the bus, which uses 8 or 16 bit protocol and can support up to 16 M bytes of memory. More than 100 different manufacturers make products for the S-100 bus.

5.2.2 *STD Bus*

The bus was launched jointly by Pro-Log and Mostek in 1978. It is designed for small 8 bit systems, and is not expandable to 16 bits. The system is built up of standard cards. Many different types are available, from over 70 vendors, a few of the boards being shown interconnected in Fig. 5.1.

The line assignment for the STD bus is shown in Table 5.2. These can be grouped into five sections, as shown. As many as five separate power supplies may be accommodated, with two separate ground returns. The auxiliary power is needed by some of the boards, such as relay drivers. The data bus consists of 8 bits, bidirectional and tri-state. The direction of data is controlled by the read–write and interrupt acknowledge lines from the processor. The 16 address lines are used for decoding by memory or input–output.

The control bus determines the flexibility of the system. A few of the commonly used lines from Table 5.2 are described here. The write to memory or I/O line is used, by the processor, to clock data from its bus into selected memory locations or output latches. The read from memory or I/O line is taken low by memory or input–output, to tell the processor that data is available on the bus for reading. Interrupt request is input to the processor from the peripheral, and the interrupt acknowledge tells the interrupting device that the processor is ready to respond to the interrupt. System reset is an ouput from the system reset circuit, which occurs on

switch on, or on any other user initiated reset. This line should be used to reset all other cards needing initialization. The clock from the processor is used in system synchronization.

5.2.3 *VME Bus*

This consists of five different bus sections: data transfer, priority interrupt, arbitration, utility, and inter-intelligence. The data transfer bus is 16 bits wide, which is expandable to 32 bits. The address bus is 24 bits, expandable to 32 bits. The address and data paths are not multiplexed, and 3 control lines determine the width of the data transfer bus.

5.2.4 *Eurobus*

This bus was designed and developed by the Ministry of Defence (UK) in conjunction with Ferranti Ltd. The aim was to have a bus which was independent of a specific product range and manufacturer. The system can operate purely as a databus, and does not need a processor to form a bus controller. The operation is asynchronous, and full handshaking exists between devices, to allow a variety of devices with different response speeds to be connected together. The address and data signals are multiplexed on to a single highway. The facility exists to link two or more Eurobuses, so a device on one bus can address a device on another.

5.2.5 *Futurebus (IEEE 896)*

This bus is intended to supersede IEEE 696 and IEEE 796. It is vendor independent and can cater for multiprocessor systems. Buses can be operated in parallel to extend the 8 bit architecture in multiples of 8 bits.

5.2.6 *CAMAC (IEEE 583)*

The computer automated measurement and control (CAMAC) bus standard defines modules and racks, and bus requirements. It has been designed for use in the nuclear industry, to transfer large volumes of data with low error rates. The equipment required is expensive, and the bus is not widely used. The CAMAC rack system can accommodate up to 24 modules. The bus has 24 address lines, 24 data input and 24 data output lines, 24 status lines, 4 sub-address lines, and 10 control, timing and command lines. It can operate at speeds up to 25 M bits per second.

5.2.7 *IEEE 488 (IEC 625)*

This was introduced in 1978 by Hewlett Packard Corp. as the HPIB, and has rapidly developed into a popular standard for interfacing instruments. The bus allows up to 15 instruments to be connected together, and these usually consist of a controller and 14 other instruments. The instruments can be listeners or talkers. Each device terminates the bus with a $3\,k\Omega$ pull-up, and a $6.2\,k\Omega$ pull-down resistor. Correct bus loading is important, and one device load should be connected every 2 m, although the overall length must not exceed 20 m. Device loads must not be separated by more than 4 m. Bus expanders can be used where more than 15 devices need to be interconnected.

IEEE 488 instruments usually have a connector for interfacing to the bus, and switches which are used to set its address. Some instruments have secondary addresses as well, which give access to different parts of the instrument.

A total of 16 active lines are used to implement the IEEE 488 bus. These are classed into three groups, as shown in Table 5.3. The transfer

Table 5.3. *IEEE 488 Bus pin designations*

	Pin	Description
Data	1	data bus
bus	2	data bus
	3	data bus
	4	data bus
	13	data bus
	14	data bus
	15	data bus
	16	data bus
Control	5	end or identify (EOI)
bus	9	interface clear (IFC)
	10	service request (SRQ)
	11	attention (ATN)
	12	shield
	17	remote enable (REN)
Transfer	6	data valid (DAV)
bus	7	not ready for data (NRFD)
	8	not data accepted (NDAC)
Grounds	18	ground
	19	ground
	20	ground
	21	ground
	22	ground
	23	ground
	24	logic ground

control bus consists of three lines, which control the transfer of data on the data bus by ensuring a fixed sequence of events. It checks that handshaking occurs, and a new data byte cannot be transmitted until the last byte has been accepted. The DAV line is set by the talker, after it has placed data on the lines. The listener sets NDAC during the process of reading this data, or NRFD if it is unable to accept data at this time.

The control bus consists of five lines. When the ATN line is activated by a controller it indicates to all the instruments, on the bus, that they must give up use of the bus and interpret the data bus as commands. The IFC line is used to initialise instruments on the bus. The SRQ line is used by an instrument to signal to the controller that it requires attention. The REN line is used by the controller to set an instrument in a programming accept mode. The EOI is used to indicate the end of a multibyte message by a talker, or with ATN this executes a polling sequence. Alternatively a line feed character can be sent to indicate an end of a message.

IEEE 488 does not define the syntax or code of messages on the bus, so instruments may use different message formats, and this needs to be accounted for in the system software.

5.3 Serial buses

Serial buses are primarily used to link controllers to their terminals, and this mode is almost exclusively used in local networks, described in Section 5.4.

5.3.1 *RS 232 Bus*

This was introduced in 1962, to connect computers to their peripherals. Each link requires a separate channel, or input–output port, since the links are point to point. If few ports are used then the serial communication system is economical, but as the number of ports increases it becomes cheaper to interface all devices to a single IEEE 488 bus.

The RS 232 bus uses a 25 way connector, although not all the pins need be used in any one system. Table 5.4 gives the designation of the most commonly used pins. Data communication to and from the terminal occurs over two lines TDATA and RDATA. The RTS line is activated by the terminal to warn the controller to prepare to receive data, and this is acknowledged by the CTS line from the controller to the terminal, to

Table 5.4. *Commonly used pins on an RS 232 bus*

Pin	Description
1	protective ground to chassis
2	data line (TDATA)
3	data line return (RDATA)
4	request to send (RTS)
5	clear to send (CTS)
6	data set ready (DSR)
7	signal ground
20	data terminal ready (DTR)

say that it is ready to receive data. DSR is used by the controller to tell the terminal that the controller is operational, and DTR is activated by the terminal to tell the controller that the terminal is operational.

The RS 232 bus operates at voltages of $+3\,V$ and $-3\,V$, which are not suitable for the long term viability of this standard. The bus also needs careful line termination for long distances.

5.3.2 *RS 422/RS 423 Buses*

The differences in bus connection between the RS 232, RS422 and RS 423 are shown in Fig. 5.2. The RS 232 bus sends data as an unbalanced signal, and uses the common ground connection between devices. RS 422 has a pair of wires to

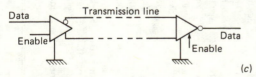

Fig. 5.2. Serial bus configurations; (*a*) RS 232, (*b*) RS 422, (*c*) RS 423.

send an unbalanced signal, and does not have a common ground. RS 423 is the most refined, using a balanced transmission signal and a common ground connection. RS 422 and RS 423 have better noise immunity, and can transmit data faster and over longer distances, that RS 232, but they use a more expensive driver chip than RS 232.

RS 422 can use a twisted cable pair for slow speed application, or it can operate from conventional coaxial cable, so that it is ideally suited for use in factories for control and instrumentation. The bus also uses a $0\,V$ and $5\,V$ level for logic 0 and logic 1 states.

5.3.3 *MIL STD 1553B Bus*

This is a high integrity, noise tolerant bus which was developed primarily for avionics systems. It has been adopted by the USA and British armed services, and by the other NATO countries. The problem of weight is overcome for avionics by use of a limited number of connections, and by time division multiplexing information from several channels onto one line.

A single two wire twisted, shielded bus is used, and data is transmitted at 1 MHz by utilising a self-clocked Manchester bi-phase format. Up to 32 data words of 16 bits can be sent in each 'packet', the data, command and status words being based on 20-bit wide blocks. Up to 31 remote terminals can share a single bus, and each terminal can support up to 30 sub-system elements.

5.4 **Local area networks (LANs)**

Local Area Networks are used to connect a variety of different terminals together over a given area. Many such systems exist at present, all of them proprietary products supported by one or more commercial concerns. Table 5.5 compares the characteristics of a few of the more popular systems.

Although LANs are generally thought of as being classified as either baseband or broadband, a much better method of differentiating between them is on the basis of the LAN topology, the LAN access scheme, and the cable media.

5.4.1 *LAN topology*

The topologies most commonly used to connect LAN stations together are bus, tree, ring, star and point-to-point. These are illustrated in

Table 5.5. *Comparison of a few commercial LANs*

Network	Company	Topology	Cable medium	Access scheme	Maximum data rate	Number of channels	Maximum number of devices on channel	Maximum distance	Electrical interface characteristics
Ethernet	Xerox Corp	bus	baseband coaxial	CSMA/CD	10 M bits/s	1	1024	2.5 km	Ethernet transceiver, four twisted pair
ARC	Datapoint Corp.	bus	baseband coaxial	token passing	2.5 M bits/s	1	255	1 km	—
Net/one	Ungermann–Bass Inc.	bus	baseband coaxial	CSMA/CD	10 M bits/s	1	250	1.5 km	RS 232 IEEE 488
Primenet	Prime Computer Inc.	ring	baseband coaxial	token passing	10 M bits/s	1	247	4 km	Prime 50 series
Z-net	Zilog Inc.	bus	baseband coaxial	CSMA/CD	0.8 M bits/s	1	255	2 km	—
Xodiac	Data General	bus	baseband coaxial	token passing	2 M bits/s	1	64	2 km	—
Sdsnet	Scientific Data Systems	tree	baseband coaxial	CSMA/CD	1 M bits/s	1	250	1 km	—
Domain	Apollo Computer Corp.	ring	baseband coaxial	token passing	12 M bits/s	1	>50 000	1 km	—
Omninet	Corvus Systems Inc.	bus	shielded twisted pair	CSMA	1 M bits/s	1	64	1 km	—
Vnet	Vector Graphic Inc.	bus	baseband coaxial	CSMA	5 M bits/s	1	127	1 km	—
Wangnet	Wang Laboratories	tree	broadband coaxial	CSMA/CD and frequency multiplexing	12 M bits/s	1	>50 000	4 km	tri-axial cable and RS 232. RS 449
Local net 20 Local net 40	Sytek Inc.	inverted rooted tree	broadband coaxial	CSMA/CD	128 k bits/s 2 M bits/s	120 5	200	50 km	RS 232
Cablenet	Amdax Corp.	bus	broadband coaxial	TDMA reservation	14 M bits/s	2	>50 000	100 km	RS 232 RS 449

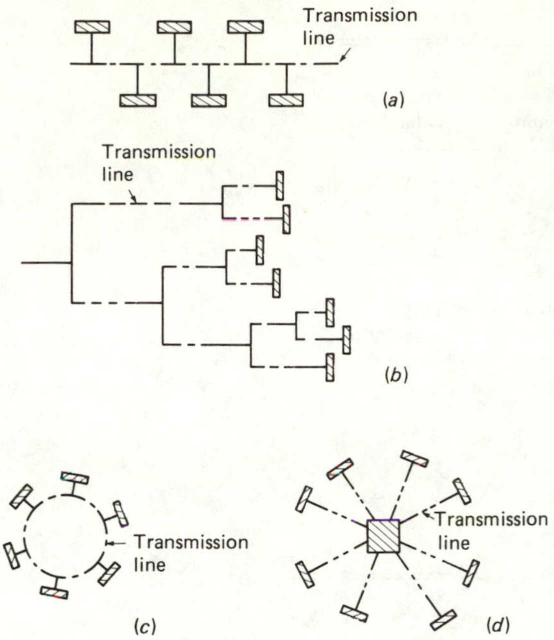

Fig. 5.3. LAN topologies; (a) bus, (b) tree, (c) ring, (d) star.

Fig. 5.3 and compared in Table 5.6. The star and point-to-point arrangements need a central controller, and are not preferred since a failure of the controller could result in a breakdown of the whole system.

A ring network can cover greater distances than a bus, because node repeaters regenerate the message. It is however more difficult to add or subtract a node from a ring.

5.4.2 *LAN access scheme*

Three main methods are used when a station wishes to gain access to a bus to transmit data. These are CSMA/CD, Token Passing and TDMA.

In CSMA/CD (Carrier Sense Multiple Access with Collision Detection) when a bus is busy all the nodes have to wait, and the nodes wishing to transmit listen to the bus to see when it clears. As soon as this occurs another station begins its transmission. If a collision occurs, due to two stations starting transmission at exactly the same time, then both stations stop transmission and wait for different (random) times before trying again. CSMA/CD is a relatively simple access system which gives good bus utilisation, unless there is heavy bus traffic when collisions occur more frequently. It is used on a linear bus, and is not suitable for real time use since it is not easy to assign priorities to individual stations connected on the bus.

In the Token Passing access scheme a software token is passed around to the individual stations, and only that station which holds the token at any time can transmit. Since the token passes in a predetermined direction, at any one time each node knows who has the token. Each station holds the token for a limited time, so that channel hogging is prevented, although priorities can be established if required. Token Passing is usually used with a ring topology.

The TDMA (Time Division Multiplexed Access) system is the easiest to implement since each node is given a fixed time slot in which it may transmit.

5.4.3 *LAN medium*

The transmission medium used for LANs establishes its cost and performance, and a few of these are compared in Table 5.7. The twisted pair cable is cheap, but is limited in speed and has poor noise immunity.

A coaxial cable can be used with baseband or broadband transmissions. A baseband cable is cheap, but can only transmit a single channel of data or digital voice. Broadband modulates the

Table 5.6. *Comparison of local area network topologies*

Topology	Connection cost	Control complexity	Expandability	Flexibility
Bus	1	1	4	4
Tree	1	1	4	5
Ring	1	3	3	3
Star	2	2	2	2
Point-to-point	3	4	1	1

1 = lowest

Table 5.7. *Comparison of local area network media*

| Parameter | Twisted pair | Coaxial cable | | Fibre-optic cable |
		Broadband	Baseband	
Cost	1	3	2	4
Component availability	4	3	2	1
Interconnect complexity	1	3	2	4
Number of nodes	tens	hundreds per channel	hundreds	two (point-to-point)
Maximum bandwidth	1 MHz	400 MHz	50 MHz	> 1000 MHz
Signal-to-noise ratio	1	2	2	3
Maximum transmission distance	100 m	300 km	3 km	> 500 km

1 = lowest

signal on an R.F. carrier, using modems, so one can get higher data rates than baseband, and multi channels using frequency division multiplexing. Standard CATV cable is usually used, with transmission speeds up to 150 M bits/s, or bandwidth up to 350 MHz, and the channels can carry data, voice or video signals.

Fibre-optic cable has a very large bandwidth, high speed, and is insensitive to electromagnetic interference. It is however costly, and the individual system components are not readily available.

Part 2: General purpose instruments

6. Voltmeters, ammeters and ohmmeters

6.1 Introduction

The measurement of voltage, current and resistance is based on the fundamental equation of voltage equal to the product of amperes and ohms. The basic instrument measures one of these parameters and derives the other two. The simplest type of meter is the moving coil instrument, which can be used for d.c. or a.c. measurements. The output is analogue, and it can be driven directly or through electronic circuits, which modify or amplify the input signal.

As the cost of electronics, principally that of memories and microprocessors, has decreased, fully electronic instruments with digital readouts have gained in popularity. They show many advantages, such as auto-calibration, auto-error correction, and auto-measurement like the automatic selection of the instrument's range. Since the measured quantity is usually analogue, these instruments incorporate analogue to digital converters, to change the signal to a digital form.

This chapter also describes comparative methods of measurement, which overcome inaccuracies of absolute measurement chiefly arising from the scale reading. These methods are mainly based on potentiometric instruments.

6.2 D.C. moving coil instruments

6.2.1 *Basic movement*

6.2.1.1 Principle of operation

The fundamental permanent magnet moving coil movement, also known as the d'Arsonval movement, consists of a coil suspended in a magnetic field, as in Fig. 6.1(*a*). When current is passed through the coil a magnetic force is exerted on it, which causes the coil to rotate against the return torque set up by the control spring. The torque T exerted on the coil in the magnetic field is given by (6.1)

$$T = BANI \qquad (6.1)$$

T is the torque in newton metres; B is the flux density in the air gap in weber/metre^2; A is the coil area in metre^2; I is the current in the coil in amperes; and N is the number of turns on the coil.

The deflection (θ) of the coil results in a restoring torque T_1 in the control springs equal to $K\theta$ and at equilibrium T_1 is equal to T so that the deflection of the coil is given by (6.2)

$$\theta = \frac{BANI}{K} \qquad (6.2)$$

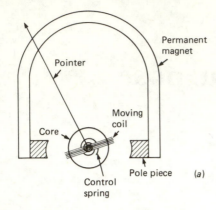

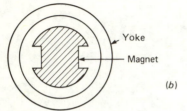

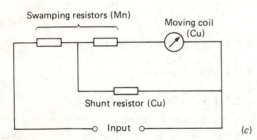

Fig. 6.1. Construction of a permanent magnet moving coil meter; (*a*) basic movement, (*b*) inverted core-magnet construction, (*c*) temperature compensation arrangement.

K is called the constant of the meter. Since in any instrument B, A, N, and K are constant the deflection of the coil is proportional to the current through it. In the design of an instrument A can usually be varied from $0.5\,\text{cm}^2$ to $3\,\text{cm}^2$, and B from $0.1\,\text{Wb/m}^2$ to $1\,\text{Wb/m}^2$ so that instruments can be made to meet a wide range of applications.

Laboratory galvanometers use a mirror and a beam of light, reflected on a distant scale, to measure the deflection of the moving coil. This gives high sensitivity, but the instrument is not portable. Industrial instruments use a relatively long pointer fixed to the coil, moving on a scale to indicate the deflection.

6.2.1.2 Construction

The permanent magnet occupies a large part of the instrument space, in order to provide an instrument with low power consumption, and low current, for full scale deflection (f.s.d.). The magnet is usually in the form of a horseshoe, with soft iron pole pieces. A soft iron cylinder is placed between the pole pieces, which gives a uniform magnetic field in the air gap between the poles.

The coil is wound on a light metal frame. The pointer is attached to the coil, but it can be adjusted to calibrate the zero position. Phosphor-bronze conducting springs, on either side of the coil, provide the force which opposes the coil's rotation. The coil is balanced by three weights. It moves in two pivot bases, supported on jewel bearings, of $0.01-0.05\,\text{mm}$ diameter at its tip, depending on the weight of the coil. The stress in the bearings, under these conditions, is of the order of $10\,\text{kg/mm}^2$, and they need to be protected against damage due to shock, often by mounting the bearings on springs. The power requirement of this movement varies from $0.02\,\text{mW}$ to $0.2\,\text{mW}$, and it has an accuracy of about $2-5\%$ of f.s.d.

Fig. 6.1(*b*) shows an alternative construction in which the magnet forms the core, and a soft iron yoke is placed around it. The yoke now serves as a magnetic shield, so that several meters can operate next to each other without interference.

Instead of using jewel bearings the moving coil can be supported by two torsion ribbons (taut-bands) under tension, on each side of the coil. This eliminates the friction of the bearings, and gives an instrument with higher sensitivity, and one which is fairly insensitive to shock. The taut-band movement gives f.s.d. for $2\,\mu\text{A}$ which is some 20 times lower than that obtained with a pivot and bearing construction.

Moving coil instruments are sensitive to temperature. As temperature increases the magnetic field strength decreases and coil resistance increases, both of which cause the meter to read low. This is partially compensated for by a decrease in spring tension with temperature. Generally the meter would read low by about $0.2\%/°\text{C}$. Temperature performance can be improved by using swamping resistors, made from manganin, as in Fig. 6.1(*c*). Manganin (Mn) has a temperature coefficient of almost zero, and copper (Cu) has a positive temperature coefficient. As the temperature increases the resistance of the copper shunt increases faster than that of the Mn–copper coil, causing a larger proportion of the total current to flow in the coil, so compensating for the other effects of temperature. The disadvantage of using swamping resistors is that they cause a reduction in the sensitivity of the movement, since higher voltages are needed to give full-scale current.

6.2.1.3 Characteristics

The sensitivity of the moving coil movement may be defined in three ways: by the current, voltage and megohm sensitivities. Current sensitivity is defined as the ratio of the deflection of the coil to the current producing the deflection, usually measured in mm/μA. Voltage sensitivity is the ratio of the deflection of the coil to the voltage producing it, measured in units of mm/mV. It is usually considered with the critical damping resistance (CDRX) across the coil. Megohm sensitivity is defined as the resistance in megohms, which needs to be connected in series with the coil, to give one unit of scale deflection, when one volt is applied to the circuit. Since the resistance of the coil is small compared to the megohms in series with it, the current to give unit deflection is $1/R$ so that megohm sensitivity is numerically equal to current sensitivity.

Dynamic characteristics of the meter are determined by the damping constant of the movement, the torque opposing the coil movement, and by the moment of inertia of the coil about its rotational axis. Ideally the meter should be critically damped, although in practice it is slightly under damped to compensate for the extra friction, which may arise in time, due to dirt or wear. Damping is provided in the instrument by three methods: friction on the coils and in the bearings; the current induced in the metal frame on which the coil is wound, or in a separate aluminium vane fixed to the shaft for this purpose; and current flow in an external resistor connected across the coil, due to voltage induced in the coil by its movement. The value of resistance which gives critical damping is called the critical damping resistance external (CDRX) for the movement.

6.2.2 *D.C. ammeter*

The ammeter uses the basic permanent magnet moving coil movement. Since the allowable coil current is very small, large current measurements require a shunt across the coil, so that only a known fraction of the total current flows through the coil. The value of the shunt resistance can be found from (6.3)

$$R_s = \frac{I_M R_M}{I - I_M} \tag{6.3}$$

R_s is the value of the shunt resistance; R_M is the internal resistance of the movement (coil); I_M is the current which causes f.s.d. of the movement; and I is the current which is required to cause f.s.d. of the ammeter (coil plus shunt). The shunt is usually placed internal to the meter for current up to about 50 A, and then it is external to the meter.

A meter can have several shunts, selected by a switch to give several current ranges, as in Fig. 6.2(*a*). The switch used is make-before-break to ensure that the movement always has a resistor across it, and so is not damaged. Fig. 6.2(*b*) shows an alternative shunt arrangement, called a universal or Ayrton shunt, which eliminates the possibility of having the meter in circuit without a shunt.

Ammeters are always connected in series in the circuit whose current is to be measured. The meter should have as low an internal resistance

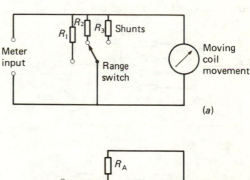

(*a*)

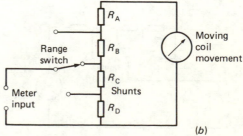

(*b*)

Fig. 6.2. Multirange voltmeter; (*a*) conventional shunt, (*b*) Ayrton shunt.

as possible, to reduce errors. The errors are highest when the meter is operating on a low current range, when the value of the shunt resistor is the largest.

6.2.3 *D.C. voltmeter*

The basic permanent magnet moving coil movement operates on a low value of current, and to turn it into a voltmeter a resistor, called a multiplier, is connected in series with the coil. This limits the current to the f.s.d. of the coil, and the value of resistance may be calculated from (6.4)

$$R_{SM} = \frac{V}{I_M} - R_M \tag{6.4}$$

R_{SM} is the series multiplier resistance; R_M is the internal resistance of the movement; I_M is the f.s.d. current of the movement; and V is the full-scale voltage of the meter. The multiplier is usually mounted inside the case for voltages up to about 500 V although for higher voltages it may be mounted externally, to avoid excessive heating inside the case.

Several multipliers can be selected in a multi-range voltmeter, as in Fig. 6.3. In the modified arrangement the first resistor (R_D) swamps the coil resistance, so that all subsequent resistors are standard, commercially available parts. The first resistor must be made to meet the special requirements of the movement.

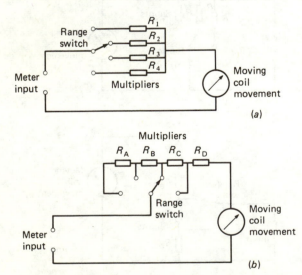

Fig. 6.3. Multirange voltmeter; (*a*) conventional multiplier, (*b*) modified multiplier.

An important parameter of a voltmeter is its sensitivity, measured in ohms per volt. Since the current through the movement must always be the same, for f.s.d. (I_M) the ratio V/R (or R/V) must be the same for each voltage range of the meter. This is stated as ohms per volt, and it defines the internal series resistance needed for each volt applied across the meter. Therefore sensitivity S is numerically equal to $1/I_M$ and (6.4) can be re-written as (6.5)

$$R_{SM} = SV - R_M \tag{6.5}$$

A sensitivity value of 20 kΩ/V is widely used for commercial d.c. voltmeters, that is a f.s.d. current of 50 μA.

Voltmeters are always connected in parallel with the circuit whose voltage is to be measured. To avoid voltmeter loading effects the resistance of the meter, on the selected scale, must be several orders larger than that of the circuit. For example consider the circuit of Fig. 6.4. Without

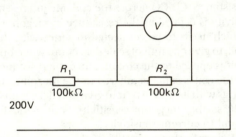

Fig. 6.4. Circuit to illustrate the loading effect of a voltmeter.

the meter connected the 'true' voltage across R_2 is 100 V. If the voltmeter has a sensitivity of 1 kΩ/V then when it is placed across R_2, and selected to its 100 V range, its resistance is 100 kΩ, giving a voltage reading across R_2 equal to 66.7 V, i.e. an error of 33.3%. If the sensitivity of the meter was 20 kΩ/V then on the 100 V range its resistance is 2 MΩ giving a voltage reading of 97.5 V across R_2, that is an error of 2.5%. Electronic voltmeters have high input impedances, and are often used where the meter loading must be minimal, as described in Section 6.2.5.

6.2.4 *Resistance measurement*

6.2.4.1 Voltmeter–ammeter method

The most direct way of measuring a load resistance is to connect a d.c. voltage across it, and

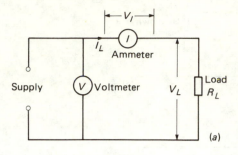

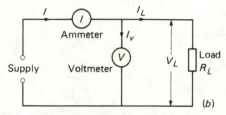

Fig. 6.5. Effect of meter positions on load resistance calculations; (a) ammeter volt drop included, (b) voltmeter current included.

to measure the voltage across the load (V_L) and the current through it (I_L). By Ohm's law the value of R_L can be found from (6.6).

$$R_L = \frac{V_L}{I_L} \qquad (6.6)$$

Fig. 6.5 shows two methods in which the voltmeter and ammeter may be connected. In Fig. 6.5(a) the voltmeter reads the volts drop across the ammeter, in addition to V_L, and in Fig. 6.5(b) the ammeter reads the current through the voltmeter, in addition to I_L. Both connections result in errors. To minimise these errors Fig. 6.5(a) should be used for high values of load resistances, when R_L is large compared to the internal resistance of the ammeter, and Fig. 6.5(b) should be used for low values of load resistance, when R_L is small compared to the internal resistance of the voltmeter.

6.2.4.2 Series ohmmeter

Fig. 6.6(a) shows an ohmmeter arrangement for measuring resistance. R_1 is a current limiting resistor, and R_2 is a zero adjust resistor. The current in the moving coil movement depends on the value of R_x. When $R_x = 0$ resistor R_2 is adjusted to give zero scale reading on the moving coil movement. This compensates for variations

in battery voltage during its operating life. Resistor R_2 can be placed in series with R_1 and adjusted for variations in E, but the calibration of the instrument is now affected to a much greater extent over the whole range of readings. When R_x is not connected the meter position indicates infinite resistance.

If R_h is the value of R_x which gives half-scale deflection of the meter (i.e. half the f.s.d current I_M), R_M is the internal resistance of the moving coil movement, and E is the battery voltage, then (6.7)–(6.9) can be derived.

$$R_h = R_1 + \frac{R_2 R_M}{R_2 + R_M} \qquad (6.7)$$

$$R_2 = \frac{I_M R_M R_h}{E - I_M R_h} \qquad (6.8)$$

$$R_1 = R_h - \frac{I_M R_M R_h}{E} \qquad (6.9)$$

6.2.4.3 Shunt ohmmeter

In the shunt ohmmeter, shown in Fig. 6.6(b), the unknown resistor is placed across the moving coil movement. Now when R_x is omitted resistor R_1 is adjusted to give maximum deflection in the movement, and this is the infinite resistance position. When R_x is zero no current flows in the movement, and this is marked as the zero ohm position. Figs. 6.6(c) and (d) show that the series type scale is calibrated right to left and the shunt type scale from left to right, and both are crowded up on one end. An on–off switch is needed for the shunt ohmmeter, to disconnect the battery when not in use.

The shunt ohmmeter is not used as frequently as the series ohmmeter. Its main use is to measure low valued resistors. If E is the battery voltage, R_M the resistance of the moving coil movement, and I_M the f.s.d. current of the movement, then the value of R_1 can be found from (6.10)

$$R_1 = \frac{E}{I_M} - R_M \qquad (6.10)$$

6.2.4.4 The megger

This instrument is used to measure very high values of resistance, and a hand driven portable d.c. generator provides the high d.c. voltage. The moving coil movement is also modified to have two coils mounted on the moving member, and

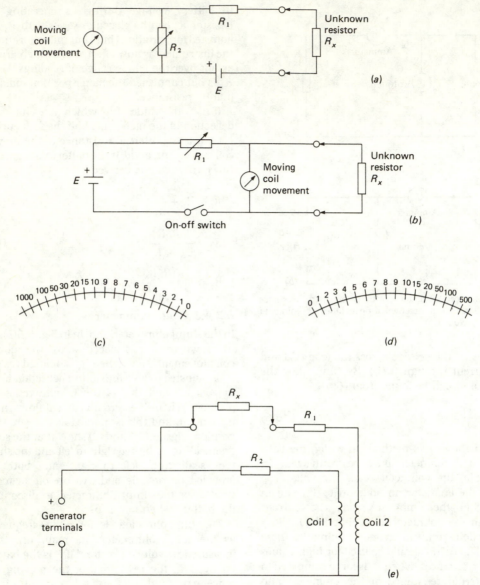

Fig. 6.6 Ohmmeters: (*a*) series, (*b*) shunt, (*c*) scale for series type, (*d*) scale for shunt type, (*e*) megger.

carrying the pointer. The arrangement is shown in Fig. 6.6(*e*). Coil 2 moves the pointer clockwise and coil 1 moves it counter clockwise. With no R_x in circuit the coil 1 current moves the pointer to the infinite resistance position. For any other value of R_x current flows in coils 1 and 2 which determine the resultant pointer position.

6.2.5 *Electronic analogue meters*

In an electronic analogue meter the signal being measured passes through one or more amplifying stages, before being applied to the moving coil movement. The amplifiers are direct coupled d.c. for low cost instruments, although for better performance, and lower drift, chopper amplifiers are usually used.

The input impedance of an electronic voltmeter is usually greater than $10\,\mathrm{M\Omega}$, irrespective of voltage range, and this is high enough to give negligible loading on most circuits. When measuring current the impedance of the meter is very low, so that loading effects are again minimised.

Resistance measurements are usually made by passing a constant current through the unknown resistor, measuring the volts drop across the resistor and amplifying it, and driving the moving coil movement. Lower voltages are used so that resistors can be measured in-circuit without risk of damage to adjoining circuitry.

Electronic meters need more power than their conventional counterparts, to drive the associated electronic circuits, so that the instruments require a long life battery, or are mains operated.

6.2.6 *Multimeters*

In a multimeter, also known as a volt-ohm-milliammeter (VOM), a function select switch is used to enable the same instrument to measure voltage, resistance or current, and a range switch selects the magnitude of the parameter being measured. Multimeters can be conventional or electronic (known as electronic multimeters or EMM), and basically consist of the separate circuits used to measure voltage, current and resistance, in one instrument.

6.3   A.C. analogue instruments

The instruments described in this section all have an analogue output, but use various types of movements to measure the a.c.

6.3.1 *Rectifier meter*

The rectifier type meter uses the same permanent magnet moving coil movement, as is used for d.c. parameter measurements. Since the movement can respond to d.c. current only, the a.c. input is first rectified, as in Fig. 6.7. Because of the inertia of the movement the meter reads the average rectified output in both instances. If R_T is the total system resistance, equal to the sum of the multiplier resistor R_1, the resistance of the series diode, and the resistance of the moving coil meter, then I_M is given by (6.11)

$$I_M = \frac{0.9\,V_r}{nR_T} \tag{6.11}$$

I_M is the f.s.d. of the moving coil; V_r is the maximum r.m.s. input voltage to be measured; and n equals 1 for full wave rectification and 2 for

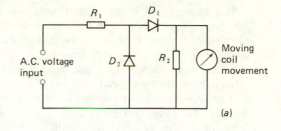

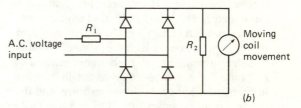

Fig. 6.7. Rectifier voltmeter; (*a*) half wave, (*b*) full wave.

half wave rectification. The value of R_T can be found from (6.11), and from this can be calculated the value of the multiplier resistor.

The scale of a rectifier instrument is calibrated in r.m.s. values, although the movement reads the mean or d.c. value. For a half wave instrument the ratio of r.m.s. to average voltage, called the form factor, is 2.22 for sine wave voltages, and it is 1.11 for a full wave instrument. If waveforms other than sinusoidal are applied to the meter than a correction factor must be applied. For example if d.c. or a square wave signal is measured on a full wave rectifier meter the scale output will give a reading which is high by a factor of 1.11, since the r.m.s. and d.c. voltages are the same for these waveforms.

In the half wave circuit of Fig. 6.7(*a*), diode D_2 is included to limit the reverse voltage across diode D_1, which would give a small leakage current through the coil movement, and result in a meter error. In both types of moving coil instruments a select switch can be used to vary the value of the multiplier resistor R_1 and so give a multirange instrument. Since the rectifier diodes have a non-linear voltage–current curve at low values of current, a small shunt resistor R_2 is placed across the moving coil movements, to move the operating point of the diodes into their linear region. A shunt resistor, however, also reduces the sensitivity of the instrument. The full wave circuit has twice the sensitivity of the half wave circuit.

The moving coil instrument usually has a small capacitor in series with the input terminal to the meter, to block d.c. if the meter is used to measure a.c. having a d.c. bias on it. The capacitor impedance would now have a slight influence on the meter reading.

A rectifier meter is commonly used to measure a.c. in the range 60 Hz–100 kHz. For high frequencies the rectifiers are built into the meter's probe, so as to minimise the pick up. The rectifier instrument is less accurate than a dynamometer, which is mainly used at power line frequencies, but is cheaper. It is also more robust than thermo meters, which are primarily used as transfer instruments at high frequencies. The resistance of the diodes connected in rectifier instruments changes with temperature, and the meter has an accuracy of ± 3 to $\pm 5\%$ of f.s.d., for sine waves at room temperature. At high frequencies the rectifiers exhibit capacitive effects which cause errors, and the meter may read as much as 0.5% low for every 1 kHz frequency increase above a few MHz.

6.3.2 *Dynamometer instrument*

The dynamometer movement is one of the oldest a.c. instrument movements. It was invented by the Siemens brothers in 1843, and it reached its present stage of development by 1910. In the construction of the instrument the permanent magnets of the d'Arsonval movement are replaced by two fixed coils, connected in series and positioned coaxially with space between them. The moving coil is placed between the fixed coils, and the current being measured flows in series through all three coils, as shown in Fig. 6.8. The moving coil is similar to that used in the d'Arsonval movement, and it has a pointer, counterbalance weights, and springs opposing its movement. Damping is provided by aluminium vanes moving in sector-shaped chambers, and the movement is surrounded by a metal shield to screen it from stray magnetic fields.

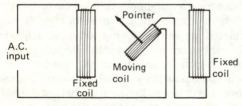

Fig. 6.8. Principle of dynamometer instrument.

The torque produced in a dynamometer movement is given by (6.1), but since the magnetic flux density B is now produced by current I the deflection of the moving coil is proportional to I^2. If the meter is used to measure d.c. the square law would give a scale which is crowded at low current values. On a.c. the instantaneous torque at any time is proportional to I^2 and is therefore always positive, so that no rectification is needed. The inertia of the movement damps the torque variations during an a.c. cycle, so that meter deflection is proportional to the mean of the squared current. Usually the instrument scale is calibrated in terms of the square root of this average current squared, so that the instrument reads r.m.s. values of the input.

The dynamometer movement requires relatively large power from the input, since the field is produced by the input current. Furthermore no iron is used in the magnetic circuit making it relatively inefficient. The low flux density affects torque, so the meter sensitivity is low. For a voltmeter the sensitivity is in the region of $10 \,\Omega/V$ to $50 \,\Omega/V$ compared to a typical sensitivity for a d'Arsonval movement of $20 \,k\Omega/V$. Because the resistance and inductance of the dynamometer coil also increase with frequency, the movement is only used at low frequencies, in the power line and lower audio range.

Multiplier resistors are used to operate the movement as a voltmeter. As an ammeter the movement can usually carry currents up to 100 mA. Above this value a shunt is placed across the moving coil only, and the thickness of the wire used in the fixed coil is increased to carry currents up to 5 A. For larger currents a current transformer is necessary to reduce the current being measured to 5 A.

Because of its low sensitivity the dynamometer movement is primarily employed to measure power, as described in Chapter 8. Its accuracy at power line frequencies is high, being better than 0.5% of f.s.d., and because of this it is sometimes used as an accurate a.c. voltmeter or ammeter for low frequency measurements.

6.3.3 *Moving iron meter*

Although moving iron meters may be of the attraction or repulsion types, the latter is the one most commonly used and is shown in Fig. 6.9. A stationary coil carries the current which is being measured. Two soft iron vanes are placed inside

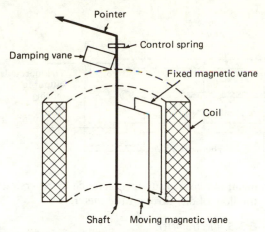

Fig. 6.9. Radial vane moving iron instrument.

the coil; one is fixed to the frame of the coil and cannot move, and the other is connected to the shaft of the instrument and can rotate. Current flowing through the coil will magnetize the vanes with the same polarity, which is independent of the instantaneous direction of the current. This causes the vanes to repel each other, with a force proportional to the value of the current. A control spring opposes the movement of the shaft, and the current is indicated by a pointer moving on a scale. An aluminium damping vane is fitted to the shaft, and rotates in a close fitting air chamber. In the attraction type a soft iron piece is connected to the pointer shaft, and this is attracted by the current in the coil. A second fixed metal vane is not used.

The radial vane instrument shown in Fig. 6.9 is the most sensitive of moving iron constructions, and has an almost linear scale. It is also possible to use a concentric vane arrangement, which has two concentric vanes, the outer one fixed and the inner one rotating. This structure is less sensitive and has a square law scale, although it is possible to modify the shape of the vanes to get scales with special characteristics.

The accuracy of the moving iron instrument is limited by the non-linearity of the magnetization curve of the iron vanes, and by hysteresis and eddy current effects in the vanes and other metal parts. The flux density in the movement is also low, giving poor current sensitivity, so that moving iron instruments are not often used in low power high impedance circuits. The advantage of the moving iron movement is that the

moving parts do not carry any current, making the system rugged and not easily damaged by overloads.

A moving iron movement can be used as a voltmeter, with the addition of a multiplier, or as an ammeter, with a shunt added. At high frequencies the stray capacitance and the impedance of the instrument circuit increase, so that when it is used as a voltmeter it will read low. It is therefore important to calibrate the instrument for the frequency at which it will be used. Commercial instruments can cover the frequency range of 25–125 Hz, although with special compensation circuits this can be extended to 2.5 kHz. Although the instrument can measure d.c. this will cause residual magnetism in the vanes, leading to errors.

6.3.4 Thermo instruments

Thermo instruments are based on the heating effect of an electric current. One such instrument is the hot wire meter, in which current flows in a fine wire (hot wire) stretched between two adjustable posts. A second wire is attached to the centre of the first wire, and passes over a pulley before being fixed to a spring, such that it pulls down on the hot wire. When current flows in the hot wire it expands, and the movement of the second wire over the pulley causes a pointer to move and indicate the magnitude of the current. Hot wire instruments have been largely replaced by thermocouple instruments, which have greater sensitivity and accuracy.

In the thermocouple movement the current being measured is passed through a small heater, which is used to heat the hot junction of a thermocouple, whose voltage output is supplied to the coil of a d'Arsonval meter. The heating effect is proportional to the square of the current, making the deflection of the instrument follow a square law, with crowding at the lower end of the scale. The thermocouple gives a d.c. voltage, which is proportional to the temperature difference between the hot and cold junctions. To ensure that this temperature difference is caused by the current being measured, and is not affected by atmospheric temperature, a construction such as Fig. 6.10 may be used. The free ends of the thermocouple are attached to the centre of copper straps, ensuring that they are at a temperature which is a mean of the ends of the heater element. The ends of the copper straps are

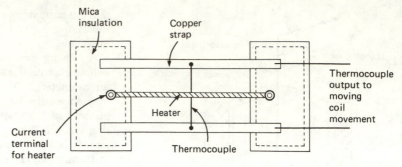

Fig. 6.10. Thermocouple probe used in a thermocouple meter.

close to, but insulated from, the ends of the heater. The temperature at the centre of the heater is greater than that at its ends, and for a given current the temperature difference between the thermocouple junctions is constant, and not affected by the ambient temperature.

Thermocouple instruments can measure currents in the range 0.5–20 A. For higher currents the heating element is usually situated outside the instrument case. For lower currents a bridge arrangement is used, with several thermocouples in series to provide a greater output and increased sensitivity. The meter has accuracies of 1% of f.s.d. at frequencies up to about 50 MHz. Above this frequency the effective resistance of the heating wire is increased, due to skin effect, reducing the accuracy. Tubular designs are used for the heating wire to reduce these errors.

Thermocouple instruments are based on the heating effect of a current, measuring r.m.s. current directly, and are not affected by the waveshape of the input. The movement may be converted to a voltmeter by using multiplier resistors and a low current thermocouple, when it has a sensitivity of 100–500 Ω/V. The instrument is not robust and it can be burnt out by large overloads. It may be used as a transfer standard by noting the scale reading when energised by the unknown a.c., and then passing d.c. until the meter gives the same scale reading as before. Errors may arise when measuring d.c., due to the Peltier heating effect at the junction between the heater and the support wires, and the Thompson heating effect along the length of the heater. These effects cause the temperature of the heater at one side to be higher than the other, resulting in reversal error. This error can be minimised by placing the thermocouple exactly at the mid point of the heater, or by taking the mean of two measurements, made by reversing the flow of d.c. through the meter.

6.3.5 *Electrometer*

The electrometer, or electrostatic voltmeter, is the only movement which measures voltage directly, rather than by the current it produces. It consists of fixed and moving semicircular metal plates separated from each other. The moving plate is attached to a pointer and controlled by a spring. When a voltage is applied to the plates, the moving plate rotates so as to align itself with the fixed plate. The amount of deflection is proportional to the square of the applied r.m.s. voltage. Therefore the meter can be used on a.c. or d.c., and its reading is not affected by the waveform of the input.

The electrostatic voltmeter can be used at frequencies up to about 100 kHz. It draws negligible current except when it is first connected into the circuit, and its plates charge up.

6.3.6 *A.C. electronic analogue meters*

The instruments described in this section use an analogue meter as the output, but this is buffered from the input by electronic circuitry which improves the performance, or measures special parameters. Two types of instruments are considered, those which measure r.m.s. voltage and those which measure peak voltage.

True r.m.s. instruments can use electronic circuits, which calculate the r.m.s. of the input voltage, and display the value on the meter. Although these instruments are expensive they can operate at high speed, which is limited only by the characteristic of the electronic circuit. An alternative approach is to make use of the heating effect of a current, similar to that used in the thermo instruments of Section 6.3.4. Fig. 6.11 shows such a system, which uses

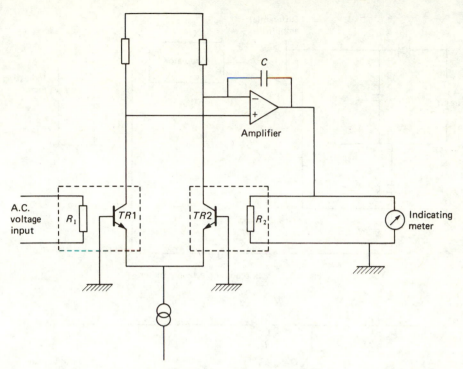

Fig. 6.11 R.M.S. reading electronic meter.

matched semiconductor junctions TR1 and TR2 heated by thick film resistors R_1 and R_2. The semiconductor junctions can be replaced by thermocouples, but then the sensitivity and response speed are decreased. Using the differential arrangement, shown in Fig. 6.11, overcomes any non-linearity problems. The voltage to the amplifier is proportional to the difference in temperature at the two semiconductor junctions. This voltage is amplified and fed back, so that equilibrium is reached when the voltage across R_2 equals that across R_1, the meter indicating this value.

In a rectifier type instrument the operating frequency is limited to below 100 kHz due to rectifier characteristics. When the meter is isolated from the a.c. side of the circuit by an amplifier, this limitation no longer applies. In a peak responding meter a capacitor is charged from the rectified a.c., and this voltage is then amplified and applied to a meter, whose scale can be calibrated to read peak or peak-to-peak voltages. The instrument can operate at frequencies over 10 GHz by putting the peak detecting circuits into the probe of the meter. The disad-

vantage of this simple circuit is that at low signal levels the diode characteristic is non-linear so that system response is poor. This is overcome by using a differential input, as in Fig. 6.12. One input to the differential amplifier is the peak of the signal which is being measured. The other input is the feedback of the voltage developed at point A. The value of this voltage is determined by the differential input to the amplifier. When the two differential inputs are the same the voltage at A is equal to that of the peak of the a.c. This is passed through another peak detecting circuit, and then read on a meter.

One of the problems with a peak responding meter is that, if the input waveform is asymmetrical, it will give a different result when the leads are reversed. This is known as turnover error, and is minimised in peak-to-peak detectors, which use an input circuit shown in Fig. 6.13.

6.4 Analogue to digital signal conversion

The majority of items which need to be measured are analogue in nature, and need to be converted to a digital signal before they can be handled by

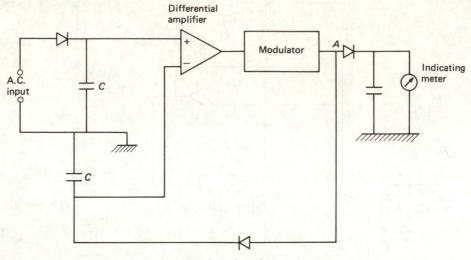

Fig. 6.12. Peak reading electronic meter.

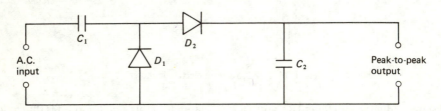

Fig. 6.13. Peak-to-peak measuring circuit.

digital instruments. These analogue to digital converters (A to D) are usually part of the measuring instrument, and are described in this section. Digital to analogue converters (D to A) are first introduced since these are often used as part of A to D converters.

6.4.1 *D to A converters*

Since a digital input can only change in discrete steps, the analogue output is also stepped, as in Fig. 6.14. The magnitude of each individual step is a function of the number of bits used to represent the digital information, and since both polarity inputs are accepted this converter is bipolar.

Fig. 6.15 illustrates three common types of converters. In Fig. 6.15(*a*) the digital input at any bit position will cause the associated switch to change over, and to connect the resistor to the reference supply V_{Ref}. This gives a current into the summing junction of the amplifier, and an output voltage whose magnitude is proportional

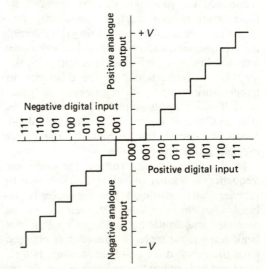

Fig. 6.14. Input–output waveforms for D to A converters.

to the junction current, and so inversely proportional to the value of the resistor in the leg. If the

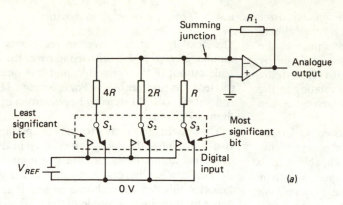

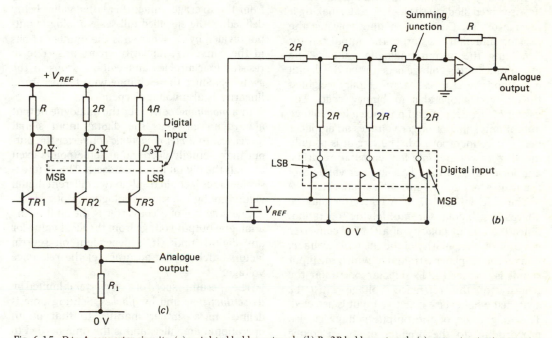

Fig. 6.15. D to A converter circuits: (*a*) weighted ladder network, (*b*) R–2R ladder network, (*c*) current output converter.

resistors are weighted in binary code, as in Fig. 6.15(*a*), then the output voltage is proportional to the binary value of the digital input. The disadvantage of this converter circuit is that its accuracy depends on the absolute accuracy of the reference voltage and of the resistors. Since resistor values double between successive legs it means that they can reach high values, making them difficult to manufacture, and to track each other closely over a wide temperature range.

The R–2R network converter, Fig. 6.15(*b*),

works on the principle that, since the input to the amplifier is at virtual earth, each node divides the current coming up the leg by two. Therefore the current at the summing junction is binary weighted, according to the number of junctions passed through. Although the R–2R ladder converter uses twice as many resistors as the binary weighted ladder network, the accuracy of the converter depends on the relative values of the resistors, and not on their absolute value. Also only two different resistor values are used, so they can be chosen to give good resistor characteristics, and to track each other closely with temperature. A further advantage of the $R-2R$ network is that the impedance seen by the operational amplifier does not vary with the digital signal input, so problems relating to the variations of the amplifier characteristics are avoided.

The D to A converters discussed so far are voltage output types. Fig. 6.15(c) shows a current output converter, which is capable of high speeds since the transistors all operate in their unsaturated mode. The diodes are connected via digital switches to ground, so that with no digital input the output voltage is negligible. A digital signal at any bit position will send the cathode of its associated diode to the logic 1 state, making it reverse biased. Current can now flow via the resistor, and its transistor, to the output resistor R_1 giving the output analogue voltage. The magnitude of this voltage depends on the digital input, and since the resistors are binary weighted the converter operates in binary code. The circuit of Fig. 6.15(c) can produce a low voltage output only, and for larger voltages an amplifier needs to be incorporated. The circuit is usually modified to compensate for temperature drifts, and for changes in the base-emitter voltages of the transistors due to current variations.

The conventional D to A converter may be altered for bipolar or multiplying operation. Since the output polarity of a D to A converter depends on the polarity of the reference voltage, and on the amplifier terminal to which the input signal is connected, for bipolar operation the polarity of the reference voltage must be switched whenever a negative input is detected. The magnitude of the output voltage is also proportional to the product of the reference voltage and the digital input. Usually this reference is fixed, but if it were to vary proportional to a second input, then the converter would be performing a multiplying function. If the reference V_{Ref} is varied proportional to an analogue signal, then the converter is a hybrid multiplier, since its output is proportional to the product of an analogue and digital signal. If V_{Ref} is obtained as an output of a previous D to A converter, which has a fixed reference input, then the two converters behave as a digital multiplier, since the two digital inputs to the converter are multiplied together.

6.4.2 *D to A converter characteristics*

The *resolution* of a converter represents the number of steps which are used to cover the full-scale output of the converter, and it is directly related to its number of bits. For example a 10 bit converter has 1024 steps and a resolution of less than 0.1 per cent.

The linearity of a converter may be specified in two ways. *Integral linearity* is measured in terms of the maximum deviation, of the output, from the best straight line drawn through it. For good linearity this deviation should be less than, or equal to, half the magnitude of the least significant bit. *Differential linearity* is measured in terms of the maximum deviation of the actual values, from the average value; this latter value being defined as the specified full-scale analogue output divided by 2^n, where n is the number of bits in the converter. Linearity errors arise due to resistor inaccuracies and voltage drops in the switches. Since these change with temperature, linearity is also voltage dependent.

In a *monotonic* converter the analogue output always increases as the digital input signal increases. In a *non-monotonic* converter the output may actually decrease at some points even though the digital input increases. This is undesirable since it could result in two different digital values giving the same analogue output.

The *accuracy* of a converter is the shift of the analogue output voltage from the ideal value for any digital input. It is dependent on several factors, such as the accuracy of the reference voltage.

The operating speed of a converter is limited by its *settling time* and by *glitches*. Settling time is defined in a similar manner to that of an operational amplifier, and is the time needed to reach to within a defined band of the final output value. This time is dependent on the types of switches used, the resistor characteristics, and the output amplifier. The output from the D to A converter is not usually regular, as shown in Fig. 6.14, but is subjected to overshoots, noise spikes, and violent dips or glitches. These glitches, illustrated in Fig. 6.16, are caused by the fact that the switches used have unequal turn on and turn off times. Therefore, for example, when switching from digital 011 to 100 it is possible for the output to pass momentarily through 000 if the switches go from 1 to 0 sooner than when

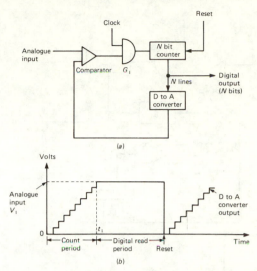

Fig. 6.17. A to D converter using a D to A converter for ramp generation; (*a*) circuit schematic, (*b*) waveforms.

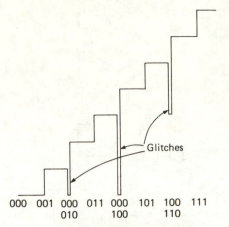

Fig. 6.16. Illustration of glitches in D to A converters.

going from 0 to 1. The size of the glitch can be reduced by using faster switches, and their effect can be smoothed out by reducing the amplifier slew rate, and by filtering. However this also reduces the overall response time. Sample and hold circuits can also be used to overcome glitch effects, since the voltage can be held until the glitch has died down. Unfortunately the acquisition time of the sample and hold circuit now reduces the speed of response of the overall converter.

The *temperature coefficient* of the converter is dependent on the stability of the reference voltage, resistors, switches and amplifier. Converters are available which have internal reference sources and amplifiers, although in some converters these must be provided externally. Where they are external components it is usual to specify the converter stability on the assumption that there is negligible change in the reference or amplifier offset with temperature.

6.4.3 *A to D converters*
The fastest method of conversion from analogue to digital is to apply the analogue signal to a bank of parallel voltage comparators, each of which is set to a different threshold. The output then feeds into a logic circuit, and gives the digital signal. This is known as flash conversion, and is very fast, although it is also expensive since it requires a considerable amount of electronic circuitry.

Fig. 6.17 illustrates an A to D converter which uses a D to A in its feedback loop. The clock steps the counter, and when its value, fed back through the D to A converter, equals that of the

analogue input, the clock is disabled until the next cycle, and the digital output can be read. The circuit shown in Fig. 6.17 is slow, since it needs to count up from zero on each reset pulse. Its speed can be increased by using an up–down counter, such that the counter is not reset to zero at the start of the count period, but it counts up or down depending on the direction in which the analogue input has changed. The device is now called a tracking or servo A to D converter.

An alternative A to D conversion method, known as dual ramp, is shown in Fig. 6.18. Initially S_1 is closed so the unknown voltage V_1 is fed into the integrator, and clock pulses are enabled by G_1. After a fixed number of clock pulses, equal to time t_1, the switch logic closes S_2 and opens S_1. At this instant in time the value of V_M is given by (6.12)

$$V_M = \frac{V_1 t_1}{\tau} \tag{6.12}$$

τ is the time constant of the integrator.

The reference input is of reverse polarity to the analogue input voltage, so the integrator output decreases to zero from V_M, at which point G_1 is disabled until the next reset cycle. The output voltage from the integrator is now zero, so the circuit performance is given by (6.13)

$$0 = V_M - \frac{V_{Ref} t_2}{\tau} \tag{6.13}$$

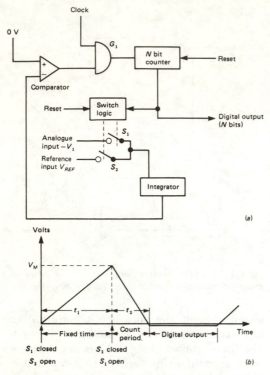

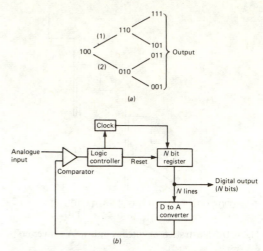

Fig. 6.19. Successive approximation A to D converter; (*a*) operating tree, (*b*) circuit diagram.

Fig. 6.18. Dual ramp A to D converter; (*a*) circuit schematic, (*b*) waveforms.

Equation (6.14) can be derived from (6.12) and (6.13)

$$V_1 = \frac{V_{\text{Ref}} t_2}{t_1} \qquad (6.14)$$

Since V_{Ref} and t_1 are constants the counter reading (t_2) indicates the value of the unknown analogue input.

Dual ramp converters give excellent accuracy, eliminate propagation errors in the circuit, and compensate for changes in clock frequency and integrator time constants, since these affect both ramps equally. The converter also compensates for comparator offset currents and voltages, since two zero crossings are involved and therefore these cancel out.

For fast analogue to digital conversion a successive approximation technique is often used, as shown in Fig. 6.19. The most significant bit of the register is initially set to logic 1 and if the output from the D to A is less than the unknown analogue input, path (1) is followed on the operating tree; otherwise path (2) is followed.

In this way the successive approximation technique works by halving the range of possible values each time, and comparing this with the input. The total number of comparisons required for any conversion is equal to the number of bits, so this technique is very fast, although it requires relatively complex logic circuitry.

Voltage to frequency conversion may also be considered as a form of A to D converter. In this the digital output consists of a series of pulses of precise width, the pulse rate being determined by the value of the analogue input voltage.

6.4.4 *A to D converter characteristics*

Many of the characteristics used to define D to A converters such as linearity, accuracy and speed, also apply to A to D converters. In addition the following three terms are used, quantization error, sampling error and electronic error.

Quantization error is the smallest analogue input to which an output signal can be approximated. It has a maximum value equal to $\pm$ LSB and is therefore dependent on the number of bits.

Sampling error, also known as *aperture error*, is defined as the difference between the digital output, and the value which would have been obtained if the input signal did not change during the conversion period. Sampling error is worse in converters having longer conversion times, and for a given conversion rate and accuracy it sets a limit to the maximum frequency of the analogue signal.

Electronic error is the error in conversion which is contributed by the various circuits through which the analogue signal needs to pass.

6.5 Digital instruments

In a digital meter the output is in the form of discrete numbers, rather than a pointer deflection on a continuous scale. The advantages of this are reduced human reading error, no parallax error and faster reading speed. Digital instruments have built in electronic sophistication, usually in the form of a microprocessor, which allows additional facilities to be incorporated. For example some instruments have a built in programmable capability, which allows them to do basic calculations such as linearizing meter readings, and displaying the modified values.

Extensive diagnostic facilities are available on some instruments, reducing repair time, whilst most modern desk-type instruments have an internal calibration check facility. Unlike older instruments the covers need not be removed for calibration; calibration constants may be entered via the instrument keyboard, and stored in its non volatile memory. Subsequent readings are then modified by the value of this constant. Many digital meters also have a bus interface, such as IEE 488, so that they may be operated as part of larger measurement systems.

The parameters for a typical multimeter are as follows:

Input range: 20 mV–1 kV; 0.2 mA–2 A; 200 mΩ–10 mΩ.

Absolute accuracy: 0.001–0.5% of maximum reading

Stability: 0.002% (24 hour period) and 0.008% (6 month period) of maximum reading

Resolution: 1 part in 10^6

Input characteristics: Resistance 10 MΩ; capacitance 40 pF

Operating time: 2 ms–1 s

Bandwidth (for a.c. volts): 100 kHz–1 MHz

Generally many larger meters can select the correct input range automatically, a feature known as autoranging. Smaller meters usually have an overload indication, which tells the operator to change the range manually. High accuracy instruments require an input resistance in the 10 GΩ range, to avoid loading the circuit being monitored. The A to D converter used in the meter usually determines its operating time, which should include the time needed to recover from an overload. Resolution indicates the smallest voltage which can be read. For example a resolution of 1 part in 10^6 enables 1 μV to be read on a 1 V input scale. Resolution depends on the number of digits in the instrument, as shown in a typical example in Table 6.1.

Digital meters all use the basic D to A and A to D circuits described in Section 6.4. Fig. 6.20, for example, shows the block diagram of a typical digital voltmeter. The input voltage is attenuated, and then amplified by a fixed gain amplifier. The value of the attenuation is varied, either manually or by autoranging, as shown in Fig. 6.20, such that the output from the signal amplifier is always within a given range. This signal is compared with the staircase waveform out of the D to A converter, and when these are equal the count clock is disabled, leaving the digital value of the analogue signal voltage in the decade counters. The frequency of the count clock is about 5 kHz. The transfer clock is much slower, about 2 Hz, and this transfers the information from the counters to the display memory

Table 6.1. *Resolution of a typical digital multimeter*

Parameter	Range	Resolution		
		$3\frac{1}{2}$ Digit	$4\frac{1}{2}$ Digit	$5\frac{1}{2}$ Digit
D.C. volts	30 V	10 mV	1 mV	100 μV
A.C. Volts (true R.M.S.)	30 V	10 mV	1 mV	100 μV
D.C. current	3 A	1 mA	100 μA	10 μA
A.C. current	3 A	1 mA	100 μA	10 μA
Resistance	3 MΩ	1 kΩ	100 Ω	10 Ω

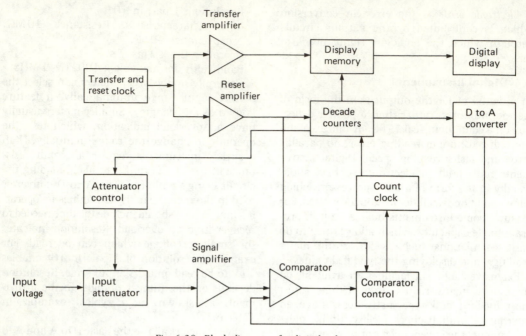

Fig. 6.20. Block diagram of a digital voltmeter.

and display, and then resets the system ready for a new count. This circuit is used in general purpose equipment.

6.6 Comparison measurements

The accuracy of a deflection meter is often limited by the accuracy of its scale markings, and this is much more noticeable for instruments having a non-linear scale. This inaccuracy is eliminated in the comparison method of measurement, which compares the unknown parameter with a reference, and the scale gives a relative indication only.

There are two principal types of comparison measurements, relative indicating and null indicating. In the null indicating type the comparison produces a null or balance, whilst in the relative indicating method the indication depends on the degree of unbalance. Null indicating instruments are more expensive, and difficult to use than relative indicating meters, but they are more accurate.

6.6.1 *Relative indicating measurements*

There are many ways in which relative measurements can be made, and Fig. 6.21 illustrates one system, known as the *equal deflection method*. The

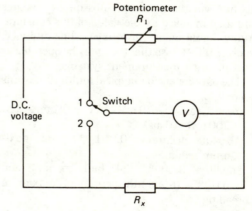

Fig. 6.21. Relative indicating method for measuring resistance.

voltmeter is switched between the potentiometer R_1 and the unknown resistor R_x, and the value of the potentiometer is adjusted to give an equal deflection on the meter in both switch positions. The value R_1 is then equal to R_x, and R_1 may be removed and measured accurately in a bridge, or it may be previously calibrated so that the value of R_x is read off. The accuracy of the voltmeter scale is no longer important, but it should have an impedance of at least ten times that of R_x.

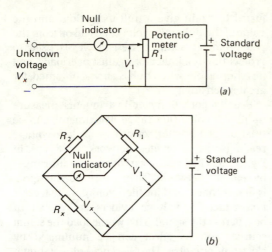

(a)

(b)

Fig. 6.22. Null indicating method of measurement; (a) potentiometer type, (b) bridge type.

6.6.2 *Null indicating measurements*

Null measurements may be considered as a special case of the equal deflection method, where the deflection is zero. Null indicating instruments are usually potentiometer types, used for voltage comparisons, or bridge types, used for impedance comparisons. Bridge instruments are described in more detail in Chapter 7, although they are compared with potentiometer types in Fig. 6.22. In the potentiometer instrument the potentiometer is adjusted until the null indicator shows zero reading. At this setting voltage V_1 equals the unknown voltage V_x and the potentiometer can be calibrated to show this voltage. At balance no current flows through the meter, so that infinite impedance is presented to the signal being measured, giving a highly accurate measurement system.

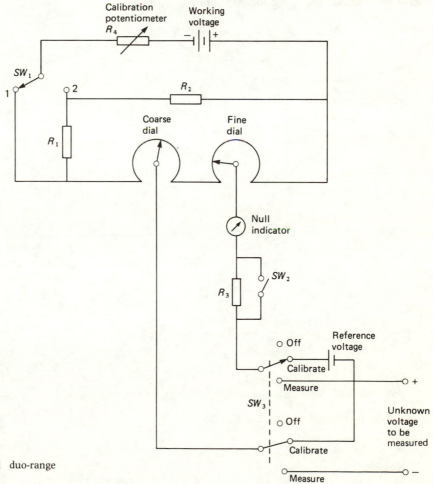

Fig. 6.23. A practical duo-range potentiometer system.

In the bridge type of null indicator, potentiometer R_1 is varied until V_1 is equal to V_x so that the indicator again shows a null. The potentiometer can be calibrated to give the value of R_x at this setting.

Fig. 6.23 shows a practical potentiometer circuit. To explain its operation first assume the switch SW_1 to be at location 1 and ignore resistors R_1 and R_2. Two potentiometers are used in series, the coarse dial having a resistance ten times that of the fine dial. The working battery causes a volt drop across the two potentiometers, and this is applied to the null indicator and switch SW_3. Initially this switch is set to the 'calibrate' position, and the potentiometer set to the value expected for the reference voltage. The calibration potentiometer R_4 is adjusted, if required, to give a null reading on the indicator. The selector switch SW_3 is then moved to the 'measure' position. The potentiometers are adjusted to again give a null indication, and the value of the unknown voltage is read off from the potentiometer settings. Resistor R_3 is used to protect the null indicator against overloads, and it is shorted out when null is almost obtained, for greater sensitivity.

Usually potentiometer instruments measure up to 1.6 volts. Duo-range instruments can be used to give greater sensitivity, at low voltage readings, by a divider network R_1, R_2 as in Fig. 6.23. This can be extended to multi-ranges by separate networks. To measure a.c. the signal is first rectified, and for high voltages the level is first reduced by voltage dividers, called a shunt box, across the signal. This would load the signal source, and its effect can be minimised by electronic circuitry. The reference voltage shown in Fig. 6.23 may take several forms, such as a standard cell or a precision zener controlled supply.

7. Bridges

7.1 Introduction

Although bridges are primarily used for component measurements, they also find applications in determining derived values such as frequency, phase angle and temperature. Bridges are introduced in this chapter, and their special applications are covered in Part 3. Bridges make measurements by comparing the unknown parameter with a known standard parameter; they are therefore usually more accurate than instruments which make absolute measurements.

7.2 Bridge principles

A bridge usually has four impedance arms, a source of power, and a detector, as shown in Fig. 7.1. If the impedance values are such that

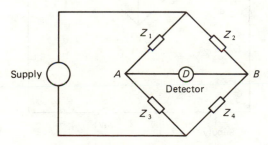

Fig. 7.1. An impedance bridge arrangement.

points A and B are at equal potential, then no current flows into the detector, and a null position is said to have been attained. Under these conditions (7.1) is followed.

$$Z_1 Z_4 = Z_2 Z_3 \tag{7.1}$$

If Z_4 is an unknown impedance then its value can be found from the null equation (7.1), as in (7.2)

$$Z_4 = Z_3 \cdot \frac{Z_2}{Z_1} \tag{7.2}$$

This shows that the null is independent of the detector impedance, since no current flows through it, and of the source voltage and impedance, so a highly stable supply is not required.

Z_3 is known as the standard arm of the bridge, and Z_1 and Z_2 are its ratio arms. To cover a wide range of unknown impedance values a commercial bridge usually has switches to change the ratio arms in decade steps, and the standard arm can be varied at each setting for balance.

If the impedances are complex then they can be denoted in polar notation as $Z/\underline{\phi}$, and now for balance the magnitudes for the impedances must be as in (7.1), and the phase angles as in (7.3)

$$\underline{/\phi_1} + \underline{/\phi_4} = \underline{/\phi_2} + \underline{/\phi_3} \tag{7.3}$$

Therefore an a.c. bridge would need two independent controls, to balance the direct and quadrature components. Complex inductors and capacitors can be represented by series or parallel resistors, and Fig. 7.2 shows the conversion between these.

In a resistance bridge the supply may be d.c. and only one component need be variable. Bridges can also measure inductance or capacitance, but not both from the same bridge arrangement. Reactive components may be expressed in series or parallel terms.

The sensitivity of a bridge is an important parameter, and is its ability to measure small changes in resistance unbalance in the bridge. It is given by the increase in detector current, for a small per-unit change of the adjustable impedance from the null position. For maximum sensitivity of the bridge of Fig. 7.1, Z_2 must equal Z_4, which at balance means that Z_1 equals Z_3. In practice this condition is rarely met, since Z_3 must be made sufficiently large, so that the resolution obtained by the smallest adjustment step gives the required precision. The source and

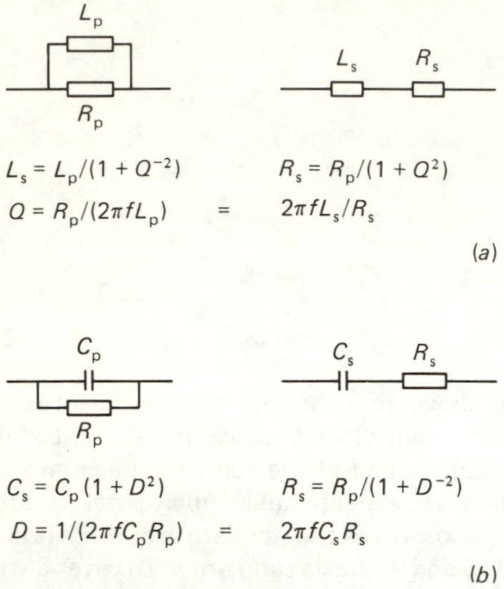

$$L_s = L_p/(1 + Q^{-2})$$
$$Q = R_p/(2\pi f L_p) \quad = \quad 2\pi f L_s/R_s$$

(a)

$$C_s = C_p(1 + D^2)$$
$$D = 1/(2\pi f C_p R_p) \quad = \quad 2\pi f C_s R_s$$

$$R_s = R_p/(1 + Q^2)$$

$$R_s = R_p/(1 + D^{-2})$$

(b)

Fig. 7.2. Conversion between series and parallel connections; (a) inductive, (b) capacitive.

detector can, however, be interchanged without unbalancing a bridge, and greatest sensitivity is obtained when the detector is connected between the junctions of the two highest impedance arms and the two lowest impedance arms. Bridge sensitivity is also proportional to the magnitude of the source voltage.

A moving coil movement may be used to measure null in a d.c. bridge. The cheapest form of detection for an a.c. bridge is headphones, although for frequencies at which the sensitivity of the ear is low radio receivers, or amplifiers and meters are used. For good sensitivity and discrimination a continuous wave source is required, together with a heterodyne detector. An amplifier feeding an oscilloscope has also been used to detect null.

The magnitude of the source voltage must not exceed the rating of devices, or cause excessive power dissipation, in the bridge arms. But the smaller the voltage the lower the bridge sensitivity, and the system is more susceptible to radio frequency interference. For a.c. bridges the line voltage may be used for low frequencies. Commercial bridges usually incorporate different frequency sources, generated from oscillators, since for a reactance bridge the sensitivity is proportional to frequency, and it can be very sharp at one end of the frequency spectrum, and flat at the other. The maximum source frequency

must be well below the self resonant frequency of the component being measured, in order to reduce measurement errors. If the balance point of the bridge is frequency sensitive then the source must have a stable frequency, and it should be free from harmonics, since a bridge which is balanced at one frequency will not have a null for the harmonics.

7.3 Resistance bridges

7.3.1 *Wheatstone bridge*

The most common form of resistance bridge is the Wheatstone bridge arrangement, shown in Fig. 7.3. R_x is the unknown resistor and R_1, R_2

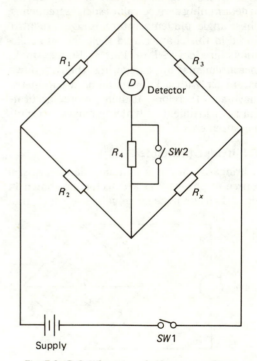

Fig. 7.3. D.C. Wheatstone bridge arrangement.

and R_3 are adjusted for zero current through detector D. At this setting the value of R_x is given by (7.4), which is similar to (7.2)

$$R_x = R_3 \cdot \frac{R_2}{R_1} \qquad (7.4)$$

R_1 and R_2 are known fixed resistors in the range $1\,\Omega$ to $1\,\text{k}\Omega$, to give ratios of R_2/R_1 between 10^{-3} and 10^3. R_3 is adjusted, in steps of $1\,\Omega$ or $0.1\,\Omega$ up to $10\,\text{k}\Omega$, to give bridge balance. In operation, R_2

and R_1 must be selected so as to give maximum bridge sensitivity, as discussed in Section 7.2. Resistor R_4 is initially maintained in circuit, to protect the detector, but it can be shorted out, for greater sensitivity, when balance is being reached.

A Wheatstone bridge is used to measure two terminal resistors in the range $1\,\Omega$–$100\,M\Omega$. The lower limit of resistance is set by the impedance of connecting leads and contacts, and for resistance values below $1\,\Omega$ the Kelvin double bridge should be used. To utilise the Wheatstone bridge to measure resistors above $100\,M\Omega$ requires high voltages, and now the effects of leakage currents to ground can result in appreciable errors. These can be minimised, and the operating range of the bridge extended to $10^{12}\,\Omega$, by the use of high sensitivity detectors, and guard terminal techniques, as described in Section 7.9.

For resistors up to $100\,M\Omega$ the Wheatstone bridge has an error of 5–100 ppm. The resistors used in the bridge are made from manganin, which has a low temperature coefficient of resistance, high stability, and a low thermoelectric e.m.f. with copper. The drift in the resistors is less than 10 ppm per year. In making measurements with the bridge it is usual to take readings with the battery terminals reversed, and to average the results, as this eliminates the effects of thermoelectric e.m.fs. The peak current through the resistors must also be kept low, to avoid resistance change due to self heating effects of current.

7.3.2 *Kelvin bridge*

The problem of using the Wheatstone bridge, shown in Fig. 7.3, to measure low valued resistors is that, since R_2 and R_x are both low resistances, considerable errors can arise due to the resistance of the interconnecting lead, depending on where the detector is connected between them. This error is avoided in the Kelvin bridge, shown in Fig. 7.4, which uses resistors R_5 and R_6 to balance out the effect of the resistance of the link (r). The bridge is called a 'double bridge' since it uses two sets of ratio arms.

R_x is the unknown resistor and R_2 is a standard resistor, of the same order of value as R_x. Both these resistors have four terminals, and R_4 is adjusted such that the current through R_2 gives a volt drop of at least 0.5 V, in order to provide adequate bridge sensitivity. At balance the value of R_x is given by (7.5)

$$R_x = R_3 \frac{R_2}{R_1} + \frac{R_6 r}{R_5 + R_6 + r} \left[\frac{R_3}{R_1} - \frac{R_5}{R_6} \right] \qquad (7.5)$$

In practice the values of R_1, R_3, R_5 and R_6 are selected such that they satisfy (7.6)

$$\frac{R_3}{R_1} = \frac{R_5}{R_6} \qquad (7.6)$$

Under these conditions (7.5) reduces to the more usual form given in (7.4). To verify that (7.6) is attained, the bridge is balanced and link r is then removed, which should not affect the balance of the bridge. Resistance r is therefore compensated for in the Kelvin bridge. The effects of lead

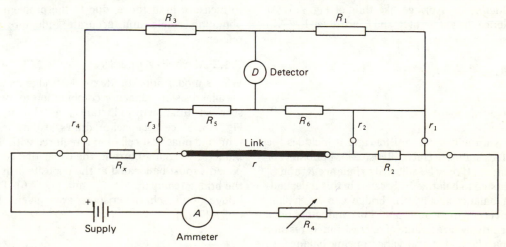

Fig. 7.4. Kelvin double bridge arrangement.

resistances r_1 to r_4 are less important, but they can be compensated for, if required, by shunting R_5 or R_6 by a large valued resistor, until the bridge is balanced with the link removed.

In operation R_x and R_2 are usually of the same order of magnitude; R_1 is varied to give the multiplier; and R_3 is varied by tap changing, for balance. R_6 is ganged to R_1, and R_5 to R_3, so that the ratios are maintained as in (7.6). To avoid the effects of thermoelectric e.m.f. the battery terminals should be reversed, as described for a Wheatstone bridge.

A Kelvin bridge gives errors of less than 0.05% for resistors in the range $10\mu\Omega$–1Ω. It can also be used to measure two terminal resistors since R_1, R_3, R_5 and R_6 form a conventional Wheatstone bridge arrangement, and the error is then less than 0.02%.

7.4 Inductance bridges

The bridges described in this section are all used to measure inductance, either by comparison against a known inductor or a known capacitor. A.C. is used as the supply to these bridges, and two components in the bridge need to be adjustable in order to balance the in-phase and quadrature phase values of the inductor. In the bridge descriptions the unknown inductor is assumed to have a self inductance of L_x, a mutual inductance of M_x and a resistance of R_x.

7.4.1 *Comparison bridge (inductance)*

The most straightforward way of measuring an unknown inductor is to compare it with a known inductor, in a bridge like that of Fig. 7.5. At balance the values of R_x and L_x are given by (7.7) and (7.8)

$$R_x = \frac{R_1 R_3}{R_2} - r \tag{7.7}$$

$$L_x = \frac{L_1 R_3}{R_2} \tag{7.8}$$

R_1 is an adjustable resistor, and it includes the self resistance of coil L_1. Resistor r is optional. For balance R_1 may be adjusted to balance R_x and L_1 adjusted to balance L_x. Because of this independent balance facility the bridge can be rapidly converged to its null state. Since inductors have a relatively large value of self resistance, resistor r may be added and varied during balance, to extend the range of resistance balance. If a fixed

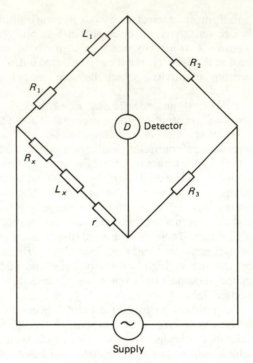

Fig. 7.5. Comparison bridge (inductive).

standard inductance L_1 is used, then bridge balance can still be obtained by adjusting R_1 and the ratio R_3/R_2, but now the two settings affect each other, so the rate of convergence is much slower, and the time to reach balance depends on the Q of the unknown inductor.

The comparison bridge is not frequently used to measure inductance, due to the problem of obtaining stable and accurate inductors as reference.

7.4.2 *Maxwell–Wien bridge*

Wien's modification of Maxwell's bridge uses a parallel resistor–capacitor combination to measure an unknown inductance, as shown in Fig. 7.6. Since a capacitor draws leading current, and phase cancellation is required with the inductor's lagging current, the capacitive components must be located at the opposite arm of the bridge (compare Fig. 7.5 and Fig. 7.6). The bridge null characteristics are given by (7.9)–(7.11).

$$R_x = \frac{R_1 R_3}{R_2} \tag{7.9}$$

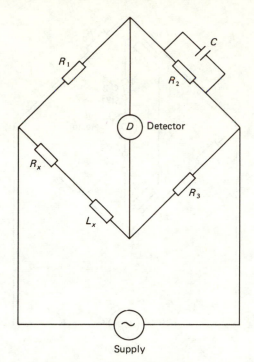

Fig. 7.6. Maxwell–Wien bridge.

$$L_x = \frac{R_1 R_3}{C} \qquad (7.10)$$

$$Q_x = \frac{\omega L_x}{R_x}$$

$$= \omega R_2 C \qquad (7.11)$$

The inductor is measured in terms of a high quality capacitor, which is more accurate and easier to make than a standard inductor, and produces a negligible external field. R_2 and C are usually adjusted for null, since this gives an independent balance for R_x and L_x. However a fixed C can be used as balance, by varying R_2 and R_1 or R_3, although obtaining a null now takes longer. Most commercial bridges do not give R_x but give the value of Q_x instead.

The Maxwell–Wien bridge is widely used for inductance measurement of coils having a Q below about 10. The upper limit on Q is due to the fact that, as shown by (7.3), the sum of the phase angles of opposite arms of a bridge must be equal for balance. Since R_1 and R_3 are both resistors, their phase angles are zero. An inductor with a high Q would have a phase angle of almost 90° lagging, and this must be balanced by

a capacitor–resistor combination of 90° leading. This means that R_2, in Fig. 7.6, would need to have an impracticably large value. This problem is overcome by the Hay bridge.

7.4.3 *Hay bridge*

The balance conditions for the Hay bridge, shown in Fig. 7.7, are given by (7.12)–(7.14)

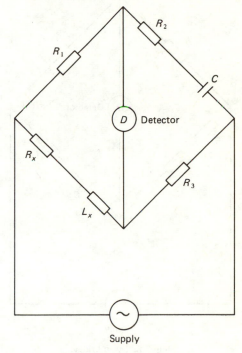

Fig. 7.7. Hay bridge.

$$R_x = \frac{R_1 R_3}{R_2 (1 + Q_x^2)} \qquad (7.12)$$

$$L_x = \frac{R_1 R_3 C}{1 + 1/Q_x^2} \qquad (7.13)$$

$$Q_x = \frac{\omega L_x}{R_x}$$

$$= 1/\omega R_2 C \qquad (7.14)$$

A resistor R_2 is used in series with the balance capacitor C, and for high Q coils R_2 can be made very small. The disadvantage of this bridge arrangement is that balance is frequency dependent, so that the instrument's dials cannot be calibrated to give inductance directly. However the Hay bridge is usually only used to measure coils having a Q greater than 10, and if the Q^2 term is ignored, in (7.13), the inductance measure is independent of frequency, and the error in reading is less than 1%.

7.4.4 *Owen bridge*

The Owen bridge arrangement is shown in Fig. 7.8 and the null equations are given by

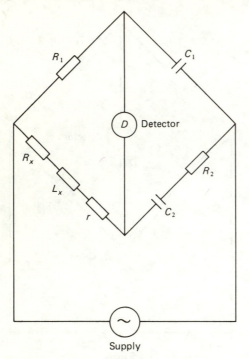

Fig. 7.8. Owen bridge.

(7.15) and (7.16)

$$R_x = \frac{R_1 C_1}{C_2} - r \qquad (7.15)$$

$$L_x = R_1 R_2 C_1 \qquad (7.16)$$

If R_2 and C_2 are adjusted for null then independent balance can be obtained for R_x and L_x, although null can also be obtained by adjusting R_1 and R_2. Resistor r is optional, and is adjusted to extend the range of resistive balance.

The Owen bridge is useful for finding incremental inductance, when a.c. is superimposed onto d.c. in an iron-cored inductor.

7.4.5 *Campbell bridge*

The Campbell bridge, shown in Fig. 7.9, is used to measure mutual inductance of a coil by

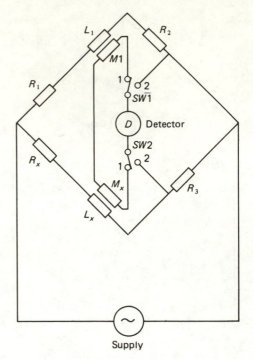

Fig. 7.9. Campbell bridge.

comparing it with a standard. The bridge conditions at null are given by (7.17)–(7.19)

$$M_x = M_1 R_3 / R_2 \qquad (7.17)$$

$$L_x = L_1 R_3 / R_2 \qquad (7.18)$$

$$R_x = R_1 R_3 / R_2 \qquad (7.19)$$

When in use the resistance and self inductance of the primary coils are first balanced, with both switches $SW1$ and $SW2$ in position 2, by adjusting L_1 and R_1. The switches are then moved to position 1 and M_1 is adjusted to balance M_x. Coupling between M_1 and M_x is prevented by careful positioning of the components.

7.5 Capacitance bridges

Capacitance bridges are used to measure the capacitance, and the series or parallel loss resistor (see Fig. 7.2), of the capacitor. Commercial instrument dials are usually calibrated in terms of the dissipation factor (D) of the capacitor, and the loss resistor may be calibrated for this. Since D is frequency sensitive the instrument's calibration is only true at one frequency, and a correction factor must be applied for other frequencies.

Three bridges are used to measure capacitance; the capacitance comparison bridge, the Schering bridge and the Wien bridge. The Wien bridge is also more commonly used to measure frequency, and it is described in Section 7.7.

7.5.1 *Comparison bridge (capacitance)*

The common form of this bridge is shown in Fig. 7.10(a), where C_1 is a standard capacitor

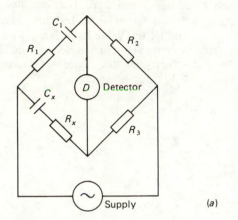

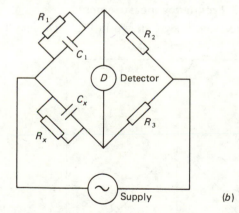

Fig. 7.10. Comparison bridge (capacitance); (a) series bridge, (b) parallel bridge.

with internal resistance R_1. An adjustable resistor is added to the capacitor arm with the smaller dissipation factor, assumed for the present to be included in R_1. The bridge null conditions are given by (7.20)–(7.22)

$$R_x = R_1 R_3 / R_2 \qquad (7.20)$$

$$C_x = C_1 R_2 / R_3 \qquad (7.21)$$

$$D = \omega C_1 R_1 \qquad (7.22)$$

R_1 and R_2 are adjusted for balance, and since these are interacting several attempts are needed to gain convergence. C_1 is usually a high accuracy standard, and is not adjustable.

For measurement of capacitors having a high D the parallel bridge configuration is preferred, since the series bridge would require a high value for resistor R_1. The bridge equations are now given by (7.20) and (7.21), and the dissipation factor is given by (7.23)

$$D = 1/\omega C_1 R_1 \qquad (7.23)$$

The Comparison bridge is not very accurate when used to measure capacitors with very low dissipation factors, and the Schering bridge is preferred.

7.5.2 *Schering bridge*

The Schering bridge is widely used to measure capacitance, and for the accurate measurement of dissipation factor. It is also used in high voltage bridges, with a high voltage standard capacitor, and shielding techniques described in Section 7.9. The Schering bridge arrangement is shown in Fig. 7.11, and the null conditions are

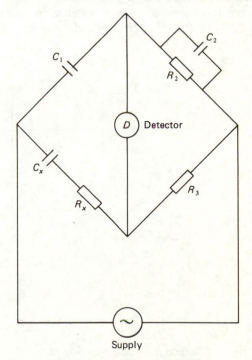

Fig. 7.11. Schering bridge.

given by (7.24)–(7.26)

$$R_x = C_2 R_3 / C_1 \tag{7.24}$$

$$C_x = C_1 R_2 / R_3 \tag{7.25}$$

$$D = \omega C_2 R_2 \tag{7.26}$$

C_1 is a standard capacitor, having a low dissipation factor, and C_2 and R_3 are adjusted for balance. Since the phase angle of C_1, R_3 is almost 90°, and that of C_x, R_x is almost 90°, the parallel combination of R_2, C_2 need only be slightly capacitive, so C_2 can be a low valued variable capacitor.

7.6 Substitution bridges

Components can be measured by substitution in a bridge. Fig. 7.12 shows a parallel substitution bridge, used to measure high valued impedances. It can measure an unknown capacitor C_x, R_x or inductor L_x, R_x. The bridge is first balanced, without the unknown components connected to terminals 1 and 2, by adjusting R_2 and C_2. Suppose the values of these components at balance are R_{2a} and C_{2a}. The unknown capacitor or inductor is then connected to terminals 1 and 2, and R_2, C_2, are adjusted to R_{2b} and C_{2b} to re-obtain balance. If R_x is greater than $10/(\omega C_{2a})$ then less than 1% error is obtained by using (7.27)–(7.29) to find the value of the unknown

components.

$$R_x = [\omega^2 C_{2a}^2 (R_{2a} - R_{2b})]^{-1} \tag{7.27}$$

$$L_x = -[\omega^2 C_x]^{-1}$$

$$= [\omega^2 (C_{2b} - C_{2a})]^{-1} \tag{7.28}$$

$$C_x = C_{2a} - C_{2b} \tag{7.29}$$

An alternative substitution bridge is shown in Fig. 7.13, in which the unknown component is connected in series into the bridge, at points 1 and 2. Initially 1 and 2 are short circuited and R_2, C_2 adjusted for balance. Let this value be R_{2a}, C_{2a}, and let the value for balance with the unknown component in position 1 and 2 be R_{2b}, C_{2b}. Then for C_{2b} greater than C_{2a} (7.30)–(7.32) are obtained:

$$R_x = \frac{R_4}{R_3}(R_{2b} - R_{2a}) \tag{7.30}$$

$$L_x = \frac{R_4}{\omega^2 R_3 C_{2a}}(1 - C_{2a}/C_{2b}) \tag{7.31}$$

$$C_x = \frac{R_3 C_{2a}}{R_4}\left(\frac{C_{2a}}{C_{2a} - C_{2b}} - 1\right) \tag{7.32}$$

7.7 Frequency measurement

The Wien bridge, shown in Fig. 7.14, may be

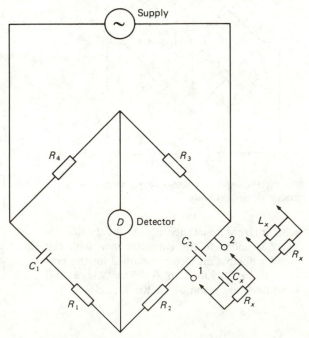

Fig. 7.12. Substitution bridge with parallel connection of unknown component.

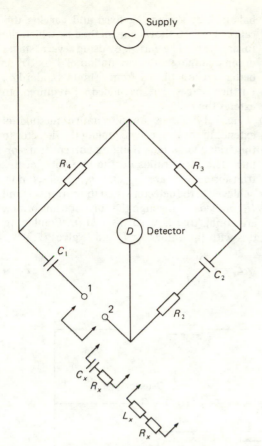

Fig. 7.13. Substitution bridge with series connection of unknown component.

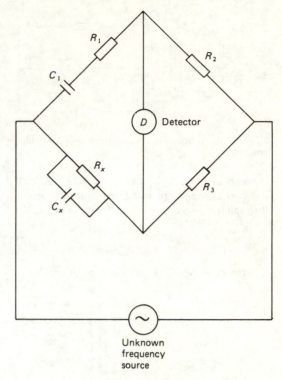

Fig. 7.14. Wien bridge.

used to measure an unknown capacitor C_x, R_x, although it is more frequently used to measure an unknown frequency by using a standard capacitor for C_x. At balance the bridge conditions are given by (7.33) and (7.34)

$$C_x/C_1 = R_2/R_3 - R_1/R_x \qquad (7.33)$$

$$C_1 C_x = 1/\omega^2 R_1 R_x \qquad (7.34)$$

These two equations can be solved to give frequency, as in (7.35)

$$f = 1/2\pi (C_1 C_x R_1 R_x)^{1/2} \qquad (7.35)$$

In a practical bridge C_1 and C_x are fixed. R_1 and R_x are known variable resistors, which are ganged to a common spindle so that $R_1 = R_x$. Also R_2 is made equal to twice the value of R_3, so that (7.35) reduces to (7.36)

$$f = 1/2\pi C_1 R_1 \qquad (7.36)$$

Therefore the bridge is balanced by a single control for R_1, and the calibration is made directly in frequency. Since the Wien bridge is frequency sensitive it will be difficult to balance if the input has harmonics, and these may need to be filtered first.

7.8 Transformer ratio arm bridge

The bridges described in the previous sections attain balance by adjustment of a standard impedance. The transformer ratio arm bridge is balanced by varying the turns ratio on a transformer. This turns ratio can be accurately determined, to better than 1 ppm, and it is not affected by temperature or environmental conditions. Therefore transformer ratio arm bridge techniques are widely used in modern instruments.

Fig. 7.15 illustrates the principle of a transformer ratio arm bridge. The turns ratio on the two portions of transformer T_1 are N_x and N_1; Z_x is the impedance to be measured, Z_1 is a fixed standard impedance, and D is the detector. If V is the a.c. supply voltage, then the voltage across the two parts of the secondary are VN_x and VN_1 giving

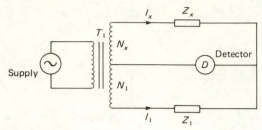

Fig. 7.15. Illustration of the transformer ratio arm bridge principle.

currents of $I_x = VN_x/Z_x$ and $I_1 = VN_1/Z_1$. At null the two currents are equal, giving zero current through the detector, so that Z_x is giving by (7.37)

$$Z_x = Z_1 \frac{N_x}{N_1} \qquad (7.37)$$

If the ratio N_x/N_1 is variable then the bridge is balanced by keeping Z_1 fixed and varying this ratio. Fig. 7.16 shows one method by which the turns ratio can be varied. By using several stages of tap changing a discrimination of 1 to 10^6 or better can be achieved, from a single standard Z_1. Further stages can be added, if required, to extend the range.

Fig. 7.17 shows an alternative method of extending the range, by coupling the detector to the bridge via a tap-changing current transformer. The turns ratios on the two parts of this transformer T_2 are n_x and n_1. Null is now achieved in the detector when the currents I_x and I_1, coupled to it through the current transformer, are equal, that is when $I_x n_x = I_1 n_1$. Combining this with the values for I_x and I_1 gives (7.38)

$$Z_x = Z_1 \frac{N_x}{N_1} \frac{n_x}{n_1} \qquad (7.38)$$

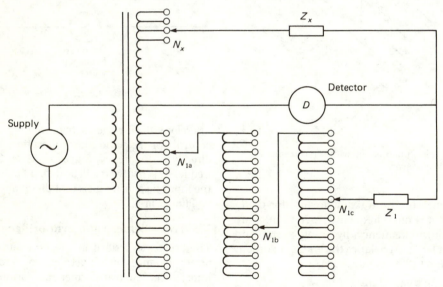

Fig. 7.16. Turns ratio adjustment in a transformer ratio arm bridge.

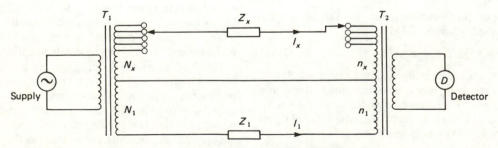

Fig. 7.17. Extending the turns ratio of a transformer ratio arm bridge by use of a current transformer.

This equation shows that the range can be extended by changing the two transformer turns ratios to achieve a null.

The transformer ratio arm bridge can be used to measure three terminal network impedances, of the type shown in Fig. 7.18(*a*). A conventional bridge, connected between A and B to measure Z_{x1}, would also measure the parallel resistors $Z_{x2} + Z_{x3}$. This network is connected into the ratio arm bridge as in Fig. 7.18(*b*). Now whereas Z_{x2} shunts the top of the transformer, it will not affect the result appreciably as long as it does not

change the transformer turns ratio, and Z_{x3} also has no effect on the final reading, since at balance no voltage appears across it. Therefore the bridge can be adjusted to read Z_{x1}.

To measure low valued resistors the transformer ratio arm bridge can be adapted to four terminal measurements, as in Fig. 7.19. In this circuit Z_x is assumed to be small, and R_a and R_b are much larger in comparison, so that $I_{x1} = VN_x/R_a$ and $I_{x2} = VZ_xN_x/R_aR_b$, where V is the supply voltage. At balance the current in the detector must be zero, so that $n_xI_{x2} = n_1I_1$. This

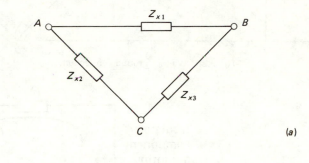

(*a*)

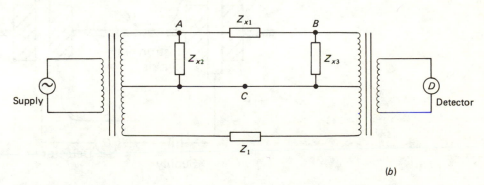

(*b*)

Fig. 7.18. Using a transformer ratio arm bridge to measure a three terminal network; (*a*) network, (*b*) bridge connection.

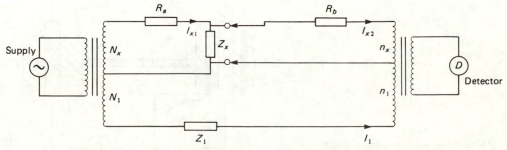

Fig. 7.19. Four terminal component measurement using a transformer ratio arm bridge.

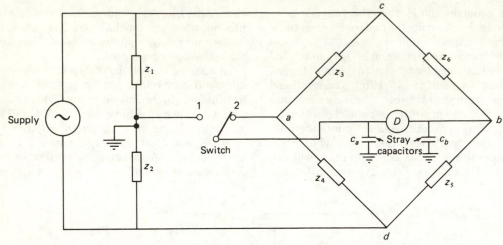

Fig. 7.20. Wagner ground connection.

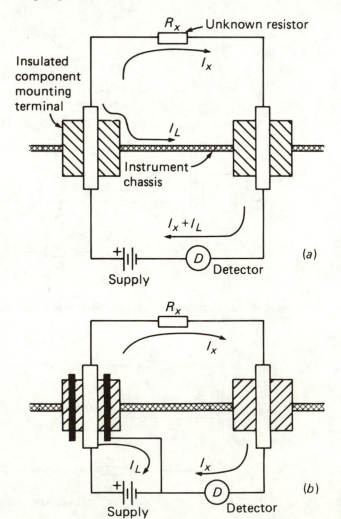

Fig. 7.21. Avoiding leakage current errors by use of a guard ring; (*a*) without guard ring, (*b*) with guard ring.

gives the value of Z_x as in (7.39)

$$Z_x = \frac{R_a R_b}{Z_1} \frac{N_1}{N_x} \frac{n_1}{n_x} \tag{7.39}$$

Since R_a and R_b are high valued they swamp the effect of contacts and connecting leads, and so reduce the error which would normally arise when measuring a low valued resistor.

7.9 Increasing bridge accuracy

Several factors affect the accuracy of bridge measurements, such as sensitivity of the detector, stability of the standard impedances used in the bridge arms, thermoelectric e.m.f. and resistance of connecting leads and contacts. These have been discussed in earlier sections, and the present section will consider the effects of leakage (stray) components, and how they can be minimised.

Stray capacitances occur to earth, and they can disturb the bridge balance and give false readings. The effect of these capacitors can be minimised by shielding all components, and the detector, and earthing the shields. This does not eliminate the stray capacitances, but keeps them at a stable value so that they can be compensated for.

Stray capacitance which occurs at the ends of a detector, as shown at C_a and C_b in Fig. 7.20, are specially troublesome, but their effect can be eliminated by the Wagner ground connection. Impedances Z_1 and Z_2 are placed across the supply oscillator, and their junction is earthed to form the Wagner ground. The switch is first set at position 1 and the bridge is balanced by adjusting its arms. The switch is then moved to position 2 and re-balanced by adjusting Z_1 or Z_2. Several iterations are usually needed before the bridge remains balanced, when switched between positions 1 and 2. In this state points a and b are at ground potential, and the effect of stray capacitors C_a and C_b on the detector reading is eliminated. Z_1 and Z_2 must be of the same type of impedance as either Z_3, Z_4 or Z_5, Z_6. Stray capacitors at ends c and d of the bridge have not been eliminated, but these load the supply and do not effect the reading of the detector.

Errors can also occur in a bridge when operating at high voltages, due to leakage currents through the component mounting terminals and the equipment frame, as in Fig. 7.21(a) The leakage current I_L adds to the true component

current in the detector, and gives a false reading. This can be eliminated by putting a guard ring around the positive terminal of the component, to collect the leakage current, and returning this to the supply, so by-passing the detector.

Problems also arise when measuring high valued resistors, due to leakage resistances to ground, as shown in Fig. 7.22(a). The effect is similar to the leakage current effect described earlier. It may be minimised by using three terminal components, the third terminal being the common return for the leakage resistors R_a and R_b. The three terminal resistor is then connected into a bridge, as in Fig. 7.22(b). Although R_a shunts R_3 it has a much greater value, so that its effect is small, and the effect of R_b is marginally to reduce the sensitivity of the detector. If a three terminal component is not used then R_a and R_b are connected across R_x, and since they are all of the same order of magnitude appreciable errors occur.

7.10 Automatic bridges

Manually balanced bridges are too slow to use in a production environment, and most bridges in current use are automatic and are microprocessor controlled. Many of these instruments measure the unknown component, and compare this with an internal standard component. Since the readings are taken within a short time of each other, using the same measurement chain, many system errors are cancelled out. The microprocessor can be used to compute the magnitude and phase of the unknown, in series or parallel form, and can also display the dissipation factor and Q.

Automatic bridges usually allow three or four terminal measurements, so that compensation for stray capacitance, and for lead and contact resistance, can be carried out. Auto trim facilities are also included to eliminate non-screened parts, such as on a jig connecting to the instrument. Some bridges have excitation level control, using an internal or external supply, to make measurements on non-linear components. Most instruments have the facility to select one of several frequencies. They can make many readings, and take their mean to eliminate errors due to noise. A typical bridge would be able to measure L, C, R, D and Q, in several frequencies within the range $100\,Hz$ to $1\,MHz$, with an accuracy of 0.1%.

Fig. 7.23 shows the arrangement of one type

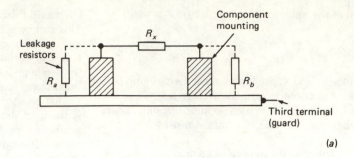

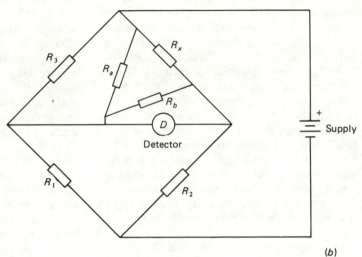

Fig. 7.22. Leakage resistor compensation in a guarded Wheatstone bridge using a three terminal resistor; (*a*) three terminal resistor, (*b*) bridge connection.

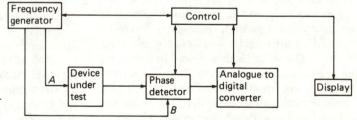

Fig. 7.23. An automatic bridge arrangement.

of automatic bridge. The crystal controlled frequency generator applies a sine wave, of an accurately adjustable frequency, to the device under test, at point *A*. In addition two other signals, having a phase shift of 90° between them, are applied to the phase detector at *B*. These signals act as a reference for the phase detector and allow the voltage and current in the device under test to be measured as in-phase and quadrature-phase vectors. These measured signals then pass

via the analogue to digital converter to the controller, which is usually a microprocessor. The value of the device under test is then computed and the results displayed.

If V_I and V_Q are the in-phase and quadrature-phase components of the voltage across the device under test, and I_I and I_Q are the in-phase and quadrature-phase components of the current through the device, then the equivalent parallel resistance R, parallel reactance X and Q

factor of the component are given by (7.40), (7.41) and (7.42).

$$R = \frac{V_I^2 + V_Q^2}{V_I I_I + V_Q I_Q} \tag{7.40}$$

$$X = \frac{V_I^2 + V_Q^2}{V_Q I_I - V_I I_Q} \tag{7.41}$$

$$Q = \frac{V_Q I_I - V_I I_Q}{V_I I_I + V_Q I_Q} \tag{7.42}$$

8. Power measurement

8.1 Introduction

Power is defined as the work performed per unit of time. Electrically this power, called true power (P_T), is obtained as the product of the voltage (V) and current (I) in a circuit, and the cosine of the phase angle (ϕ) between them $(VI\cos\phi)$. There are two other types of power, apparent power P_A (VI) and reactive power P_R $(VI\sin\phi)$. The three types of power are connected by (8.1).

$$P_A = (P_T{}^2 + P_R{}^2)^{1/2} \qquad (8.1)$$

In this chapter the concept of power and power factor are first introduced, followed by a description of the electrodynamometer meter, which is used to measure power and power factor at low frequencies. Instruments and techniques used to measure high frequency and pulsed power are then described.

8.2 Power and power factor

The reactive power, given in (8.1), should be minimised, and electricity supply authorities apply penalties for poor power factor loads. Fig. 8.1 illustrates an a.c. circuit and from this it is seen that reactive power can be eliminated by making $V_C = V_L$, which is known as power factor correction.

At low frequencies power is usually calculated from the current and voltage measurements. At higher frequencies, greater than about 1 MHz, it is more convenient, and accurate, to measure power, and to calculate voltage and current from this. At frequencies above 1 GHz, voltage and current cease to be useful, and power is almost exclusively the measured parameter.

Power varies instantaneously, in an a.c. circuit, since voltage and current vary. Meters measure the average or d.c. power, which for r.f. work means an average over many cycles. The

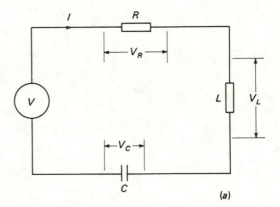

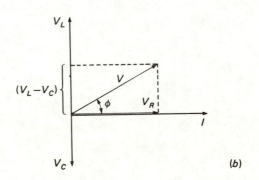

Fig. 8.1. Illustration of voltage and current in an a.c. circuit; (a) circuit schematic, (b) vector diagram.

average period used depends on the type of signal. For a CW signal the average power is many periods of the highest frequency. In the case of an amplitude modulated wave the power must be averaged over several modulation cycles, and for a pulse modulated wave the power is averaged over many pulse repetitions.

Power in a series of pulses is given by the *pulse power* (P_p), as illustrated in Fig. 8.2(a). This is measured by averaging the power over the pulse

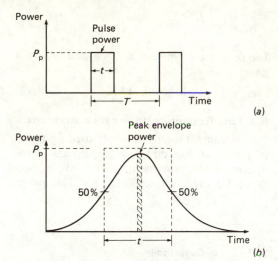

Fig. 8.2. Pulse power and peak envelope power; (*a*) pulse power in a rectangular pulse train, (*b*) peak envelope power in a Gaussian wave.

width (P_{av}) and then using (8.2).

$$P_p = \frac{P_{av}}{\text{duty cycle}} \tag{8.2}$$

The duty cycle is equal to the pulse width divided by the pulse period.

Pulse power is not a suitable unit for measuring power at microwave frequencies, or where the pulse is non-rectangular so that pulse width is difficult to measure. Such a pulse is shown in Fig. 8.2(*b*), where pulse power calculates *t* as the 50% points, and calculates a value of P_p from (8.2), which is too high. In these instances peak envelope power is a better measurement parameter, found by making the averaging time very short, much shorter than the maximum frequency component of the modulation waveform; but it must still be long enough to cover several r.f. cycles. For a perfect rectangular pulse the peak envelope power equals the pulse power, and for a CW signal these two are also equal to the average power. The most common measure of power is average power.

Power measurement comparisons are frequently done in decibels (dB), which is one tenth of a bel. For example if P_2 is the input to an amplifier and P_1 is the output, then the gain of the amplifier is given by (8.3)

$$G(\text{dB}) = 10 \log_{10} \frac{P_1}{P_2} \tag{8.3}$$

The decibel is a convenient scale to use in power measurements, since it gives a more compact range of numbers, and to find the gain of several cascaded stages one can add the individual gains, rather than multiply them.

The reference level of power, used in communication work, is usually taken as 1 mW dissipated in a 600 ohm resistor. This is obtained by a voltage of 0.775 V across a 600 ohm load. The comparative power is known as dBm, which is the same as dB but with the denominator always equal to 1 mW, as in (8.4).

$$G(\text{dBm}) = 10 \log_{10} \frac{P_1}{(1 \text{ mW})} \tag{8.4}$$

dBm is used to measure absolute power. As an example suppose that the power to an amplifier is measured as 2 dBm, and the output is 20 dBm. Then the gain of the amplifier is 18 dBm, which gives a power ratio of 63.10, from (8.3), and a voltage ratio of 7.94, from (8.5). dBm can also have negative values. For example if a signal of + 5 dBm is fed into an attenuator having an attenuation of 20 dBm, then the output signal would be − 15 dBm, that is 15 dB below 1 mW.

Voltage or current can also be compared on a decibel scale, since they are related to power. If the reference and the unknown operate into identical impedances, then the ratios of V_1 and V_2 or of I_1 and I_2 are given by (8.5) and (8.6)

$$R_{db} = 20 \log_{10} \frac{V_1}{V_2} \tag{8.5}$$

$$R_{db} = 20 \log_{10} \frac{I_1}{I_2} \tag{8.6}$$

If the voltages and currents operate into unequal impedances of magnitude Z_1 and Z_2, and of power factors $\cos \phi_1$ and $\cos \phi_2$, then the voltage and current ratios are given by (8.7) and (8.8)

$$R_{db} = 20 \log_{10} \frac{V_1}{V_2} + 10 \log_{10} \frac{Z_2}{Z_1}$$
$$+ 10 \log_{10} \frac{\cos \phi_1}{\cos \phi_2} \tag{8.7}$$

$$R_{db} = 20 \log_{10} \frac{I_1}{I_2} + 10 \log_{10} \frac{Z_1}{Z_2}$$
$$+ 10 \log_{10} \frac{\cos \phi_1}{\cos \phi_2} \tag{8.8}$$

The decibel is used on the briggsian or base to 10 scale. An alternative measure of comparative power, the neper, is used on the napierian or base to e scale. In nepers the power ratio is given by (8.9), and the voltage and current ratios, when feeding into identical impedances, are given by (8.10) and (8.11)

$$R_{np} = \tfrac{1}{2}\log_e \frac{P_1}{P_2} \tag{8.9}$$

$$R_{np} = \log_e \frac{V_1}{V_2} \tag{8.10}$$

$$R_{np} = \log_e \frac{I_1}{I_2} \tag{8.11}$$

The conversion between decibels and nepers is given by (8.12).

$$\text{nepers} = \text{decibels} \times 0.1151 \tag{8.12}$$

8.3 Low frequency power measurement

The instrument most commonly used for measurements at line and low frequencies is the dynamometer movement, described in Section 6.3.2 and illustrated in Fig. 6.8. This instru-

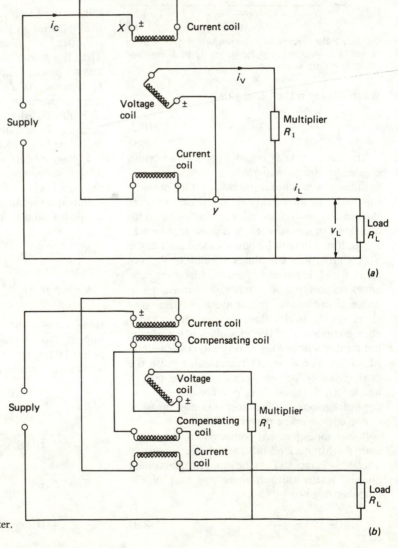

Fig. 8.3. Dynamometer wattmeter.
(*a*) ordinary, (*b*) compensated.

ment is capable of measuring relatively large power levels, and its modifications to read true power, reactive power and power factor are described in this section.

8.3.1 *True power measurement*

8.3.1.1 Single phase circuits

Basically the same electrodynamometer construction is used to measure watts, as was used for voltage and current measurements, but the meter connections are different. Fig. 8.3(*a*) shows the arrangement for measuring single phase power. The fixed coils carry the load line current, and are connected in series. The moving coil, which is made from thinner wire than the fixed coils, is placed in the magnetic field of the fixed coils. It is connected, via a multiplier resistor R_1, to measure the voltage across the load. If the currents and voltages shown in Fig. 8.3(*a*) are assumed to be instantaneous values then, over a period T, the average power (P_A) to the load is given by (8.13).

$$P_A = \frac{1}{T} \int_0^T i_L V_L \, dt \qquad (8.13)$$

The average deflection (θ_A) of the wattmeter is given by (8.14).

$$\theta_A = K \frac{1}{T} \int_0^T i_c i_v \, dt \qquad (8.14)$$

The value of i_v is given by (8.15), where R is the resistance of the moving (voltage) coil plus R_1.

$$i_v = \frac{V_L}{R} \qquad (8.15)$$

If i_v is assumed to be small then i_c is equal to i_L and the value of θ_A can be derived from (8.14) and (8.15), as in (8.16).

$$\theta_A = K_1 \frac{1}{T} \int_0^T i_L V_L \, dt \qquad (8.16)$$

Comparing (8.13) and (8.16) it is seen that the deflection of the meter is proportional to the average power in the load, so the meter can be calibrated to read watts.

The dynamometer wattmeter can measure a.c. or d.c. power, and it is not affected by the waveform of the voltage or current. It has an accuracy of better than 0.25% of f.s.d., from d.c. to 2.5 kHz. It can be designed to have multi ranges, and to work as a voltmeter or an ammeter. The current carrying capability of the meter is important since, if it is used in low power factor circuits, the power reading may be within the meter's scale, but the current could be very large.

The dynamometer wattmeter dissipates some power to maintain its magnetic field, but this is small compared to the power taken by the load. Power is also dissipated in the coil windings, and in the multiplier resistor. The value of the power taken by a wattmeter can be measured by connecting it as shown in Fig. 8.4.

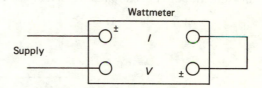

Fig. 8.4. Connection to measure the power loss in a wattmeter. I = current coil terminals (two in series). V = voltage coil terminals.

In the arrangement shown in Fig. 8.3(*a*) the current coil measures the current in the voltage coil (i_v), in addition to that in the load. If the voltage coil connection is moved from Y to X this current error is avoided, but now the voltage coil measures the volts drop across the current coil in addition to that across the load (V_L). Therefore, in both cases, the wattmeter reads higher than the true power loss. To minimise this error, connection should be made at Y for high load current and low load voltage, and at X for high load voltage and low load current.

Fig. 8.3(*b*) shows a compensation method to overcome the error caused by the voltage coil current. The current coil now has two windings, each having the same number of turns. The compensating coil carries the same current as the voltage coil, and it is wound such that its flux opposes that produced by the current coil; therefore it cancels out the effect of the voltage coil current.

8.3.1.2 Polyphase circuits

Power in a polyphase circuit is measured by multiple wattmeters, whose readings are summed. The number of wattmeters needed to measure the power is one less than the number of wires in the polyphase system, as long as there is a common line for all the potential circuits (Blondel's theorem).

Fig. 8.5(*a*) shows the two wattmeter method of measuring power in a three phase delta connected load. It can be shown that the total power consumed in the circuit is equal to the sum of the two wattmeter readings, and is not dependent on whether the load is balanced, or on the waveform or power factor of the load. Fig. 8.5(*b*) shows the three wattmeter method of measuring power in a three phase four wire star connected load. Each wattmeter is connected to measure the power in one phase, so that the sum of the three readings equals the total power in the load.

8.3.2 *Reactive power measurement*

The unit of reactive power is the VAR (volt–ampere–reactive), and meters which measure reactive power are called varmeters. If V, I and $\cos \phi$ are parameters of a circuit, then $VI \cos \phi$ is real or true power, and $VI \sin \phi$ or $VI \cos (\phi - 90°)$ is reactive power. Therefore VARs can be indicated by a wattmeter, in which phase shifting is used to measure reactive power. In a single phase circuit this can be done by using suitable R, L, C components, although in a three

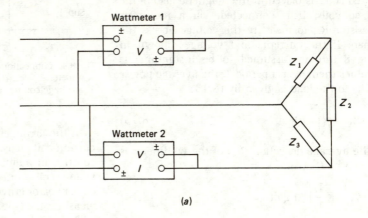

(*a*)

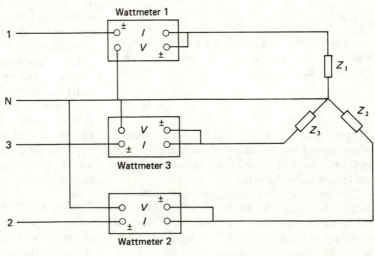

Fig. 8.5. Measurement of power in three phase circuits; (*a*) two wattmeter method, (*b*) three wattmeter method.

(*b*)

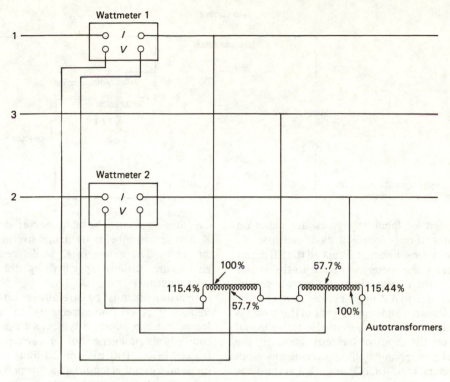

Fig. 8.6. Measurement of reactive power in a three phase circuit.

phase circuit it is more convenient to use two autotransformers, connected as shown in Fig. 8.6. Line 3 is connected to the junction of the two autotransformers, and lines 1 and 2 are connected to the 100% taps. The voltage coils of the two wattmeters are connected between the 57.7% tap and 115.4% tap of the opposite transformers. This produces a voltage equal to the line voltage, but shifted by 90°, for each wattmeter, so each meter measures reactive power, and their readings are added to find the total reactive power of the load.

8.3.3 *Power factor measurement*

The electrodynamometer movement can be modified to form a power factor meter. In this two moving coils are used, which are placed at right angles to each other. One coil is connected in series with an inductor, and the second in series with a non inductive resistor. The two current coils are connected in series with the supply line so that the current through it is in phase with the line current, as in Fig. 8.7. For a

unity power factor load the current in coil $CL1$ is therefore in phase with the line current, and the movement will rotate such that the plane of this coil is parallel to that of the fixed coils. This corresponds to 1.00 power factor on the meter scale. If the load power factor is zero then the movement will rotate such that coil $CL2$ is aligned with the fixed coil. This is the 0.00 power factor position. For any intermediate load power factor the torque produced by the two moving coils will give a deflection between the unity and zero positions.

8.4 **High frequency power measurement**

8.4.1 *Instruments and parameters*

Instruments used to measure power at r.f. and microwave frequencies are of two types; absorption power meters, which contain their own load, and through-line power meters, where the load is remote from the meter. Absorption types are more accurate and usually contain a 50 ohm load for r.f. work. The main instruments cur-

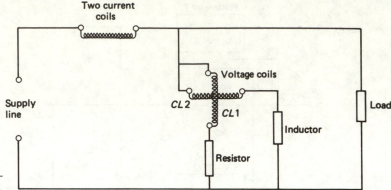

Fig. 8.7. Connection of a power factor meter in a single phase circuit.

rently used for laboratory work are based on thermistor, thermocouple or diode sensors.

A power sensor should accept all the r.f. power input, since the meter connected to the sensor measures the power which has been accepted. However, due to mismatch between impedances of the r.f. source and sensor, some of the power is reflected back, and not dissipated in the sensor. If P_i denotes the input or incident power, P_r the reflected power, and P_d the power dissipated in the sensor, then (8.17) and (8.18) may be derived

$$P_i = P_r + P_d \qquad (8.17)$$

$$P_r = \rho_r^2 P_i \qquad (8.18)$$

where ρ_r is called the *reflection coefficient* of the sensor. Ideally the reflection coefficient should be zero, although it is acceptable to have a value of 0.05 i.e. 5% reflection.

R.F. power may also be dissipated in parts of the instrument other than the sensing element. Examples of such losses include dissipation in the conducting walls of the power sensor; leakage into instrumentation; and power radiated into space. These losses are not measured by the power meter, and should be minimised. The amount of loss is given by the *effective efficiency* (η_e) of the sensor, which should ideally be 1.00 (100%).

The *calibration factor* (K_b) of the sensor connects together the reflection coefficient and effective efficiency, and is given by (8.19)

$$K_b = \eta_e(1 - \rho_r^2) \qquad (8.19)$$

A calibration factor of 0.90 (90%) means that the meter will read 10% below the value of the incident power P_i. Power meters usually have a

dial, to allow them to be set to the correct value of K_b for the sensor being used, and so compensate for errors. The values of K_b at different frequencies are usually specified by the sensor manufacturer.

Instruments must be carefully set up for high frequency power measurements, to minimise losses. Fig. 8.8 shows a typical set up. The r.f. source is a signal generator, or a sweeper which is used below 100 mW (or 20 dBm) only. The input attenuator is required for low power work, below 1 mW (or 0 dBm), and sometimes in the medium power range, 1 mW–10 W (or 0–+ 40 dBm) to allow the input level to be set at a safe value for the device under test. The circulator is used to channel reflections back to the load, and prevent them reaching the source. Circulators are generally used at high power levels, above 10 W, to protect the source.

Directional couplers monitor the input and output to the device under test. At low power levels the indicator may be placed in series with the load, but for high power it is placed off the coupler. At high microwave powers the power level at each component in the test set up is also very important, and must be limited to the safe value for that device.

8.4.2 *Voltmeter based instruments*

For frequencies below about 100 MHz the voltage V_s across a standard resistor R_s may be measured, and the power calculated from (8.20)

$$P = V_s^2/R_s \qquad (8.20)$$

Fig. 8.9 shows such an absorption meter. The value of the resistor must not change over the frequency and power range. Generally this

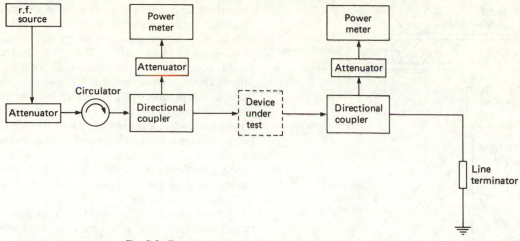

Fig. 8.8. Test set up to measure power at high frequencies.

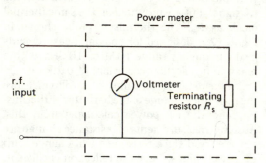

Fig. 8.9. Absorption meter which indicates power by measuring voltage.

power measurement technique is not frequently used.

8.4.3 *Calorimeter*

Calorimeters are used to measure high power in a standards laboratory environment, rather than in industry. It consists of a load resistor, enclosed in a heat insulator, and immersed in a fluid or in air. The fluid may be static, or may flow in and out of the calorimeter at a known rate. The temperatures at the inlet and outlet are measured. If r is the rate of flow of the coolant in cc./s; d is the density of the coolant in g/cc.; s is the specific heat of the coolant; T_i is the temperature of the coolant at the input and T_o at the output; then the power dissipated in the calorimeter P_i is given by (8.21)

$$P = \frac{(T_o - T_i)rds}{0.2389} \text{ watts} \qquad (8.21)$$

Substitution methods may be used in calorimetric measurements. For example after the r.f. measurement has been completed d.c. power is adjusted to the calorimeter, to give the same value of $(T_o - T_i)$, under the same flow conditions as used for the r.f. power measurement. The value of the d.c. power is now measured, and equals that of the r.f. power.

8.4.4 *Bolometers*

A bolometer is a bridge with a barretter or thermistor in one arm to detect the r.f. power. A barretter is a thin wire, usually made from platinum, with a positive temperature coefficient of resistance. The wire is very thin and short, to give a reasonable change of resistance for small power dissipations. The barretter is biased to an operating resistance between 50 ohms and 400 ohms, usually 200 ohms for efficient operation. It needs to be run close to its burn out level, which means that it can be accidentally destroyed by overloads. Barretters have now largely been replaced by thermistors.

The thermistor is a semiconductor with a negative temperature coefficient of resistance; it was introduced in Section 4.9.3. For r.f. power measurements the thermistor is made as a small bead, of 0.5 mm diameter, with a wire lead of about 0.3 mm, as in Fig. 8.10. All the resistance of the thermistor is effectively concentrated in the bead. This resistance changes in a non-linear fashion, as in Fig. 8.11, and will vary from one thermistor to the next in a batch.

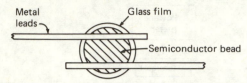

Fig. 8.10. Thermistor construction for high frequency power measurement applications.

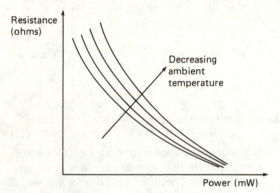

Fig. 8.11. Thermistor characteristics.

The thermistor is enclosed in a special mount, usually a coaxial or waveguide structure, so that it is compatible with the lines used at r.f. and microwave. The mount must match the impedance of the transmission line over the operating frequency range, and the resistive and capacitive losses within the mount should be small, enabling most of the r.f. power to be dissipated in the thermistor. The mount must also be a good insulator, to prevent leakage from the thermistor; it must be resistant to shock and vibration, and it should be capable of shielding the thermistor from stray power fields.

Fig. 8.12 shows a coaxial thermistor mount arrangement, used with an H.P. 478A instrument. Four matched thermistors are fixed to a thermally conductive mount. R.F. power is dissipated in thermistor pair R_m, which are connected in series as far as the low frequency measurement bridge is concerned (terminal A). For r.f. these two resistors are in parallel since capacitor C_2 acts as a by-pass capacitor. The bias adjusts each resistor to 100 ohms, giving a 50 ohm termination for the r.f. input. Capacitor C_1 is an r.f. coupling capacitor.

Thermistor pair R_c is used for compensating against temperature variations. These are biased to 100 ohms each by a separate bridge, at terminal B, and are electrically isolated from the r.f. signal, but are mounted on the same thermal block as R_m. The thermistor probe shown in Fig. 8.12 is designed for coaxial measurements in the frequency range 10 MHz–10 GHz, with a maximum reflection coefficient of 0.2.

A Wheatstone bridge may be used to measure the resistance change of the barretter or thermistor, and the r.f. power calculated from this. Absolute measurements of resistance change are not preferred since resistance will change with r.f. power, and so vary the reflection coefficient. This limits the range of the instrument to about 2 mW. Therefore other techniques are used to measure r.f. power. In the balance bridge technique a d.c. or low frequency bias is applied to the bridge, without any r.f. power input. The bridge is balanced and r.f. power is then applied, which

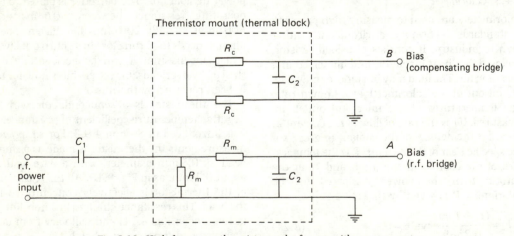

Fig. 8.12. High frequency thermistor probe for use with a power meter.

heats the sensor and unbalances the bridge. The d.c. bias is then reduced to return the resistance to its original value, and so balance the bridge. The decrease in the bias d.c. power is measured, and is equal to the r.f. power.

Fig. 8.13 shows an alternative bridge arrangement using two thermistors. The thermistors R_3, R_4, and heating resistors R_h, are matched. Any unbalance in the bridge is de-

tected, and applied to R_4 so as to bring the bridge back to balance. The power indicator reads the unbalance power, which is equal to that of the r.f. source.

A modern power meter, such as the H.P. 432A illustrated in Fig. 8.14, uses two bridges and a probe, like the one shown in Fig. 8.12. D.C. bias voltages V_{rf} and V_c are used to keep the bridge in balance. The change in the resistance of either

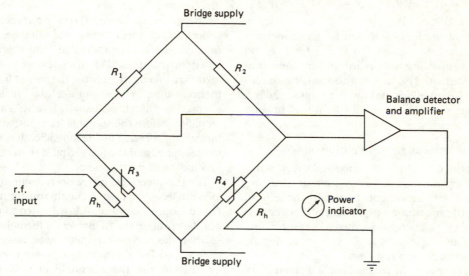

Fig. 8.13. Simplified arrangement of a thermistor power measurement bridge.

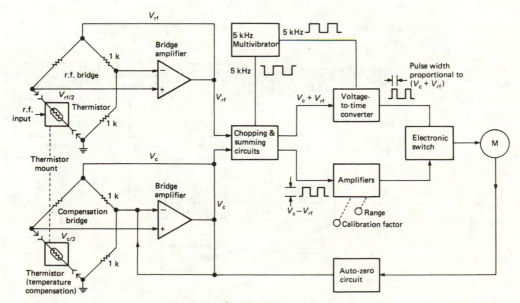

Fig. 8.14. The H.P. 432A power meter.

thermistor causes an unbalance in the bridge, which automatically compensates to bring the resistance back to its initial value. During set up V_c is made equal to V_{rf}, with no r.f. power input. If r.f. power P_r is now applied to the meter, it can be shown that the value of this power is given by (8.22), which is the value measured by the meter, where R is the thermistor resistance at balance.

$$P_r = \frac{1}{4R}(V_c - V_{rf})(V_c + V_{rf}) \qquad (8.22)$$

The advantage of a thermistor based meter is that it operates at a high signal level, so no special shielding is needed in an industrial environment. The meter works on the fundamental assumption that the same amount of r.f. and d.c. power cause equal amounts of heating in the thermistor.

8.4.5 *Thermocouple based instruments*

The principle of the thermocouple was introduced in Section 4.9.2. For r.f. power measurements the thermocouple is usually made from bismuth and antimony. The hot junction is heated by r.f. energy dissipated in a resistor, and thin film techniques are used to make the thermocouple and resistors.

The output of the thermocouple is given in μV per °C of temperature difference between the hot and cold junctions. It is also dependent on the thermal resistance, measured in °C/mW, so that the sensitivity of the overall system is in $\mu V/mW$ of r.f. power. Typical sensitivities of the sensor vary from 100 to 200 $\mu V/mW$, and since the thermoelectric voltage depends on temperature difference, rather than absolute temperature, it is almost independent of the ambient temperature, as long as this does not cause a temperature gradient between the two junctions.

Thermocouple based r.f. power meters have a wider measurement range, and are more accurate, rugged and reproducible than thermistor based instruments. Fig. 8.15 shows a typical meter arrangement. Because the output from the thermocouple is low voltage d.c. it is first chopped and amplified, usually locally in a probe amplifier, before the signal is transmitted to the measuring meter. Further amplification occurs in the meter, and the d.c. signal is then detected and indicated on the meter.

The thermocouple meter operates in an open loop mode, as opposed to the thermistor meter's closed loop, where feedback was used to cancel out the effects of the r.f. power on the bolometer, so that the thermistor meter was largely self calibrating. The thermocouple meter is affected by drift in the thermocouple and associated electronics, so most instruments have a built in

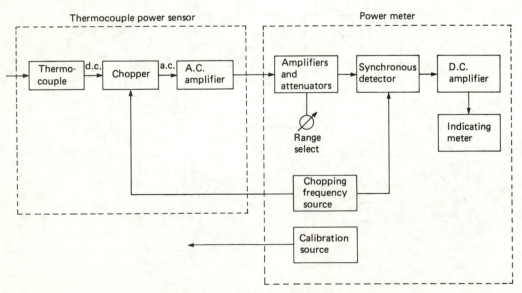

Fig. 8.15. A thermocouple based meter for measuring high frequency power.

calibration power source, of known frequency and power, which can be connected in place of the r.f. input into the probe, to calibrate the meter.

8.4.6 *Diode sensor based instruments*

A diode detector can be used to measure r.f. power levels down to 100 pW, over the frequency range 10 MHz–20 GHz. It is faster than a thermistor or thermocouple probe, but is also less accurate.

Fig. 8.16 shows a typical diode probe arrangement. The diode used has a barrier Schottky construction, with a very low 'knee' and a characteristic as shown in Fig. 8.17. Provided

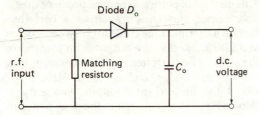

Fig. 8.16. A diode probe for measuring high frequency power.

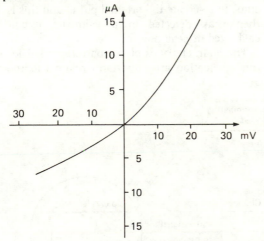

Fig. 8.17. Characteristic of a low barrier Schottky diode.

the probe is operated on the square law portion of the diode characteristic, the voltage on the capacitor C_0 is proportional to the average r.f. power input. At high power level the operating point may move into the linear portion of the diode curve, and the probe now records average power for sinusoidal inputs only.

For maximum power transfer to the diode the

resistance of the diode, at low r.f. voltages, should be matched to the source resistance. The resistance of the diode is temperature dependent, and at the origin it is given by (8.23).

$$R_0 = [\alpha I_s]^{-1} \tag{8.23}$$

α is equal to $q(nkT)^{-1}$; q is the electron charge; n a correction constant of about 1.1; k is Boltzmann's constant; T is absolute temperature; I_s is the diode saturation current. Since R_0 varies with temperature, the sensitivity and reflection coefficient of the probe will also vary. To reduce this effect a matching resistor, of about 50 ohms, is used to terminate the source.

The same low level power meter may be used to measure the output from a diode probe, as was used for a thermocouple probe. The output from the diode sensor is about 50 nV for a 100 μW input signal, so chopper amplification needs to be built into the probe, and care must be taken to prevent leakage and thermoelectric effects from influencing the results.

8.5 Pulse power measurement

The power measurement techniques described so far have all been concerned with CW power, i.e. power which is continuously applied to a meter. To measure power which is present as a pulse, different techniques must be used, and the power is measured as pulse power or peak envelope power, as described in Section 8.2. Four methods for measuring power in a pulse are described in this section. They are the average power per duty cycle method, the d.c./.pulse power comparison method, the integration–differentiation method, and the sample and hold method.

8.5.1 *Average power per duty cycle method*

This method of power measurement is shown in Fig. 8.18. The r.f. power source is passed via a

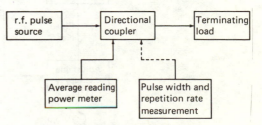

Fig. 8.18. Average power per duty cycle method of measuring pulse power.

directional coupler to a terminating load. An average reading power meter is connected to measure the mean power of the pulses, and this is then replaced by instruments to measure the pulse width and pulse repetition rate. These last readings give the pulse duty cycle, and the power is then found from (8.2).

8.5.2 *D.C./pulse power comparison method*

This method is illustrated in Fig. 8.19. The input r.f. power pulse is split by the power divider. Part of this power goes to a diode peak detector, which gives a d.c. signal proportional to the peak value of the r.f. pulse. This pulse is shown on an oscilloscope. The diode in the detector is forward biased to bring its operating point to a suitable impedance, and so make its response to power almost linear.

The voltage developed by the diode is connected to one terminal of a mechanical chopper. The other terminal of the chopper is supplied with a variable d.c. voltage. With proper synchronization both traces can be seen on the oscilloscope. Initially, with no pulse input, the two traces overlap at zero meter reading. A null control, on the front panel of the instrument, allows the d.c. bias on the diode to be effectively erased from the video output. This control also compensates for long term drift in the diode.

When making a measurement the r.f. pulse is applied, and the d.c. reference voltage is then adjusted until its level is equal to the peak pulse amplitude. This value can be read off the d.c. meter, which is calibrated to read power. To calibrate the test set-up a known r.f. C.W. source is connected to the input, and the terminating load is replaced by a C.W. reading power meter. The power output of the diode detector can now be monitored simultaneously as the power meter reading.

The d.c./pulse power comparison method can measure pulse power in the range 50 MHz–2 GHz, with pulses having a minimum pulse duration of $0.25\,\mu s$. The accuracy is better than $\pm 1.0\,dB$ for pulse repetition rates up to 2 MHz.

8.5.3 *Integration–differentiation method*

This method of pulse power measurement uses a slow responding sensor, such as a barretter, having a thermal time constant between $100\,\mu s$ and $200\,\mu s$, so that the output is an integral of the pulse. The barretter is put in one arm of a Wheatstone bridge, and a square wave input would give an integrated output, with a rising slope which is proportional to the peak power of the pulse. This integrated signal is then amplified and differentiated, using active or passive circuits, to re-create the original pulse, and this is then peak detected in a voltmeter circuit calibrated in peak power.

The characteristics of the barretter used for this application are important, relative to the

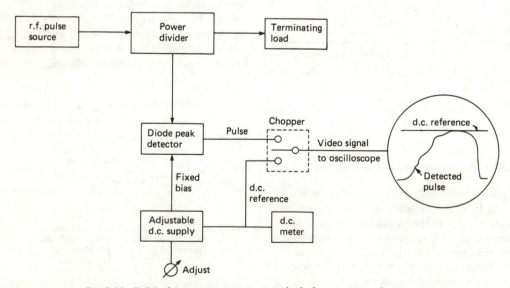

Fig. 8.19. D.C./pulse power comparison method of measuring pulse power.

pulse being measured. If the pulse is too narrow the barretter will not heat up enough to give a signal level, which is appreciably greater than the noise level of the amplifiers in the system. Also if the pulse width of the signal is close to the barretter's time constant the integration will be less accurate. The maximum power capability of the system is limited by the barretter's rating, and the barretter is easily destroyed by overloads. Usually the pulse width of the power being measured should be within a band 0.25–$10 \mu s$, with a pulse repetition rate between 100 and 10 000 pulses per second, and a power level up to 300 mW peak. The accuracy of measurement is then better than 0.8 dB.

8.5.4 *Sample and hold method*

This method depends on the ability of a diode detector to make a power measurement quickly. A small portion of the input envelope, between 50 ns and 100 ns, is sampled and stored on the capacitor, and then amplified and metered. The

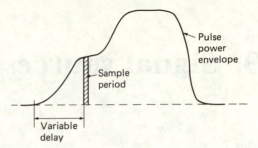

Fig. 8.20. Illustration of the sample and hold method of pulse power measurement.

delay time of the sample can be varied, as in Fig. 8.20, so as to sample different parts of the wave, and so find the peak envelope power on the meter. Alternatively the profile of the whole pulse can be plotted, and the pulse power found from this, as before. With the sampling rates used this method can measure the power of pulses down to $0.2 \mu s$ duration.

9. Signal sources

9.1 Introduction

Instruments which generate electrical signals are widely used in a variety of applications; some of these applications, such as measuring the frequency response of amplifiers, and the alignment of radio receivers, will be described in later chapters. The signal generating instruments are of different types, and have been called by different names. The demarcation between the various modern day instruments is blurred, and an instrument can generally perform several different functions.

Sine wave generators, both in the audio and r.f. frequency range, are called oscillators. These instruments are based on circuits which generate a high accuracy fixed frequency output. Pulse and square wave generators also deliver waveforms of excellent accuracy. The square wave has a 50% duty cycle, whereas the duty cycle for a pulse generator can usually be varied.

The term signal generator is often applied to an oscillator which has the capability of being modulated. Function generators provide a variety of output waveshapes, such as sine, square, pulse and triangular. They usually have a lower specification than that of a corresponding dedicated instrument.

9.2 Oscillator circuits

There are many different circuits used in commercial sine wave oscillators. Generally they fall into two groups: low frequency oscillators, covering the frequency range from a fraction of a hertz to 1 MHz, and high frequency oscillators in the range 100 kHz to greater than 500 MHz.

All oscillators can be considered to be made up of three sections; the oscillatory circuit which determines the frequency; an amplifier to provide the output; and a feedback circuit, to divert some of the amplified output back to the input, and so compensate for losses in the oscillating circuit. If the amplifier gain in G, and the feedback ratio is α, then for oscillations to occur two conditions, known as Barkhausen conditions, must be met. First the loop gain, equal to $G\alpha$, must be unity, and secondly the phase shift between the input and feedback voltages, called the loop phase shift, must be zero. An amplifier with an odd number of stages will give a $180°$ phase shift, therefore the network used in the feedback loop must also give a $180°$ phase shift, at the desired oscillation frequency.

In this section the parameters used to measure the performance of an oscillator are first discussed, followed by a description of four types of frequently used oscillator circuits. The first two, called the Hartley and the Colpitts oscillators, use an $L–C$ tuned circuit to produce the desired sine wave frequency, and they are used primarily in the high frequency range. The other two circuits, called the Wien bridge and the Phase Shift oscillators, use $R–C$ circuits to determine the oscillator frequency, and they are used mainly at lower frequencies. Oscillator techniques, using frequency synthesis, are described in Section 9.4.4.

9.2.1 *Oscillator parameters*

Several parameters need to be considered when measuring the performance of an oscillator, or when selecting one for an application. The *frequency range* of the instrument must cover the desired values, and laboratory instruments are available in the range of 0.00005 Hz to over 50 MHz, but not from the same instrument. The *power output* or *voltage output* must also be sufficient to meet the application needs.

The *output impedance* of the oscillator is an important consideration. Some oscillators have a low output impedance, so they can be converted to the required value by adding series impedances. Other oscillators use transformer coupling, giving a balanced and isolated output. Many oscillators need to work into 600 ohm impedance systems, and have 600 ohm output attenuators.

The *dial resolution* and *dial accuracy* of an oscillator specifies how closely it can be set to the desired frequency, and also specifies the accuracy of the output frequency and amplitude compared to the dial setting.

The *frequency stability* of the oscillator is a measure of its ability to maintain a selected frequency over a given period of time. This stability is primarily affected by component ageing, temperature drifts, and the effects of the changes in the power supply. Frequency stability is improved by careful choice of the components used in the design of the oscillator circuit. *Amplitude stability* is the variation of the oscillator's amplitude with time, at a fixed setting. It is improved by using negative feedback techniques in oscillator design.

The distortion of the oscillator, usually specified as *total harmonic distortion (THD)*, indicates the purity of the output waveform with reference to a sine wave. A distorted waveform puts harmonics into the system being tested. When the oscillator is used for distortion measurements its signal distortion should be many orders below that of the system being tested.

9.2.2 *L–C oscillators*

An *L–C* circuit resonates at a frequency given by (9.1)

$$f = \frac{1}{2\pi\sqrt{(LC)}} \qquad (9.1)$$

In an oscillator this signal frequency is amplified and fed to the output, and some of the signal is fed back from the output to compensate for losses in the *L–C* circuit. *L–C* circuits are not suitable for producing low frequencies since they would require large values of inductance. Two types of *L–C* oscillators are introduced in this section: the Hartley oscillator and the Colpitts oscillator.

9.2.2.1 Hartley oscillator

Many versions exist of the same oscillator circuits, and two of these are illustrated for the Hartley oscillator in Fig. 9.1. Capacitor *C* shunts the tapped inductor *L*, and together they form

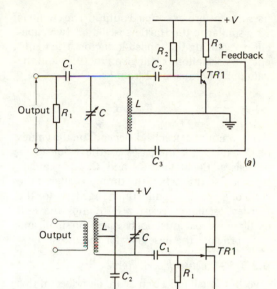

Fig. 9.1. Hartley oscillator circuits; (*a*) using a bipolar transistor and *R–C* coupled output, (*b*) using a FET transistor and transformer coupled output.

the *L–C* tank. Feedback is provided via an *R–C* circuit. In Figure 9.1(*a*) the output is *R–C* coupled, whereas in 9.1(*b*) it is coupled via a winding on the inductor. In both versions the transistors provide a 180° phase shift, and a further 180° shift between output and feedback is achieved by the tapping on the inductor *L*. Capacitor *C* is usually adjusted to vary the frequency of oscillation.

9.2.2.2 Colpitts oscillator

Fig. 9.2 shows one version of this circuit, which

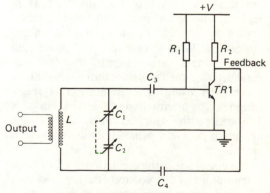

Fig. 9.2. Colpitts oscillator.

uses a transformer coupled output. The circuit is very similar to the Hartley oscillator; two capacitors replacing the tapped inductor. The frequency of oscillation is still given by (9.1) and the value of C is calculated from (9.2)

$$C = \frac{C_1 C_2}{C_1 + C_2} \qquad (9.2)$$

The amount of feedback depends on the values of C_1 and C_2; it increases as the value of C_1 decreases. Capacitors C_1 and C_2 are ganged together, so that when the tuning is altered the ratio of C_1 to C_2 remains unchanged. As for the Hartley oscillator, resistor R_2 damps the L–C oscillations, so it must not be made too small, and $R_1 C_3$ provides the base bias for the transistor.

9.2.3 *R–C network oscillators*

Two R–C oscillator circuits are described in this section, the Wien Bridge oscillator and the Phase Shift oscillator.

9.2.3.1 Wien bridge oscillator

The Wien bridge oscillator is mainly used to produce frequencies in the audio range, with an upper limit of about 100 kHz. It is simple in design, has an excellent frequency stability, and low output distortion.

The principle of the Wien bridge was introduced in Section 7.7, and illustrated in Fig. 7.14. The frequency of the bridge, at balance, is given by (7.35), and for $C_1 = C_x = C$ and $R_1 = R_x = R$ this reduces to (9.3)

$$f = \frac{1}{2\pi \sqrt{(CR)}} \qquad (9.3)$$

The Wien bridge oscillator uses the bridge in the feedback circuit of the amplifier, as in Fig. 9.3. The series and parallel R–C combinations, together with R_1 and R_2 form the original bridge. The Wien circuit has zero phase shift between input and output, therefore the amplifier used must also have zero phase shift, for example by using an even number of stages within it. The attenuation of the Wien circuit is 3, therefore the amplifier must also have a minimum gain of 3, which is easily obtained in practice.

The amplitude of the oscillations in the circuit is determined by the product of the amplifier gain (G) and the attenuation ratio (α). It can be varied

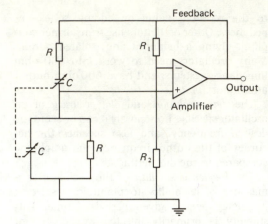

Fig. 9.3. Wien bridge oscillator.

by changing α, usually by controlling the value of R_2. If R_2 has a positive temperature coefficient of resistance, then the circuit can be made to compensate automatically for amplitude changes. The same effect can be achieved by making R_1 from material having a negative temperature coefficient of resistance. The frequency of oscillations is continuously varied by changing C, and it can be stepped through ranges by changing R.

9.2.3.2 Phase shift oscillator

The phase shift oscillator, shown in Fig. 9.4, is capable of operating over a wide frequency range from a few hertz to many kilohertz. A single stage transistor is used, giving a phase shift of 180°, so the three stage R–C circuit must produce a further 180° shift. Usually each stage is arranged to give a shift of 60°, and at this setting the

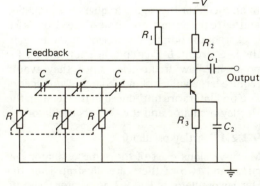

Fig. 9.4. Phase-shift oscillator.

frequency of oscillation is given by (9.4)

$$f = \frac{\sqrt{6}}{2\pi RC} \qquad (9.4)$$

The overall feedback attenuation of the network is now 29, so the amplifier must have a gain at least equal to this value. In operation C is varied for continuous frequency adjustment, and R for range selection.

9.3 Pulse and square wave circuits

A square wave is a special case of a pulse, and has a 50% duty cycle. In a pulse generator the width of the pulse remains constant whilst the repetition rate is varied, whereas in a square wave generator the width changes to keep the duty cycle constant at 50%. Pulses are used in power measurements, where the pulse width can be set to a short duration to prevent over heating. Square waves are preferred for testing low frequency systems, such as audio.

9.3.1 *Pulse parameters*

The main parameters of interest are illustrated in Fig. 9.5. The *base line* is referred to the d.c. level and is the line at which the pulse starts and finishes. The shift of this line from zero volts, or the expected value, is called the *base line offset*. The *amplitude* of the pulse is measured from the base line to the steady state pulse value.

The pulse *rise time* is the time needed for the pulse to go from 10% to 90% of its amplitude, and the *fall time* is the time for the trailing edge to go from 90% to 10%. These times are also called

leading edge and trailing edge *transition times*. The *linearity* of the pulse is the *deviation* of an edge, from the straight line drawn through the 10% and 90% points, expressed as a per cent of the pulse amplitude.

The pulse *preshoot* is the deviation prior to reaching the base line at the start of the pulse. The *overshoot* is the maximum height immediately following the leading edge. *Ringing* is the positive and negative peak distortion, excluding overshoot. *Settling time* is the time needed for the pulse ringing to be within a specified percentage of the pulse amplitude, measured from the 90% point of the leading edge. Pulse *droop* or *sag* is the fall in pulse amplitude with time. Pulse *rounding* is the curvature of the pulse at the leading and trailing edges.

The *width* of the pulse is measured, in units of time, between the 50% points on the leading and trailing edges. The pulse *period* is the time between equal points on the waveform. The pulse *repetition rate* is a measure of how frequently a pulse occurs. It is equal to the reciprocal of the pulse period, and is measured in units of frequency. The *duty cycle* of the pulse is the ratio of its width to its period, usually expressed as a percentage. Pulse *jitter* is a measure of short term instability of one event with respect to another, for example instability in the starting time, or the pulse width, or the pulse amplitude. It is usually expressed as a percentage of the main parameter.

9.3.2 *Pulse and square wave circuits*

Pulse circuits are of two types; passive, which shape an input waveform, usually a sine wave,

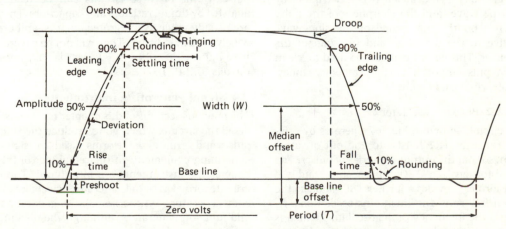

Fig. 9.5. Illustration of critical pulse parameters.

to give a pulse output; and active, such as a blocking oscillator, or a multivibrator. These are briefly described in this section.

9.3.2.1 Pulse shaping circuits

Fig. 9.6 illustrates a circuit in which a sine wave,

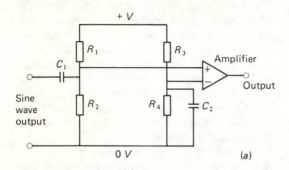

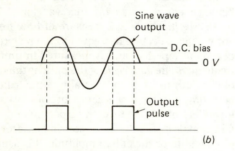

Fig. 9.6. Generating a pulse from a sine wave; (*a*) circuit diagram, (*b*) waveforms.

from an oscillator, is converted into a pulse train. The amplifier is saturated between two limits, by the sine wave, and the sharpness of the pulse depends on the magnitude of the sine wave relative to the d.c. bias, and the gain of the amplifier. This circuit can also be used to 'clean up' a pulse input having overshoots, under-shoots, rings and so on.

9.3.2.2 Blocking oscillator

The circuit, shown in Fig. 9.7 operates by driving transistor *TR1* hard into and out of saturation, so that the output is a series of pulses. The transistor gives a 180° phase shift, and the transformer provides a further 180° shift, so the circuit free runs and maintains oscillations. R_1 adjusts the frequency of the oscillator, which is given by (9.5), where R is the total resistance in

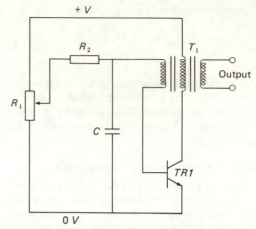

Fig. 9.7. A blocking oscillator pulse generator.

series with capacitor C and the transformer has a step down ratio of 1:N.

$$f = \frac{N+1}{RC} \tag{9.5}$$

The width of the pulse is determined by the characteristics of the transformer, mainly the self inductance and self capacitance of the primary.

9.3.2.3 Multivibrators

There are three types of multivibrators: bistables or flip flops, which provide an output pulse for every two input trigger signals; monostables which give a timed pulse out for a trigger input; and astable or free running multivibrators, which give a string of pulses. The period and pulse widths of these circuits can be readily adjusted by timing resistors and capacitors. They are usually made as integrated circuits, and are widely used in industry. They will not be covered here but are described in F.F. Mazda, *Integrated Circuits*, C.U.P. 1978.

9.4 Signal generating instruments

The previous sections of this chapter have considered the circuits which go to produce the sine, pulse and square waveforms used in signal generating equipment. In the remainder of this chapter the instruments are introduced. These are categorised as signal generators, swept frequency generators, synthesizers, pulse generators, and function generators, although the dividing lines between these groups are often blurred.

9.4.1 *Signal generators*

These instruments usually produce a fixed frequency sine wave, whose output can be frequency or amplitude modulated by another signal. The instruments cover a frequency range of 0.001 Hz–50 GHz, but not from the same device.

Fig. 9.8 shows a signal generator circuit. Frequency modulation is achieved by varying the voltage across a variable capacitance diode in the tuning circuit of the oscillator. This gives a system with low output distortion, for modulation depths below 1% of the carrier frequency. Above this modulation level the waveform applied to the tuning diode needs to be deliberately distorted, in order to compensate for its non-linear characteristics. During frequency modulation manual or automatic methods may be used to keep the amplitude of the output constant.

Amplitude modulation is most conveniently done by varying the supply voltage to the oscillator. This method is, however, only suitable for small modulation depths, up to about 50%. It also gives phase modulation due to the effect on the components used within the oscillator circuit. Feedback can be used to reduce output distortion, as shown in Fig. 9.8, by detecting the output to obtain the modulation envelope, comparing this with the amplitude modulation input, and then amplifying and feeding back the difference as the modulation signal. This technique is known as envelope feedback.

The output amplifier shown in Figure 9.8 provides the required signal level, and buffers the oscillator from changes in the load impedance. The attenuator is needed to give low level output signals.

The heterodyne principle is sometimes used, as in Fig. 9.9, to give a continuously variable, wide frequency range output from a single instrument. The signal quality is good, but the frequency stability is poor, especially at low frequencies. This is because the output is a difference frequency $(f_2 - f_1)$ so a slight change in one of the frequencies, when they are close to each other, can give a wide drift in the output. The output also has a considerable amount of noise and spurious signals.

Multiplier and divider techniques can be used to extend the frequency range available from a single instrument. In the multiplier generator,

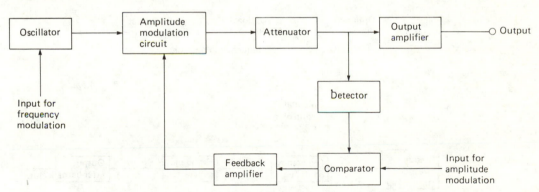

Fig. 9.8. A signal generator using envelope feedback for amplitude modulation.

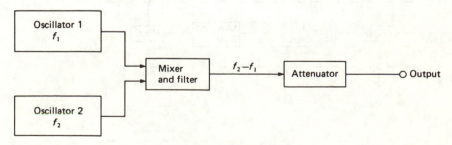

Fig. 9.9. A heterodyne oscillator.

shown in Fig. 9.10, the output from the fixed frequency oscillator is fed through a series of tuned multipliers, each having a non-linear amplifier which produces harmonics. The output from each stage is fed to a tuned filter which selects the high frequency output. Frequency modulation may be applied to the master oscillator, and amplitude modulation is achieved by varying the d.c. supply to the last multiplier stage. The disadvantage of the multiplier generator is that it produces a large amount of spurious signal around the desired frequency.

The divider signal generator uses a high frequency master oscillator, whose output is divided down by a series of electronic stages. The outputs from the divider stages are square waves, which need to be filtered to produce sine waveforms. Frequency modulation is applied to the main oscillator, if required, and amplitude modulation is obtained by diode modulators in the output amplifier. The divider generator does not produce any spurious signal at frequencies below the required frequency.

9.4.2 *Swept frequency generator*

A swept frequency generator, or sweeper, is a special type of signal generator in which the output frequency is cyclically swept through a

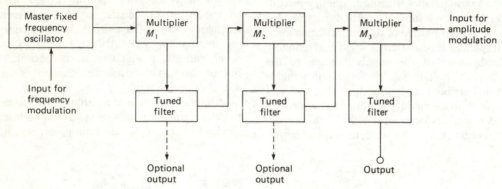

Fig. 9.10. A multiplier frequency generator.

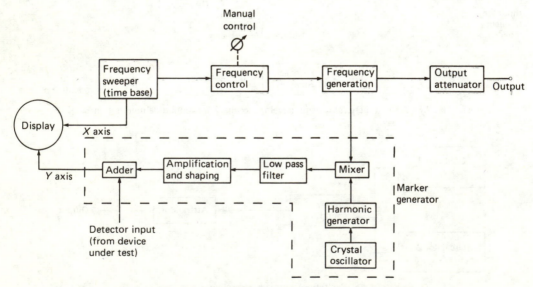

Fig. 9.11. Block diagram of a sweep generator with marker facility.

range of frequencies. The instrument may have a display built into it, to show the variation of amplitude with frequency, or it may provide signals for use with an external oscilloscope.

Fig. 9.11 shows the block diagram of a typical sweep generator. The time base is usually adjustable, to give output sweep times in the range from 10 ms to greater than 100 s. It is also often possible to control the sweep manually from the front panel of the instrument. The time base is frequently a triangular or sawtooth waveform.

Two modes are used to set the swept frequency range. (i) The stop–start or $(f_1 - f_2)$ mode. In this mode the stop and start frequencies are set from the front panel, and the instrument sweeps between these limits. This mode is used for wide sweep widths. (ii) Delta frequency mode (Δf). In this mode the centre frequency, and the maximum excursion about this frequency, are set from the front panel. It is used for narrow sweep widths.

The frequency range of the swept frequency generator usually extends over three bands, 0.001 Hz–100 kHz (low frequency to audio), 100 kHz–1 500 MHz (r.f. range), and 1–200 GHz (microwave range). Three approaches may be used to cover a wide band of swept frequencies from a single instrument. (i) Manually switching between different frequency oscillators. The problem occurs when the frequency range needed overlaps two bands. (ii) Stacked switching. In this the bands are automatically selected by electronic switches, so that one can sweep the whole instrument range as one continuous band. (iii) Heterodyne control. In this method, illustrated in Fig. 9.12, two high frequency signals are mixed to give a lower difference frequency output, as one continuous band. If, for example, generator 1 can be varied over a frequency range 3–5 GHz (i.e. a change of 2:3), and generator 2 has a fixed frequency output of 3 GHz, then the output can vary from 1 to 2000 MHz (i.e. a change of 2000:1). Since the tuning range needed for this wide output variation is small, the instrument has good linearity. However a non-linear mixer will produce many other frequencies in addition to the required one.

Output level control is used in a swept frequency generator to keep the output amplitude to the value set on the front dial. Usually an r.f. detector monitors the r.f. amplitude of the output. This is compared with a signal corresponding to the required amplitude, and the error is fed back to an electronic attenuator circuit, to keep the output constant.

The graticule on the display is not accurate enough to be able to interpret the frequency being displayed, and some swept frequency generators have the facility for putting a series of accurately known frequency markers onto the

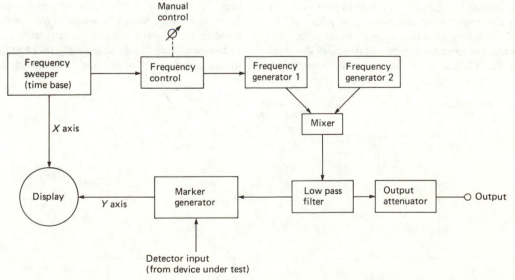

Fig. 9.12. Block diagram of a sweep generator using a heterodyne frequency generator.

screen, to act as a reference. One method of generating the marker frequencies, as shown in Fig. 9.11, is to pass the output from a stable crystal oscillator through a harmonic generator, which gives a series of narrow pulses, spaced at harmonic intervals. These are mixed with a sample of the r.f. output from the sweeper. The mixer will produce a series of low difference frequency output bursts, called birdies, as the sweeper frequency approaches and passes through each harmonic frequency. The birdies are shaped and amplified, and then combined with the signal received back from the device under test (via a detector), before being fed to the display. This gives a composite picture of the frequency response characteristic, of the device being tested, and the frequency calibration markers. An alternative method, of generating a marker, is to pass a single fixed frequency signal to the mixer, to give a spot of accurately known frequency on the screen.

9.4.2.1 Sweeper errors

Several errors occur in swept frequency generator measurements, due to imperfections in the output of the generator. Any harmonics produced in the frequency source will be swept along with the fundamental, but at a faster rate. For example the nth harmonic will be swept at n times the rate of the fundamental. These harmonics may be within the pass band of the device being tested, for part of the sweep, and out of the pass band for the rest. This would give a change in the response of the device, which could be falsely interpreted as a change in its characteristics. For example Fig. 9.13 illustrates the effect of the nth harmonic on a test to measure the

flatness response of a low pass filter. A bump is generated when the nth harmonic is present, and several overlapping harmonics will produce a series of bumps. Generally only the 2nd and 3rd harmonics are of concern, since the higher harmonics have a much lower amplitude.

The accuracy of the output, and the change in its amplitude with frequency, is another important consideration. The amplitude varies with the sweep rate, since the amplitude feedback correction circuits have less time to work at faster sweep rate. The effect of this error can be minimised by sweeping at the lowest rate acceptable for the device being measured, and by making readings with respect to the markers, since these are also affected by a change in amplitude.

The output from a sweeper is not linear, which means that the spacing between markers is unequal, so that one cannot easily estimate a value at a frequency between two markers. The error is usually very small, but if an exact reading is required then the sweeper is set to its manual mode. Its frequency can then be changed until the desired frequency appears on the display, and this can be read with an external accurate frequency meter.

9.4.3 *Random noise generators*

A random noise generator produces a signal output whose instantaneous amplitude varies at random, and has no periodic frequency component. Generators are available which cover frequency bands from low audio to microwave. Random noise generators have many uses, such as testing of radio and radar for signal reception in the presence of noise, and intermodulation

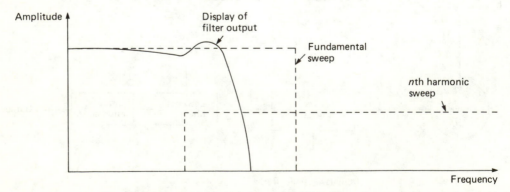

Fig. 9.13. Effect of sweep generator harmonics on a low pass filter response measurement.

and crosstalk tests in communication systems.

Two parameters are of interest in random noise generators:

The noise–power spectrum. Usually one is only interested in noise which has a bandwidth greater than that of the circuit under test. Within this frequency range the noise power spectral density can follow one of three curves, as shown in Fig. 9.14. White noise has a constant power

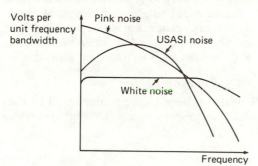

Fig. 9.14. Noise spectrum.

spectral density from 20 Hz to 25 kHz. Pink noise has higher amplitudes at lower frequencies (its amplitude varies inversely as the square root of frequency), and is given its name because of the similarity of its spectrum with that of red light. It is used in bandwidth analysis. The spectrum of USASI noise is approximately equal to the energy distribution of speech and music frequencies, and it is used for testing audio systems.

Probability distribution function (p.d.f.). The power spectrum of a noise signal generator gives its frequency content, but does not characterise its waveshape; this is done by its p.d.f. The p.d.f. is the statistical calculation of the portion of time which the signal spends at the various amplitudes. Usually, for naturally occurring random signals, this function follows a Gaussian curve.

Random noise can be generated by using a gas discharge tube or a zener diode generator. A semiconductor noise diode gives an output frequency in the band from 50 kHz to 200 kHz. This can be amplified and modulated to produce an audio frequency band. The signal is then passed through a filter to get one of the spectra shown in Fig. 9.14.

The disadvantage of the random noise generation method is that sometimes the total output

is subject to long term variations, giving an unpredictable power spectrum. An alternative method is to generate a pseudo-random noise, such as by clocking a long shift register having several feedback loops. The noise pattern will now be repeatable, but within a cycle the signal behaves as a truly random noise. The output from the shift register is fed through a low pass filter to give the pseudo-random analogue noise signal.

9.4.4 *Frequency synthesizer*

Frequency generators are of two types. (i) Free running, in which the output can be tuned continuously, over a frequency range, by mechanical or electronic methods. These are the types described so far, and they have good overall performance, although their frequency, accuracy and stability are relatively poor. (ii) Synthesizer, which has its output derived from a fixed frequency, highly stable oscillator, and covers the range in a series of steps.

Synthesizers are of two types, direct and indirect. The direct synthesizer uses a stable crystal oscillator, followed by a series of harmonic multipliers and mixers, to provide the range of different output frequencies. The indirect frequency synthesizer uses a phase locked loop to give an output which is a fraction of that of a stable crystal oscillator. The synthesizer technique provides a generator with excellent frequency accuracy and stability, but careful design is needed to overcome the problem of spurious signals in the output.

Fig. 9.15 shows the basic indirect frequency synthesizer. The output from the phase detector is a signal proportional to $f_1 \pm f_2$. The low pass filter selects the signal, which is then amplified and fed to the voltage controlled oscillator (VCO). The output from the VCO is fed to a programmable divider, and then back to the phase detector. The system drives f_2 towards f_1 until they are equal, and at this stage there is still a phase difference between the two frequencies, which produces sufficient d.c. signal to keep f_2 locked to f_1. The output frequency is Nf_1 and this can be changed by varying N, usually through a keyboard from the front panel of the instrument.

The switching speed, between frequency settings, of the frequency generator depends on the time needed for f_2 to lock to f_1 and on the settling time. For large frequency differences the lock time can be long. It is minimised by setting the VCO close to its final value, usually under microprocessor control, and then letting the phase lock loop pull it to its exact value. The settling time is the time, after lock, for the output to reach within a given range of the final output setting; it is typically a few milliseconds.

The problem with the basic system shown in Fig. 9.15 is that for a wide frequency range, with small incremental steps, a low reference frequency, and a VCO which can cover a wide range, are required. The programmable divider also needs a large divider ratio, and this produces noise within the loop bandwidth. A multiloop frequency synthesizer, as shown in Fig. 9.16, overcomes these disadvantages. The low frequency synthesizer has a small range, and can operate with small steps, whilst the high frequency synthesizer operates over a wide range with wide steps. Combining these two signals, in a mixer, gives a system which can work over a wide frequency range, with small steps.

Fig. 9.17 shows an alternative system, which is often called a decadic synthesizer since the fixed dividers have a ratio of 10. The range of ratios needed from each programmable divider is now much less. The fixed divider frequency synthesizer uses a high reference frequency, allowing wide loop bandwidth, and giving a faster lock time and better noise performance.

Commercial synthesizers are available covering frequency ranges from 10 kHz to 3000 MHz, and from 0.01 GHz to 30 GHz; and with modulation facilities for use as a signal generator.

9.4.5 *Function generators*

Function generators are characterised by their ability to produce several different waveforms

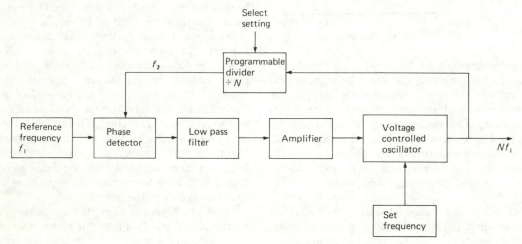

Fig. 9.15. An indirect frequency synthesizer.

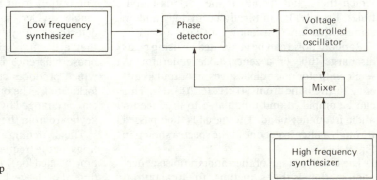

Fig. 9.16. An indirect multiloop frequency synthesizer.

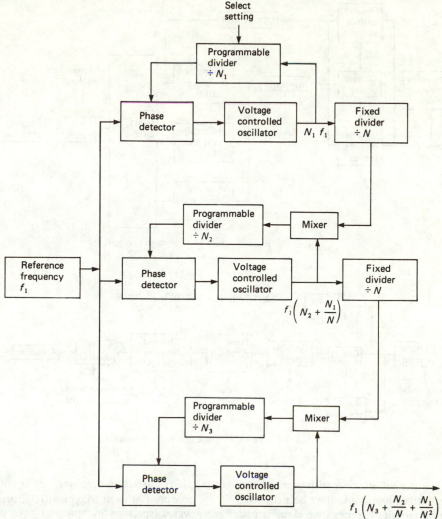

Fig. 9.17. A fixed divider (decadic) indirect frequency synthesizer.

from the same instrument, and by their flexibility. The three basic types of outputs produced are sine, square and triangular wave, over a frequency range of 1–50 MHz, although some function generators also have the capability of producing pulses. The performance of the output from a function generator is usually lower than that from a dedicated instrument.

Fig. 9.18 shows the basic parts of a function generator. A triangular voltage V is generated by integrating the positive and negative current source according to (9.6)

$$V = \frac{1}{C} \int i \, dt \qquad (9.6)$$

The positive and negative sources are switched in and out when the required output level is reached, and the frequency is determined by the amount of current supplied by the sources. Range change is accomplished by switching different capacitors into the integrating circuit. The frequency is selected by a dial on the front panel of the instrument, or remotely via an IEEE 488 interface.

The triangular wave goes through a squaring circuit to produce a square wave output, and it can go through shaping circuits to give a sine wave. The function select circuit choses one of these waveforms, passing it through the level control and output amplifier. Sawtooth wave-

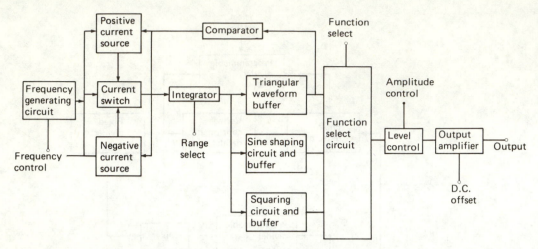

Fig. 9.18. Block diagram of a function generator.

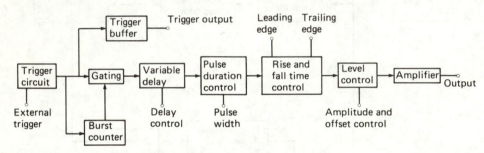

Fig. 9.19. Block diagram of a pulse generator.

forms are triangular waves with unequal positive and negative slopes, and may be generated by the same circuit. The negative slope is usually very steep, to give a fast flyback.

A pulse generating circuit, as in Fig. 9.19, differs from that used for general function generation, although it may be included in the same instrument. The trigger pulses are gated, and then fed through circuits which give control over the pulse delay, width, rise and fall times, amplitude and offset. The output amplifier used is a switching amplifier, to give fast rise and fall times, whereas it is a linear amplifier in the case of a function generator. The range of a pulse generator is also much wider, and it can give a very narrow pulse, at a low repetition rate, since separate circuits are used for the pulse duration and period. Pulses can vary from 5 μs to 5 days.

Function generators operate in several modes.

(i) Triggered, where one cycle of output is produced at each trigger. (ii) N-burns, where N cycles (specified by the user) are produced at each trigger. (iii) Gate, where a continuous output is produced, as long as the signal, at a gate input to the instrument, is above a given value. (iv) Sweep mode. Function generators with sweep capability sometimes have a second generator built into the first instrument, to provide the sweep. The sweep function may be linear, or a choice of linear and logarithmic. (v) Modulation mode, where the output of one generator is modulated by another using AM, FM and phase modulation. (vi) Phase locked mode, where the function generator is phase locked to a signal. Two generators could therefore be locked to each other, and their outputs phase shifted with respect to each other. Some function generators also have a variable

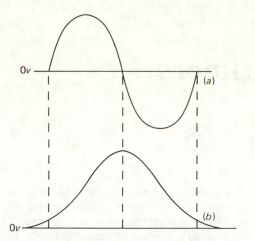

Fig. 9.20. Generating a haversine wave from a sine wave; (*a*) sine wave. (*b*) haversine wave.

start–stop phase control capability. This allows, for example, a sine wave to be changed to a haversine wave, as in Fig. 9.20.

9.4.5.1 Arbitrary waveform generator

Arbitrary waveform generators are a form of function generator where the user can store and generate his own waveshape. These instruments are very flexible, but are limited to low frequency ranges. Arbitrary waveforms are generated in two ways:

(i) By representing the required waveform as a series of points on a grid, and storing this in random access memory within the instrument. This can then be readout, as the waveform is produced, with linear interpolation between the points. A typical system would store 1000 time points and 4000 amplitude points, giving a frequency of about 5 kHz. The number of points used on the grid is usually variable; the greater the number of points the slower the speed, but the higher the resolution.

(ii) By storing a sequence of vectors in memory, and using them as tangents to build up the waveform. The vectors are joined on end to produce the waveform, but can be changed in length and orientation over wide limits. In a typical system 200 vectors may be selected during waveform generation, with a minimum interval of 0.1 ms; giving a maximum output frequency of about 1 kHz.

The vector method of arbitrary waveform generation is more accurate than the grid method, since many vectors may be used in the curved areas of a waveform, and fewer where the curve is straight. The vector method is also slower since each vector needs to be generated as a separate ramp.

10. Counters and timers

10.1 Introduction

Frequency and time interval are both fundamental parameters, which need to be measured in the field of electronics. Counters, which are used to make these measurements, are of many different types, some general purpose and others specialised for a specific application; the operating principle and basic components are however the same in all types.

This chapter first introduces the principles of counters and the principal specifications which need to be considered in their selection. The different operating modes for counters are then described, followed by a description of counter modifications needed to make measurements at high frequency. The chapter concludes with a description of the errors which can arise when using counters, and how these can be minimised.

10.2 Counter fundamentals

A basic block diagram of a counter is shown in Fig. 10.1. The input signals may not be clean square waves, and may have too high or too low an amplitude. The signal conditioner circuitry attenuates or amplifies the signal; converts d.c. to a.c. so as to remove any undesirable d.c. bias on the signal; varies the slope of the signal; sharpens the signal by feeding it through a Schmitt trigger, which also introduces hysteresis and makes the system less sensitive to noise. Some counters enable the hysteresis band and the point of trigger on the waveform to be controlled.

The time base consists of a stable oscillator, usually crystal controlled, which is temperature compensated or temperature controlled in an oven. The frequency of the time base element is usually 1, 5 or 10 MHz.

The input signal, and time base, are fed through programmable dividers, which can provide independent control on these sources, to meet the needs of the various measurement modes, as described in Section 10.3. These modes are usually selected by a mode select switch on the front panel of the instrument.

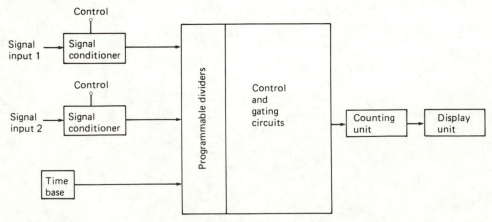

Fig. 10.1. Basic block diagram of a counter.

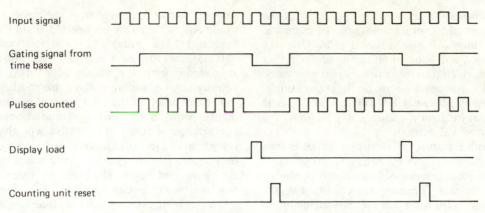

Input signal

Gating signal from
time base

Pulses counted

Display load

Counting unit reset

Fig. 10.2. Example of a frequency measurement cycle.

The control and gating systems consist of integrated circuits such as flip-flops, registers, counters and gates, and most modern instruments incorporate microprocessors (F.F. Mazda *Integrated Circuits*, C.U.P., 1978). These circuits determine the signal path through the instrument in order to achieve the required measurement mode.

The counting unit measures the number of pulses coming past the gating circuit, and this is shown on the display unit. Many different technologies have been used for the display, such as light emitting diode, liquid crystal, vacuum fluorescent and cathode ray tube.

The basic principle of frequency, period or time interval measurement is to compare the unknown quantity with a known quantity, using gating. For example Fig. 10.2 shows the measurement cycle for frequency. The gate is enabled for the period of the time base pulse, and the pulses passed through (6 in the illustration) are counted in the counting unit, and then loaded onto the display unit. The counting unit is subsequently cleared, ready for the next cycle. Knowing the period of the time base signal (say 1 ms in this case) the frequency can be calculated and displayed (6 kHz).

10.3 Measurement modes

Counters can be used in several modes to measure different parameters, the main ones being frequency, period, ratio, time interval and totalising. All these measurement modes use the same basic blocks, as shown in Fig. 10.1, but interconnect them in different ways within the control and gating structure.

10.3.1 *Frequency measurement*

This is probably the most frequently used application for a counter, and its operation was illustrated in Fig. 10.2. The arrangement of the gating structure is shown in Fig. 10.3(*a*), where

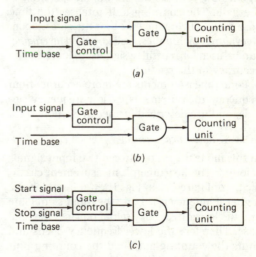

(a)

(b)

(c)

Fig. 10.3. Measurement modes for counters: (*a*) frequency measurement (Adapt for ratio measurement.), (*b*) period measurement, (*c*) time interval measurement. (Adapt for totalising.)

the time base controls the on and off operation of the gate. If the time base opens the gate for time *t*, and during this time the counting unit totals *n* pulses of the input signal, then the frequency of this signal is given by n/t. For example if 10 000 pulses are totalled in 0.1 ms then the frequency is 100 kHz.

Several variations are available in commercial

instruments. To increase the accuracy of the measurement several readings can be taken and their values averaged. This is called *frequency averaging*, and is based on the assumption that factors which cause inaccuracy are random, and so will average to zero on a large sample. Averaging obviously takes longer to calculate the final frequency, and can only be used on a repetitive waveform.

Another feature on a counter, called *normalised counting* allows the displayed frequency to be equal to the measured frequency multiplied by some constant. The value of this constant may be set up from the front panel of the instrument, or be programmed into the instrument's memory.

10.3.2 *Period measurement*

This is the inverse of frequency measurement, and is the time for the signal to complete one cycle. The gating arrangement, within the instrument, is shown in Fig. 10.3(*b*), and the input signal now controls the on and off operation of the gate. The counting unit totalises the time base pulses during the time that the gate is open. Averaging techniques, called *period averaging*, can be used for repetitive signals, to increase the accuracy of the reading.

Period measurements are more accurate than frequency measurements for low frequency signals, and are described further in Section 10.5.

10.3.3 *Ratio measurement*

In this mode the ratio between two input signals is found. The instrument's measurement circuitry is configured as in Fig. 10.3(*a*) in which the higher frequency is fed into the input signal line and the lower frequency signal replaces the time base. Therefore the lower frequency opens and shuts the counting gate, and the counting unit totalises the pulses from the higher frequency signal during this time. If the gate is open for one cycle of the lower frequency the counting unit gives the frequency ratio directly, although the average of several cycles may be taken to increase accuracy.

10.3.4 *Time interval measurement*

In this measurement mode, shown in Fig. 10.3(*c*), the gate is controlled by separate start and stop signals, and the counting unit totalises the pulses from the time base during this period, to give the elapsed time. The start and stop signals can be derived from various system conditions, which can be selected within the instrument. For example if the start signal is obtained from the rising edge of a pulse and the stop signal from the falling edge, then the counter gives a reading of the pulse width.

The resolution of the time interval measurement depends on the speed of the time base clock. For example a speed of 500 MHz will give a resolution of 2 ns. Time interval techniques can be used on repetitive signals to give resolutions in the picosecond range. The speed at which the stop and start circuits operate are important when measuring these small time intervals, and it is also important that the start and stop channels have equal delays.

10.3.5 *Totalising measurement*

The circuit used is similar to Fig. 10.3(*c*) except that the time base is replaced by the signal pulses which are to be totalised. The counting unit will now measure the number of pulses which occur between the start and stop signals. These signals may be derived from the instrument's front panel, or from certain system conditions, as in Section 10.3.4.

A variation of the totalising measurement mode is the *preset counter*. In this a count value is loaded into the instrument, and an output is obtained only when the number of signal pulses totalised by the instrument exceeds this preset value. Preset counters are widely used in process industries for applications such as limit setting.

10.4 Counter specification

There are two basic types of specifications: input specification, which describes the counter's input signal conditioning, and operating mode specification, which specifies how the counter will perform in its various operating modes.

10.4.1 *Input specifications*

Seven parameters are described in this section, although several more may be included in the instrument's data sheets.

(i) *Sensitivity.* This is the minimum input signal level, at a given frequency, which can be counted by the instrument. Sensitivity is usually specified as an r.m.s. value of a sinusoidal input. If the input is in the form of pulses then the minimum sensitivity is $2\sqrt{2}$ of the specified sinusoidal value. Therefore for the waveform

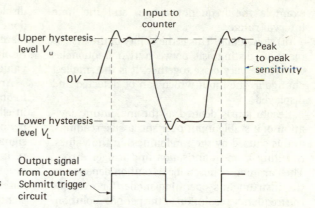

Fig. 10.4. Illustration of the effect of the counter's Schmitt trigger circuit.

shown in Fig. 10.4 the sine wave sensitivity is $(V_u - V_L)/2\sqrt{2}$.

Sensitivity is mainly determined by the gain of the input amplifier circuit, and by the hysteresis of the Schmitt trigger circuit. It is frequency sensitive, and so may be specified over several frequency bands. Peak to peak sensitivity is the difference between two hysteresis levels, as in Fig. 10.4. The pulse will register on the counter only when it crosses both levels, so the effects of noise and pulse ringing are minimised.

(ii) *Range.* This is the range of frequencies over which the input amplifier is specified. It usually varies depending on whether a.c. or d.c. coupling is used, since the lower frequency is limited for a.c. coupling.

(iii) *Trigger level.* If a pulse train goes positive or negative only, as in Fig. 10.5(a), then it may

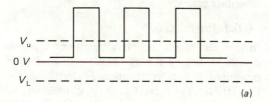

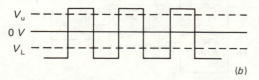

Fig. 10.5. Illustration of trigger level shifting; (a) no level shift, (b) positive level shift.

not cut both the hysteresis levels of the instrument, and will not be counted. Most counters have the facility for shifting the trigger level, either by a fixed amount or by a continuously variable amount, as in Fig. 10.5(b), so as to register the pulses. The trigger level specification is usually the voltage value at the centre of the hysteresis band.

(iv) *A.C. and d.c. coupling.* Signals having a d.c. bias may not register on the counter if the signal does not cross both hysteresis levels. The effect is similar to that shown in Fig. 10.5(a) and can be overcome by trigger level offset. However the d.c. bias can be removed by selecting a.c. coupling in the instrument, for sinusoidal or symmetrical signals, which cause the waveform to swing about the zero volts line.

(v) *Signal operating range.* This is the range of input signal voltage amplitudes over which the counter will work.

(vi) *Dynamic range.* This is the maximum peak to peak signal range, centered around the middle of the trigger level range, which the input amplifier can accept. If this range is exceeded the amplifier may saturate, and give errors.

(vii) *Damage level.* This is the maximum voltage which the input amplifier can stand without damage. The value of dynamic range may vary with the attenuation level setting.

10.4.2 *Operating mode specifications*

Four parameters in this group are described in this section.

(i) *Range.* This is the minimum to maximum value of the parameter which can be measured and displayed. The parameter selected depends on the mode of operation of the instrument, for example the frequency range, and the time interval range.

(ii) *Resolution.* This is the ability of the instrument to distinguish between two frequencies which are very close together. It is the smallest change in the input which can be detected and displayed.

Resolution is defined as the maximum deviation of a stable input over successive readings, and is caused by the randomness in the values, resulting from quantisation and trigger errors. The parameter chosen for resolution depends on the instrument's operating mode, for example the frequency resolution or the period resolution.

(iii) *Accuracy.* This is a measure of how close the measured value of the parameter is to the true value. Differences arise due to systematic errors, (e.g. channel delay error, time base error and trigger level error), or random error (e.g. quantisation error and time base short term stability). These errors are described in more detail in Section 10.7.

(iv) *Minimum dead time.* This is used for time interval measurement. It is the minimum time between the 'stop' signal, on a previous time interval measurement, and the 'start' signal on the current time interval measurement.

10.5 Low frequency measurement

The problem of measuring a low frequency signal is that a long measurement time is needed, to give good resolution. For example, in the conventional frequency measurement method, to measure an input frequency of 0.1 Hz would need 10 s to yield a resolution of 1 Hz. Measuring a 100 Hz signal over 1 s would give a resolution of 1 in 100.

Two methods are used to measure low frequencies. In the first the input frequency is multiplied by a factor of 100 or 1000 before going through the counting stage. The output is then divided down by the same factor before being

displayed. Frequency multiplication is usually done in a phase locked loop multiplier (F.F. Mazda, *Integrated Circuits*, C.U.P., 1978), and the resolution of the instrument is improved by a value equal to the multiplication factor.

An alternative technique for measuring low frequencies is to use a reciprocal counter, which in effect measures period. The reciprocal of the period is then taken to give the frequency of the signal. Fig. 10.6 shows the block diagram of a reciprocal counter. Two counting units are used, one to count the signal and the other the time base. The time base counter is gated by the signal, so that the number of time base pulses, for one cycle of the signal, is noted. Knowing the period of the time base the period of the signal can therefore be calculated, by the arithmetic unit, and displayed. Period averaging techniques can also be used where the gate is open for several cycles of the signal, and the reading of the time counter divided by that of the signal counter, before being multiplied by the period of the time base. If the input signal is 1 Hz, and the clock frequency is 1 MHz, then the reciprocal counter has a theoretical resolution of 1 in 10^6 for a measurement time of 1 s.

10.6 High frequency measurement

The speed of the logic used in the manufacture of a modern counter usually limits its direct measurement ability to below 1 GHz. As component techniques improve this limit will be exceeded, but at present some form of down-conversion is used to measure frequencies above 500 MHz, as described below.

10.6.1 *Prescaling*

In this technique the high frequency input is divided down by a ratio of 2 or 4, and then measured by a conventional circuit as in Fig. 10.3(*a*). The correct frequency value may be displayed by multiplying the counted frequency

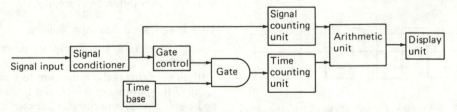

Fig. 10.6. Block diagram of a reciprocal counter.

by the divider ratio, or by scaling down the clock frequency by the same ratio, that is increasing the time for which the gate is open.

In the prescaler technique only the relatively simple prescaler circuitry needs to operate at high frequency. It can be used for frequencies up to about 2 GHz.

10.6.2 Heterodyne down-converter

In the heterodyne down-converter counter the high frequency input signal is fed into a mixer which beats it with a high stability internal signal. This internal signal is obtained by multiplying the time base frequency, and then passing it through a harmonic generator, which produces a comb line of frequencies spaced at intervals of f_t (Fig. 10.7), where f_t is typically between 10 MHz and 500 MHz. One of these frequencies, Nf_t is then selected by the tunable filter and sent to the mixer. It is the difference frequency $(f_i - Nf_t)$, from the amplifier, which is counted.

The value of N is chosen automatically by the instrument. It starts at a value of $N = 1$ and then steps through the values, in the tunable filter, until the output from the amplifier is within the bandwidth of the instrument. The value of N is then noted and Nf_t is added to the counted value $(f_i - Nf_t)$ in the display, to give the input frequency f_i.

Automatic gain control in the amplifier pro-vides a degree of noise immunity, by ensuring that only the strongest signal, within the pass band of the amplifier, is detected.

10.6.3 Transfer oscillator down-converter

In this system, illustrated in Fig. 10.8, a low frequency oscillator is phase locked to the microwave input frequency. The lower frequency, f_i/N can now be measured, using the conventional counter circuitry, and is usually in the range 10–200 MHz. A second phase locked loop is used to find the value of N, and this feeds a divider which slows down the time base by the same value, so that the counting unit records the value of f_i.

10.6.4 Harmonic heterodyne down-converter

As its name implies this is a mixture of the heterodyne and transfer oscillator techniques. The high frequency input f_i is mixed with a variable lower frequency Nf_s from an internal frequency synthesiser, and the difference frequency $f_i - Nf_s$ is measured, as is done in a heterodyne converter. The ratio N is determined by a phase locked loop method, as in the transfer oscillator converter, and the result is found by adding Nf_s to the value of the counter, before the result is displayed.

10.6.5 Comparison of techniques

Table 10.1 summarises some of the basic dif-

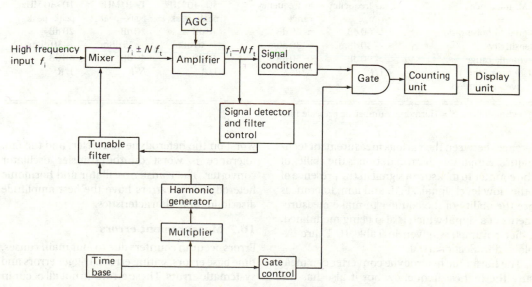

Fig. 10.7. Block diagram of a heterodyne down-converter counter.

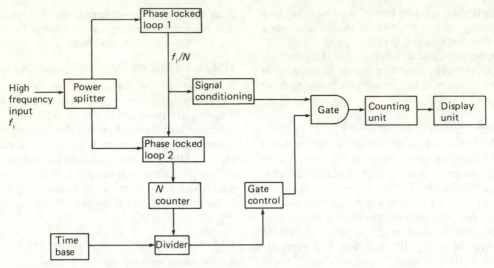

Fig. 10.8. Block diagram of a transfer oscillator down-converter counter.

Table 10.1 *Comparison of high frequency counter techniques*

Parameter	Direct gating	Prescaler	Heterodyne	Transfer oscillator	Harmonic heterodyne
Frequency range	700 MHz	2 GHz	25 GHz	30 GHz	40 GHz
Acquisition time	immediate	immediate	150 ms	150 ms	400 ms
Amplitude discrimination	accepts all in frequency range	accepts all in frequency range	3–30 dB	2–10 db	2–10 db
A.M. tolerance	> 90%	> 90%	< 50%	> 90%	> 90%
F.M. tolerance	as frequency range	as frequency range	30–40 MHz peak–peak	1–10 MHz peak–peak	10–50 MHz peak–peak
Signal to noise ratio	> 60 dB	> 60 dB	40 dB	20 dB	20 dB
Sensitivity	− 30 dBm	− 30 dBm	− 30 dBm	− 35 dBm	− 30 dBm
Dynamic range	> 90 dB	> 50 dB	40 dB	40 dB	40 dB
Gate time for a given resolution*	1/R	N/R	1/R	N/R	1/R

*R = Resolution. N = Harmonic number or prescale ratio.

ferences between the various measurement techniques. Amplitude discrimination is the ability of the counter to measure a signal in the presence of other low level signals. F.M. and a.m. tolerances are the ability of the counter to make measurements of a signal which is also being modulated. Other parameters, given in Table 10.1, are explained in Section 10.4.

The harmonic heterodyne converter can measure the highest frequency, but it also has the longest acquisition time. The a.m. tolerance is

worst on the heterodyne converter, and the f.m. tolerance is worst on the transfer oscillator converter. The transfer oscillator and harmonic heterodyne converters have the best amplitude discrimination characteristics.

10.7 Measurement errors

Errors occur in counters due to four main causes, time base errors, gating errors, trigger errors and systematic errors. These errors do not all occur in every measurement mode, and the modes in

Table 10.2. *Types of errors and the measurement modes in which they occur*

Type of error	Measurement mode		
	Frequency	Period	Time Interval
Time base	√	√	√
Gating	√	√	√
Trigger	—	√	√
Systematic	—	—	√

which they occur are given in Table 10.2. Frequency measurements, therefore, are only affected by time base and gating errors; and systematic errors are only important when making time interval measurements.

10.7.1 *Time base error*

This error is caused by the change in frequency of the internal time base oscillator. There are several causes for time base error.

(i) Initial error, due to faulty set up during calibration. Counters are usually calibrated by comparing them with a standard frequency broadcast.

(ii) Short term stability error. This gives momentary frequency changes due to effects such as shock, vibration and voltage transients. Their effect can be minimised by taking measurements over a long time, and then averaging the result.

(iii) Long term stability error. This is mainly due to ageing of the crystal oscillator causing a frequency drift. Ageing is most pronounced during the early life of a crystal, and it can be minimised by operating the device in a temperature controlled oven.

Time base error is constant regardless of frequency. If the time base is supposed to be 1 MHz but is really 1 000 010 Hz then the instrument error is 0.001%, and this error will occur whether a 1 kHz or 100 kHz signal is being measured.

10.7.2 *Gating error*

This is also known as ± 1 count error, and is illustrated in Fig. 10.9. For the signal shown in Fig. 10.9(a) the counter will record five pulses during the gating period, whereas for a slightly phase shifted waveform, as in Fig. 10.9(b) this reading is six pulses.

Gating error gives a ± 1 count error in the least significant digit of a reading. The error is inversely proportional to the frequency being measured and to the gate time, and at low frequencies it can be very large. It is better therefore to take period measurements at low frequencies. Generally if f_T is the frequency of the internal time base of the instrument, then for signal frequencies below $(f_T)^{1/2}$ period measurements should be made, and for input frequencies above this value the frequency mode should be used. The error at this frequency is given by 100 $(f_T)^{-1/2}$ per cent. Gating error is also minimised by using the averaging measurement mode.

10.7.3 *Trigger error*

Trigger error arises primarily due to noise on the input signal or on the input channel of the counter. For period and time measurements the input signal opens and shuts the control gate of the counter, and noise on this signal can give a false gate operating time.

Fig. 10.10(a) illustrates the effect of noise on the input signal, giving a false count due to the waveform in Fig. 10.10(b). Everytime a noise

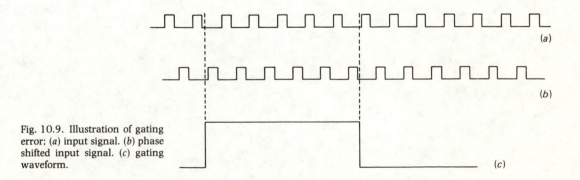

Fig. 10.9. Illustration of gating error; (a) input signal. (b) phase shifted input signal. (c) gating waveform.

spike crosses both the upper and lower trigger levels a false count is registered. If the trigger levels are spaced wider apart when making this measurement, as at *CD*, then the effect of the signal noise is nulled, as in Fig. 10.10(*c*).

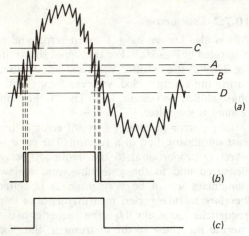

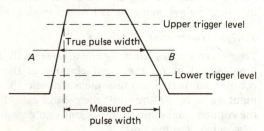

Fig. 10.10. Illustration of trigger error due to signal noise giving high count; (*a*) input signal, (*b*) counter waveform with trigger level set at AB, (*c*) counter waveform with trigger level set at CD.

Fig. 10.11. Illustration of trigger error giving false period measurement due to asymmetrical pulse rise and fall times.

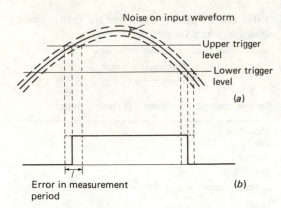

Fig. 10.12. Illustration of trigger error giving uncertain period of measurement; (*a*) input waveform, (*b*) counter measurement waveform.

Fig. 10.11 shows the effect of a signal having different rise and fall times; the measured pulse width is greater than the actual width. The error can be minimised by narrowing the trigger band so that it almost coincides with the centre line *AB*.

Noise on a shallow input signal waveform can also cause jitter in the measurement period, as in Fig. 10.12. Its effect can be reduced by sharpening up the waveform edges, and by narrowing the hysteresis band, so that the waveform passes quickly through the band.

10.7.4. *Systematic error*

Systematic error occurs in time interval measurements. It is mainly caused by mismatch between the propagation delays and rise times in the stop and start channels within the instrument, and by delay differences in external cabling used to connect the signals to the instrument.

11. Signal analysis

11.1 Introduction

The analysis of electrical signals is required for many different applications. Instruments which do this include spectrum analysers, wave analysers, distortion analysers, audio analysers and modulation analysers. The last three instruments are primarily used in audio or telecommunication applications, and are described in Chapters 14 and 15. In this chapter the principles of wave and spectrum analysers are described, and the measurement of signal noise.

All signal analysis instruments measure the basic frequency properties of a signal, but they use different techniques to do so. A spectrum analyser sweeps the signal frequency band, and displays a plot of amplitude versus frequency. It has an operating range of about 0.02 Hz–250 GHz. A wave analyser is a voltmeter which can be accurately tuned to measure the amplitude of a single frequency, within a band of about 10 Hz–40 MHz.

Distortion analysers operate over a range of 5 Hz–1 MHz, and give a measure of the energy present in a signal outside a specified frequency band. They therefore tune out the fundamental signal and give an indication of the harmonics. An audio analyser is similar to a distortion analyser but can measure additional functions, such as noise. Modulation analysers tune to the required signal and recover the whole a.m., f.m. or phase modulation envelope for display or analysis.

11.2 Wave analyser

The wave analyser has also been called a frequency selective voltmeter, a selective level voltmeter, and a carrier frequency voltmeter. It is used to measure the amplitude of a single frequency in a complex wave, such as measuring amplitude in the presence of noise.

There are two types of wave analysers, depending on the frequency ranges. For measurements in the audio frequency range the signal is attenuated to the required level, and amplified. It then goes through a narrow pass band filter, which is tuned to select the required frequency. This frequency is amplified and fed to the output indicator. The indicator usually consists of an analogue meter, which gives the amplitude of the selected frequency, and a mechanical or electronic digital readout, which indicates the frequency selected.

The instrument has a very narrow bandwidth of about one per cent of the selected frequency.

For frequency measurements in the megahertz range a heterodyne wave analyser construction is used, as shown in Fig. 11.1. The input signal is fed through an attenuator and amplifier before being mixed with a local oscillator. The frequency of this oscillator is adjusted to give a fixed frequency output which is in the pass band of the i.f. amplifier. This signal is then mixed with a second crystal controlled oscillator, whose frequency is such that the output from the mixer is centred on zero frequency. The subsequent active filter has a controllable bandwidth, and passes the selected component of the frequency to the indicating meter.

Good frequency stability in a wave analyser is obtained by using frequency synthesisers, which have high accuracy and resolution, or by automatic frequency control (AFC). In an AFC system the local oscillator locks to the signal, and so eliminates the drift between them.

11.3 Spectrum analyser

A spectrum analyser is a wave analyser whose measurement frequency is swept electronically, and displayed on a cathode ray tube. In this section the principle of operation of spectrum

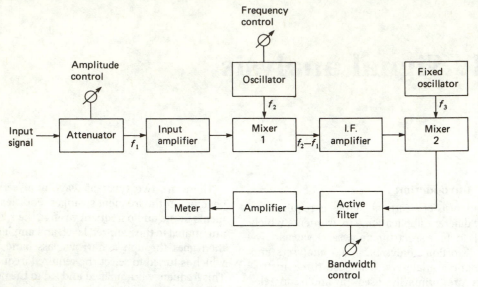

Fig. 11.1. Block diagram of a heterodyne wave analyser.

analysers, and their characteristics, are described. Applications of spectrum analysers are covered under the relevant chapters in Part 3 of the book.

Fig. 11.2 illustrates the basic difference between an oscilloscope trace and spectrum analyser trace. The oscilloscope shows the amplitude of the signal plotted to a time base, and is therefore said to work in the time domain. The spectrum analyser shows the amplitude of the signal plotted to a frequency base, and is therefore said to work in the frequency domain. The instrument breaks the signal down into its individual frequency components, and displays these on the cathode ray tube (CRT) screen as a series of vertical lines. The length of each line represents the amplitude of the signal, and the position on the screen represents its frequency. The advantage of the spectrum analyser over an oscilloscope is evident from Fig. 11.2. The oscilloscope shows the signal to be almost sinusoidal, but on the spectrum analyser each component of the spectrum is seen individually, rather than summed in the time domain, so it is more accurate in measuring the distortion.

There are two principal types of spectrum analysers, those operating in real time and those operating in a swept tuned mode. These are described in the sections 11.3.2 and 11.3.3.

11.3.1 *Spectrum analyser characteristics*

A few of the basic parameters used to determine the performance of a spectrum analyser are described in this section. They are primarily related to a swept superheterodyne analyser (see Section 11.3.3) which is the most popular analyser in current use.

Resolution

The resolution of an analyser is its ability to distinguish between neighbouring signals, and it

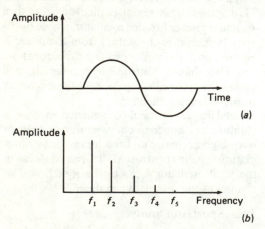

Fig. 11.2. Difference between an oscilloscope display and a spectrum analyser display; (*a*) amplitude–time plot of oscilloscope, (*b*) amplitude–frequency plot of spectrum analyser.

is determined by the instrument's narrowest i.f. bandwidth. If this is 1 kHz, for example, then this is the closest any two signals can come, and still be resolved. Other limitations are the analyser's residual f.m., which affects its usable i.f. bandwidth, and noise sidebands, which occur above the skirt of the i.f. filter, and so reduce its out of band rejection. The i.f. bandwidth cannot be made too narrow since this increases its time constant, and therefore the scan time of the analyser.

The ratio of the 60 dB bandwidth to the 3 dB bandwidth, both in hertz, is called the shape factor of the filter, and the smaller this factor the higher the resolution of the analyser. Synchronously tuned filters are usually used in analysers, because of their phase linearity, and this limits the shape factor. Stagger tuned filters can be used, which have better shape factors, but they also have phase discontinuities at band edges, and produce ringing when sweeping frequencies through them, as is done in a spectrum analyser.

Sensitivity

Sensitivity is the ability of the analyser to detect small signals, and it is limited by the noise generated internally in the instrument. Noise can be thermal and non thermal. Thermal noise is caused by temperature effects, and is proportional to temperature and the bandwidth of the system. Therefore decreasing the bandwidth will give a corresponding noise improvement. Non thermal noise is due to mismatches of impedance within the analyser, and switching noise in the electronic devices. Noise adds to the signal amplitude, and one method of specifying sensitivity is as the input signal level, which is equal to the internal noise level.

To measure low signal levels a low pass filter can be used after the detection stage, which averages the internal noise and so isolates the input signal more effectively. Other techniques which may be used to minimise the effects of noise are phase locking, and the synthesis of the local oscillator.

Stability

Analyser stability is clearly an important parameter, and it is essential that the analyser is more stable than the signal it is measuring. The stability of a spectrum analyser is determined primarily by the frequency stability of its local oscillator. Stability can be improved by phase locking the oscillator to a tooth of a comb generated by a crystal controlled oscillator.

Input signal level

Very large input signal levels can damage the input circuitry of the spectrum analyser, and this is called the *maximum input level* of the analyser.

For signals below the maximum level the analyser compresses the input signal. This results in the signal amplitude, displayed on the CRT, not being a good measure of the actual input signal level. The *linear input level* of the analyser is the signal level which gives less than 1 dB gain compression.

All signals give some distortion in the analyser, usually due to non-linear effects of the input mixer. Input attenuators are frequently used before the mixer to limit this distortion. The *optimum input signal level* of the analyser is the value, at a specified attenuator setting, which keeps the internally generated distortion below a defined value.

Dynamic range

This is the ratio of the largest to the smallest signal which can be displayed simultaneously, without appreciable distortion. The dynamic range can be improved by reducing the signal level at the mixer, but now the sensitivity suffers. To have good dynamic range the display range must be large enough, no spurious responses should occur, and the sensitivity must be sufficient to eliminate noise from the displayed amplitude range.

Frequency response

The frequency response is basically a measure of the amplitude linearity of the analyser over its frequency range. It is very important to have this linearity since a spectrum analyser primarily compares amplitudes at different frequencies. Good frequency response requires a mixer with a conversion loss which is independent of frequency, and which has a flat response.

11.3.2 *Real time spectrum analyser*

Real time analysers can be either multichannelled, or based on the Fourier transform.

Multichannel analyser

The multichannel real time analyser uses a

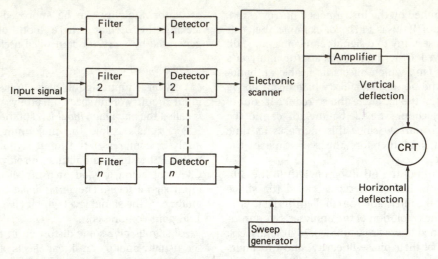

Fig. 11.3. Real time multichannel spectrum analyser.

series of fixed band pass filters, as in Fig. 11.3, whose cut off slopes intersect at their 3 dB response points. The input signal is applied to each filter at the same time, and the individual frequency components are selected in parallel by the filters. The electronic scanner selects the output of each detector in turn, and this is fed to the vertical deflection plates of a CRT, in synchronism with the signal which operates the scanner, and which also controls the horizontal deflection plates. This spaces the amplitudes of the signals on the CRT, on a frequency grid.

The resolution of the multichannel real time analyser is limited by the bandwidth of the filters, and the frequency range is defined by the band covered by all the filters, that is the number of filters and their individual bandwidths. The analyser is relatively expensive, since it requires several filters, and it is not very flexible, due to the fixed resolution of the filters. Its advantage is that, since all the filters see the signal at the same time, a transient can be readily caught by the instrument. The multichannel analyser also has a better performance than swept tuned instruments for d.c. and low audio frequencies, since the filter pass band can be made very narrow for good resolution. If this is done in a swept tuned instrument then the sweep speed would need to be reduced.

Fourier transform analyser

The Fourier transform analyser determines the magnitude and phase of each frequency component from a series of time domain samples of the input signal. It works on the mathematical principle of the Fourier transform, which breaks a waveform down into the sum of its sinusoidals. From a given time domain of the signal the transform calculates the amplitude of the individual sines and cosines, and plots these in a frequency domain. The mathematical statement of the Fourier transform is given by (11.1)

$$S(f) = \int_{-\infty}^{+\infty} X(t)\exp(-j2\pi ft)dt \qquad (11.1)$$

where $S(f)$ is the frequency domain function and $X(t)$ is the time domain function.

The Fourier transform is based on continuous data, which stretches $-\infty$ to $+\infty$. In most instruments the signal is measured as a series of samples in finite time, and the discrete Fourier transform has been developed to handle this. It is stated mathematically as a sum of the samples, in (11.2)

$$S(n) = \frac{1}{N}\sum_{m=0}^{N-1} X(m)\exp(-j2\pi nm/N) \qquad (11.2)$$

where $n = 0, 1, 2, 3, \ldots, (N-1)$

The fast Fourier transform, on which modern spectrum analysers are based, is a version of the discrete Fourier transform, which allows faster calculations. It takes advantage of the fact that, when integral values of n and m are put into (11.2), several terms generated have identical values. These predictable identical terms can be

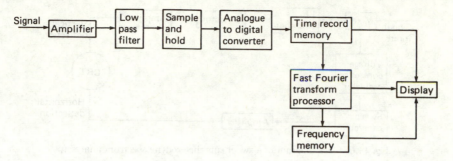

Fig. 11.4. Block diagram of a spectrum analyser using a fast Fourier transform.

combined, so reducing the number of calculations needed compared with the conventional Fourier transform.

Fig. 11.4 shows the block diagram of a typical spectrum analyser based on the fast Fourier transform. The input signal is amplified, and then filtered to remove out of band components. The characteristics of this low pass filter are important, since it prevents aliasing due to signal sampling. The signal is then sampled, in the sample and hold circuit, at regular intervals, and converted to digital form by the analogue to digital converter. The digitised samples are stored in memory until the required sample size, known as a time record, has been obtained. The processor then performs a fast Fourier transform, on the time domain points, to get the frequency domain results, which are saved in the frequency memory. These can be further analysed, or displayed on a CRT display, or on a plotter.

The fast Fourier transform spectrum analyser has an operating frequency range from d.c. to about 100 kHz. It is faster at low frequencies than the swept spectrum analyser, since it can simultaneously calculate all frequency values from one set of observed inputs.

11.3.3 *Swept tuned logic analysers*

There are two principal types of swept tuned analysers, swept trf and swept superheterodyne, and these are described in the following sections.

Swept trf spectrum analyser

The swept trf (tuned radio frequency) analyser, shown in Fig. 11.5, uses a voltage tuned bandpass filter which has a very narrow pass band. The filter is driven from a sweep generator which provides a sawtooth voltage waveform. This causes the centre frequency of the filter to be swept from a minimum to a maximum frequency. A component of the input signal frequency is passed by the filter only when its pass band is tuned to it, so the input circuit selects each frequency in turn. The signal amplitude from the filter is detected, amplified and then fed to the vertical deflection plates of the cathode ray tube. The frequency corresponding to the amplitude of the signal at that time is fed to the horizontal plates of the CRT, by the sweep generator, giving an amplitude–frequency plot of the input signal.

A swept trf spectrum analyser is relatively inexpensive, and is the instrument most often used for microwave measurements since broadband tuned filters are readily available. Its disadvantages are poor sensitivity, and poor resolution, which is determined by the filter band-

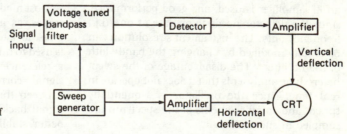

Fig. 11.5. Block diagram of a swept trf spectrum analyser.

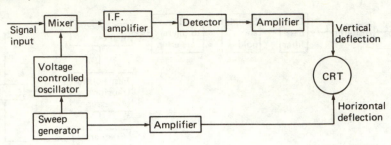

Fig. 11.6. Block diagram of a swept superheterodyne spectrum analyser.

width and which varies with frequency for a tuned filter.

Swept superheterodyne spectrum analyser

The swept superheterodyne analyser differs from the swept trf analyser in that the signal spectrum is swept through a fixed band bass filter, rather than sweeping the filter through the spectrum. It may be considered to be a narrow band receiver which is electronically tuned over a frequency range.

A block diagram of the swept superheterodyne spectrum analyser is shown in Fig. 11.6. It is similar to the swept trf analyser in which the tuned filter is replaced by a voltage controlled oscillator, mixer and i.f. amplifier. The voltage controlled oscillator is swept between a minimum and maximum frequency, and the difference between this and each element of the input frequency is fed to the i.f. amplifier. The i.f. amplifier passes one intermediate frequency only, and if this is present from the mixer output it drives the vertical plates of the CRT whilst the sweep generator drives the horizontal plates. For example if the i.f. amplifier passes 100 kHz signals, and the voltage controlled oscillator sweeps between 200 kHz and 300 kHz, then frequencies in the input signal between 100 kHz and 200 kHz will be registered by the analyser.

The swept heterodyne is the most popular spectrum analyser. It has high sensitivity, since an i.f. amplifier is used, and good performance, since the detector needs to operate at one (i.f.) frequency only. The instrument's resolution can be readily modified by changing the bandwidth of the i.f. filters. The disadvantage of the swept heterodyne analyser is that it does not operate in real time, since the oscillator is sequentially tuned, and small segments of the spectrum are sampled in succession.

11.3.4 *Spectrum analyser developments*

The basic spectrum analyser can be modified for special applications. To measure very high frequencies would need a high frequency local oscillator. For example to measure a 0–2 GHz input signal requires a 2 GHz i.f. amplifier stage, and a local oscillator of 2–4 GHz. For very high input frequencies it is possible to generate harmonics from the local oscillator in the mixer, and to use this. There are several problems with the use of harmonic mixers, such as image responses, multiple responses and spurious responses. These can be overcome by utilising a bandpass filter which tracks the voltage controlled oscillator frequency, so that the analyser sees an input signal only when it is tuned to that signal frequency. This filter is called a tracking preselector, and tracks automatically. The preselector reduces the sensitivity and flatness of the system, although its effect is small, but it can be removed from the circuit when these factors are critical.

Modern spectrum analysers use microprocessor control, and have an IEEE 488 bus to interface to a remote controller. They incorporate features such as automatic peak signal search; digital signal averaging, which allows low level signals to be measured in the presence of noise, without needing a slow sweep; electronic control of display grids; and automatic selection of optimum resolution, bandwidth, and sweep speed.

Most modern spectrum analysers use multiple stages, as in Fig. 11.7, and possibly also pilot phase lock techniques. In this a second i.f. system is used to mix signals from a local oscillator with a reference signal, and then to feed an error signal, corresponding to the phase difference between this pilot i.f. signal and the reference signal, back to the first stage. Such a system gives better stability and tuning accuracy, and less

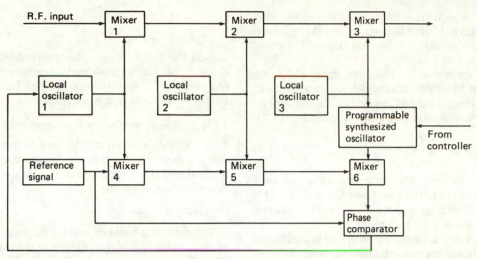

Fig. 11.7. Swept superheterodyne spectrum analyser using three stages and a pilot signal phase-lock technique.

noise. A typical specification would be: 10 Hz resolution at 2 GHz; a frequency range of 100 Hz–2 GHz; 10^{-9} per day frequency stability; amplitude measurement range of $+30$ to -140 dBm; and amplitude accuracy of ± 2.5 dB.

11.4 Noise measurement

In this section the terms used to describe noise are first introduced. This is followed by a description of the methods employed to measure noise, specially those methods which use a noise meter. Finally the errors which can arise when measuring noise are highlighted.

11.4.1 *Noise definitions*

Noise is the electrical interference in a system, which degrades its performance from the predicted value, and often masks the signal being measured. System sensitivity is affected both by the noise which is present in the signal being measured, at the input of the instrument, and by the noise which is generated internally within the instrument.

Four terms are commonly used to define noise.

Noise factor. This defines the amount of noise generated in an instrument, and is given by the ratio of the signal to noise ratios at the input and the output of the instrument. Therefore if a signal passing through an instrument has a signal to noise ratio of SN_i at the input to the instrument,

and a signal to noise ratio of SN_o at the output of the instrument, then its noise factor NF_a is given by (11.3), as a dimensionless ratio.

$$NF_a = SN_i/SN_o \qquad (11.3)$$

Noise figure. This is numerically equal to the logarithm of the noise factor, and is measured in decibels. Therefore the noise figure NF_i (in dB) is given by (11.4).

$$NF_i = \log(SN_i/SN_o) \qquad (11.4)$$

The noise figure can also be defined in terms of the gain and bandwidth of the measurement system. For instance if N_o is the noise generated at the output of a system, of gain G and bandwidth B, when there is no noise in the signal which is present at the input of the system, then the noise figure of the system is given by (11.5).

$$NF_i = N_o/kTBG \qquad (11.5)$$

In this equation k is Boltzmann's constant $(1.374 \times 10^{-23}$ joule/K) and T is the absolute temperature (290 K).

Alternatively if a system is fed from a noise source having a relative excess power of E; N_o is the noise power output from the system with the noise source off (cold); N_i is the noise power output with the noise source on (fired); then the noise figure of the system is given in dB, by (11.6).

$$NF_i = 10 \log E - 10 \log \left(\frac{N_i}{N_o} - 1 \right) \qquad (11.6)$$

The first term, $10 \log E$, is called the excess noise ratio (ENR) of the noise source, and (11.6) is used very frequently in noise measurement systems.

Noise temperature.   The noise figure (NF_i) of a system in dB can also be defined in terms of its noise temperature (T) in kelvin and the conversion between the two terms is given by (11.7)

$$NF_i = 10 \log \left(1 + \frac{T}{290} \right) \qquad (11.7)$$

Both noise figure and noise temperature are used to define a system's performance and (11.7) is often plotted on a spiral chart to aid in conversion between the two terms.

The noise figure of a system can be calculated in terms of the noise temperatures in the measurement instrumentation. For example if T_c is the temperature of the noise source when it is cold; T_F is its temperature when fired; N_o and N_i are the noise power outputs from the system with the noise source cold and fired; Y is the Y-factor and is numerically equal to N_i/N_o; then the noise figure of the system is given by (11.8)

$$NF_i = \frac{(T_F/290 - 1) - Y(T_c/290 - 1)}{Y - 1} \qquad (11.8)$$

The excess noise ratio of the noise source is given by (11.9)

$$\text{ENR} = 10 \log (T_F/290 - 1) \qquad (11.9)$$

If the cold temperature of the noise source (T_c) is assumed to be about $290 \, \text{K}$, then (11.8) gives the value of noise figure approximately, in dB, as in (11.10)

$$NF_i = \text{ENR} - 10 \log (Y - 1) \qquad (11.10)$$

The error in using (11.10) is about $0.1 \, \text{dB}$, and is caused by the fact that T_C is not exactly $290 \, \text{K}$.

Cascaded noise.   This gives a measure of the noise in a cascaded system. If two amplifiers of gain G_1 and G_2 are cascaded together, and NF_1 and NF_2 are the noise figures of the first and second stages, then the cascaded noise figure is

given by (11.11)

$$NF_c = NF_1 + (NF_2 - 1)/G_1 \qquad (11.11)$$

Usually the second stage noise is unwanted and therefore results in a noise measurement error of $(NF_2 - 1)/G_1$. It may be compensated for if the values of NF_2 and G_1 can be measured.

11.4.2 *Manual methods of measuring noise*

Two methods are commonly used to measure noise. These are known as the twice power method and the Y-factor method, and they are described in this section.

Twice power method

This method is based on making the value of N_i equal to twice that of N_o in (11.6), so the value of noise figure reduces to the ENR of the noise source. Fig. 11.8 shows a system which can be used for this. Switch SW is initially closed to take the 3 dB pad out of circuit. The noise source is switched off (cold) and the attenuator adjusted to give a convenient reading of N_o on the power detector. The attenuator has no effect on the input terminator noise power since the signal termination and attenuator are at the same temperature, so the receiver sees a matched input.

The noise source is now turned on (fired) to give N_i. Switch SW is opened, to insert the 3 dB pad into the circuit, and the attenuator again adjusted to give the same detector reading as before. The value of N_i, as output from the system under test, is therefore given by (11.12)

$$3 \, \text{dB} = \log \frac{N_i}{N_o} \qquad (11.12)$$

This gives $N_i = 2N_o$. Therefore, from (11.6), the noise figure of the system under these conditions is equal to the attenuated excess noise ratio of the source. This can be found as the noise figure of the noise source less the setting of the attenuator. An argon source is usually used which has an ENR of 15.2 dB at 4 GHz and 17 °C, varying by about 0.3 dB over a frequency range of 10 MHz–20 GHz.

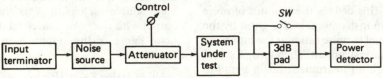

Fig. 11.8. Twice power method of measuring noise figure.

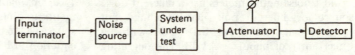

Fig. 11.9. *Y*-factor method of measuring noise figure.

The twice power method of measuring noise is reliable, simple and accurate. But it is not convenient to use because of the need to switch the 3 dB pad in and out of circuit, and to make adjustments of the attenuator.

Y-factor method

The *Y*-factor is equal to the ratio N_i/N_o and the *Y*-factor method consists of measuring this ratio and inserting it into (11.6) to give the noise figure, knowing the ENR of the noise source.

Fig. 11.9 shows an arrangement for measuring noise by the *Y*-factor method. With the noise source cold the attenuator is adjusted to give a convenient reading for N_o on the meter. The noise source is then fired and the attenuator increased to bring the detector reading back to its original value. The increase in the value of the attenuator is equal to the *Y*-factor of the system in dB. This can be converted to a power ratio and substituted in (11.6), along with the ENR of the noise source, to get the noise figure of the system under test.

The *Y*-factor method is more popular than the twice power method for two main reasons. (i) The insertion loss of the attenuator does not contribute to the noise figure of the system under test. (ii) The signal into the system under test is kept at a constant level, so that errors, such as those due to non-linearities, are minimised.

11.4.3 *Noise figure meter*

Several different principles are used in automatic noise figure meters, the most popular being the *Y*-factor method. In the example shown in Fig. 11.10, the *Y*-factor (ratio N_i/N_o) is measured and the instrument then calculates the noise figure using (11.6). The gating signal turns the noise source on and off at a low frequency, so the signal which arrives at the input to the i.f. amplifier is alternately with the noise source cold and fired. The i.f. amplifier is also gated, after a delay, which allows the noise source to settle to the required value before it is passed on to the square law detector. This detector converts the input power to an output voltage level.

When the noise source is fired the signal N_i, after amplification, is passed by the switch, which is also controlled by the gating source, to an AGC integrator. This provides a controlling voltage which regulates the gain of the i.f. amplifier. The system has a long time constant so the gain is controlled even when the noise source is cold and the N_o signal is passing through it. Since during this period the N_i signal is kept at a constant level by the AGC, the value of N_o which is detected is really the pulse ratio of N_i/N_o. This is directed by the switch to the meter processor where the noise figure can be calculated, from (11.6), knowing the ENR of the noise source. This noise figure is then displayed.

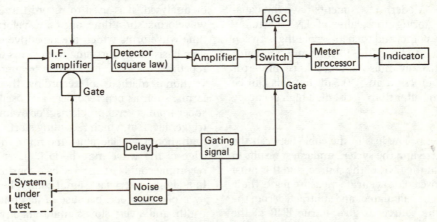

Fig. 11.10. Block diagram of a noise figure meter.

Most noise figure meters use a different noise source for different noise level readings, but cover a band of about 0.1–30 dB with an uncertainty of about ± 0.1 dB. The operating frequency band is about 10 MHz–18 GHz.

11.4.3 *Errors in noise measurements*

Several errors arise in noise measurements, a few of these being described in this section.

Mismatch.   This results in power loss, and since noise is random the loss is unpredictable and is difficult to account for. Matching at the noise source is critical. However the gain of the amplifiers changes due to change in the source impedance, as the source switches between its cold and fired states.

Mismatch effects can be minimised by using a coupler of 20 dB or so, but the test signal then needs to have a power at least 20 dB above the noise level for good results. The maximum noise value is set by the requirement to prevent the amplifier from limiting. An alternative solution to the mismatch problem is to buffer the amplifier from the noise source by an isolator or circulator. This reduces the effective impedance variations as seen by the amplifier. Although the insertion loss in the buffer adds to the noise figure of the system under test, this is usually below 0.5 dB in value.

Accuracy of ENR of noise source.   Inaccuracies in the ENR of the noise source result in a proportionate error in the noise figure measured, and this should be compensated for. The variation of ENR depends on the type of noise source and the frequency. Modern instruments have the capability of storing the value of ENR at many frequency points and then interpolating between the points to obtain the ENR at any frequency setting. Usually noise sources have an error variation of less than ± 0.5 dB but this can be reduced to better than ± 0.1 dB if high accuracy is needed.

Cable losses.   The loss in the cable between the noise source and the system under test results in changes in the ENR of the source, and this must be minimised. For accurate measurements these losses can be calculated, and subtracted from the ENR of the source to give a true ENR of the system under test.

T_c *approximation errors.*   Most simple instruments assume the cold temperature of the noise source (T_c) to be 290 K, and use the approximate equation (11.10) to calculate the noise figure. This can result in errors of about 0.1 dB in noise figure measurements below 1 dB, since the value of T_c is usually about 296 K, or room temperature. Modern instruments, using microprocessor control, can calculate the value of the noise figure from (11.8), using the actual value of T_c, and so avoid this error.

Cascaded noise error.   This is the error caused by the second stage noise figure of the instrument. A calibration step can be used to measure an instrument's own noise, followed by the measurement of the gain of the first stage. The noise figure of the system under test can then be calculated from (11.11).

Interference.   Special precautions, such as shielding connections and using screened rooms, need to be taken to minimise interference from outside sources, since the voltages involved in noise measurements are very low.

Noise meter errors.   These errors are caused by hardware factors such as the variation of the detector from a square law, ageing effects and meter tracking. Usually the accuracy is within ± 0.5 dB but, for low noise measurement, meters can be obtained with an accuracy of ± 0.2 dB.

11.5 Waveform recorder

A waveform recorder is primarily used to capture a fast or transient waveform, which is stored digitally and can subsequently be manipulated or analysed. It is used to record three types of waveforms; short time, one off waveforms; short time waveforms which are repetitive on a random basis; and short time waveforms which are repetitive at a low predictable frequency. Conventional methods of recording these waveforms, such as pen recorders and oscilloscopes, suffer from several problems. For example if the trigger level, at which recording starts, is set too high then a considerable amount of the leading edge of the wave may be lost. If set too low, recording could be triggered by a spurious signal. In addition a very fast signal cannot be recorded by a pen recorder, because of its limited bandwidth, and a very slow signal is inconvenient to view on an oscilloscope.

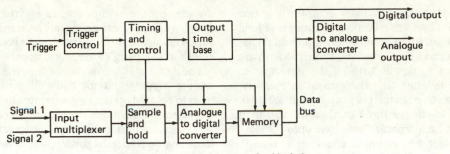

Fig. 11.11. Waveform recorder block diagram.

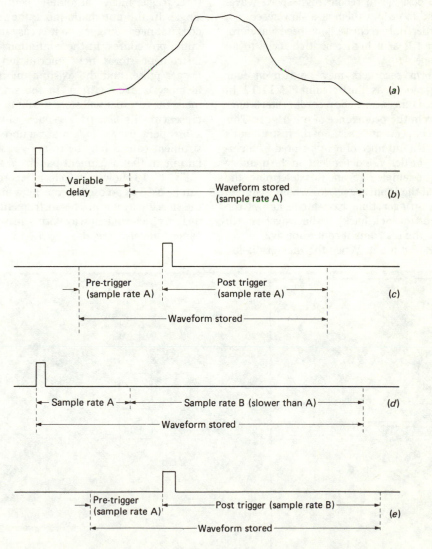

Fig. 11.12. Waveform recorder operating modes: (a) waveform being recorded, (b) delayed time based mode, (c) pre-trigger time base mode, (d) dual time base mode, (e) pre-trigger dual time base mode.

Fig. 11.11 shows a block diagram of a two channel waveform recorder. A sample and hold circuit is used to sample the input analogue signal at short time intervals. These readings are converted to digital form by the analogue to digital converter, and then stored in memory. The data stored in memory can be manipulated, such as to change the X or Y scales, and then read out by a separate time base onto the data bus. A digital to analogue converter is used to recreate the original analogue waveform. It is therefore possible to record high speed waveforms, and to output them at a slow speed to a pen recorder, or to record a slow speed waveform and output it at a high repetitive speed to an oscilloscope.

Waveform recorders may be run in four operating modes, as illustrated in Fig. 11.12. In the delayed time base mode recording starts after a delay from the occurrence of the trigger. This delay can be accurately set from the instrument's controls. The number of points stored in memory, after the delay, is dependent on the memory size and the sample rate, and this determines the amount of the input waveform which is stored.

The pre-trigger time base mode allows the instrument to 'look back' at the waveform. In this mode the instrument continuously samples and stores the input. When the memory is full storage continues by discarding the oldest sampled data. Once the trigger is detected, stored data is discarded up to the pre-trigger point only, and then sampling of the input stops.

The dual time base modes, in both the normal and pre-trigger situations, allow different sample rates to be used on the two separate parts of the waveform. This enables the maximum coverage of a signal whilst preserving the precision of recording on a selected portion of the waveform.

Most waveform recorders can also operate in three trigger modes, automatic, continuous and single. In the automatic mode the instrument does not need a trigger, so it is easy to set up its initial procedures. In the continuous mode the instrument stores new information on each trigger pulse, and the existing information in memory is over written. In the single trigger mode the recorder will store the input waveform present on its first trigger pulse only, and all subsequent trigger pulses are ignored. The instrument can usually be reset by a signal or a button on the instrument panel.

Fig. 11.13 shows a waveform recorder, which can be used as an oscilloscope plug-in to display the stored waveform. This instrument can store up to six independent waveforms, using the front panel controls. (See also Fig. 12.13.)

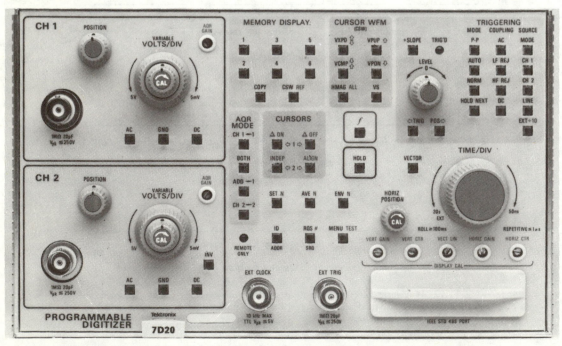

Fig. 11.13. The Tektronix model 7D20 waveform recorder (courtesy Tektronix UK Ltd).

12. Oscilloscopes

12.1 Introduction

The oscilloscope, in common with the multi-meter, is probably the most frequently used instrument in an electronics environment. It is much more versatile than a meter, since it can show the shape of a waveform, as well as give a reading of its amplitude. The oscilloscope works on the principle of a beam of electrons writing onto a phosphor coated glass screen. The beam causes the screen to fluoresce, the beam being deflected over the area of the screen by X and Y deflection plates. The brightness of the beam can also be controlled by a Z control, so the system is very similar to a television tube. However for an oscilloscope the X control is usually time, so the display consists of the amplitude of the wave against time.

Oscilloscopes have evolved continuously, and instruments are now available which can measure frequencies up to 1 GHz, and observer events as small as 20 ps in duration. Many features are available with some oscilloscopes, such as built in digital multi meters and counters. The oscilloscopes are also getting smarter, and many are microprocessor controlled. They have the ability to calculate several features, such as the rise time or pulse width of the measured waveform, and to display these values with the waveform. The instruments are also easier to use and internal routines often guide the user, and display a warning if there is an error in a setting.

Many oscilloscopes are now available with IEEE 488 bus capabilities, so that they can be used as part of a measurement test bed, with the instrument's controls being set remotely, and the readings digitised and extracted for recording or further analysis.

In this chapter the basic structure of an oscilloscope is first described. It consists primarily of a cathode ray tube and its associated controls. The construction and operation of a cathode ray tube is then introduced. Although most instruments are monochromatic, colour oscilloscopes are finding increasing applications. The circuits used to control the cathode ray tube are described, and these include the controls for the X and Y deflection systems, and for the trigger system.

There have been several developments of the basic oscilloscope, for enhanced applications. Most modern day instruments are capable of accepting two or more input signals, and displaying these simultaneously. This may be done by using a single split beam, or by using a multiple beam tube. For high speed operations sampling oscilloscopes are used, which employ time samples to extend the range of a conventional oscilloscope to measure very high frequency signals, up to about 20 GHz. They can only detect a repetitive waveform, and work on the principle of taking a sample once per cycle, over several cycles, each sample point being shifted from the previous point. The complete picture of the waveform is stored, and can be displayed as a stationary signal.

Storage oscilloscopes can be used to capture transient signals, and then to display them for periods which vary from a few minutes to several years. An analogue storage system uses a modified form of the standard cathode ray tube to store the trace. The digital storage oscilloscope first converts the analogue signal into digital form, and stores this in digital memory. The signal can then be called up for display on the screen as often as required.

Because of the variety of oscilloscopes available, they need to be carefully selected for a given application. The chapter therefore includes sec-

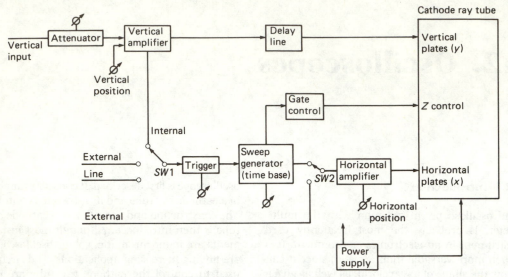

Fig. 12.1. Block diagram of a basic oscilloscope.

tions on the specification of oscilloscopes, and errors which arise in their use. It concludes with descriptions of oscilloscope accessories, such as cameras and probes, and of a few of the applications of an oscilloscope.

12.2 The basic oscilloscope

A block diagram of a basic oscilloscope is shown in Fig. 12.1. The heart of any oscilloscope is the cathode ray tube (CRT) which displays the signal being observed. The basic CRT consists of a cathode which emits a beam of electrons. This beam is regulated by a control grid, and is then focused and accelerated past two pairs of plates, which can deflect the beam in the horizontal (X) or vertical (Y) directions.

If no voltage is applied to either deflection plate of the CRT, then the beam of electrons would strike the centre of the CRT, and produce a dot. The extent to which a beam is deflected by any of the plates is proportional to the voltage on that plate. Suppose, for instance, that a voltage of 50 V across the X plates gives a deflection of 1 cm, and that the screen has a width of 20 cm. If a continuous sawtooth voltage is now applied across the X plates, having a peak of 1000 V, then the electron beam would trace a straight horizontal line across the screen. During the flyback period of the horizontal wave the beam would need to be switched off, and then restarted at the beginning of the next saw tooth.

If a repetitive voltage, as shown in Fig. 12.2, is applied to the vertical plates, at the same time as the sawtooth wave at the horizontal plates, then the screen will display the waveform shown. Since this display is repeated every cycle of the sawtooth waveform, it will appear to be stationary on the screen. The amplitude of the display is determined by the voltage on the vertical plates. This can be reduced, if too large, by means of the adjustable attenuator.

The sweep time of the horizontal voltage is also adjustable. If it is doubled, as shown in Fig. 12.3, then two cycles of the input waveform will be displayed on the screen of the CRT.

The magnitude of the input signal can vary from a few millivolts to many hundreds of volts. This wide range of signals is handled by a variable input attenuator, which can typically cover an attenuation ratio of up to about 1500, plus a vertical amplifier having a gain of about 2000. For very high voltages an external attenuator can be used, usually built into the oscilloscope's probes (Section 12.10). The frequency response of the vertical amplifier is important, and it determines the bandwidth of the oscilloscope. For measuring pulses the rise time of the oscilloscope is the important parameter. The relationship between bandwidth and rise time is given approximately by (12.1)

$$\text{Bandwidth (MHz)} = \frac{0.35}{\text{Rise time } (\mu s)} \qquad (12.1)$$

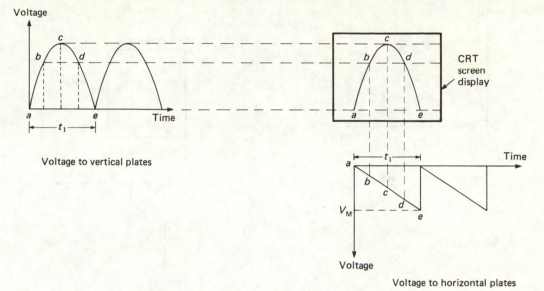

Fig. 12.2. Oscilloscope trace at one time base setting.

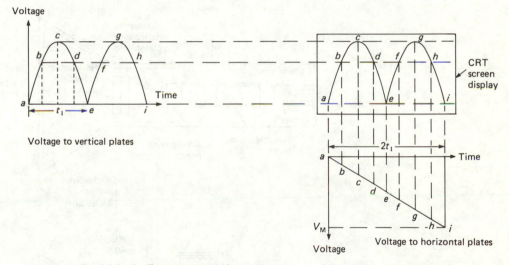

Fig. 12.3. Oscilloscope trace with time base twice as long as that in Fig. 12.2.

The trigger circuit synchronises the sweep generator such that it is a fixed multiple of the input, to give a stationary trace. Therefore in order to display half a cycle of the input trace, as in Fig. 12.4, the sweep generator frequency is doubled, but it is always triggered at the start of the input waveform. The trigger point on the waveform can be controlled from the instrument's front panel. The oscilloscope can also be triggered from an external source, or the

mains frequency, by changing the position of switch SW1 (Fig. 12.1).

The sweep time of the generator is usually variable from a few nanoseconds per division to several seconds per division. The linearity of the sweep is an important consideration. During the sweep the gate control ensures that the beam is on only during the linear portion of the sweep. An external signal can be selected by switch SW2, to feed into the horizontal amplifier. This is

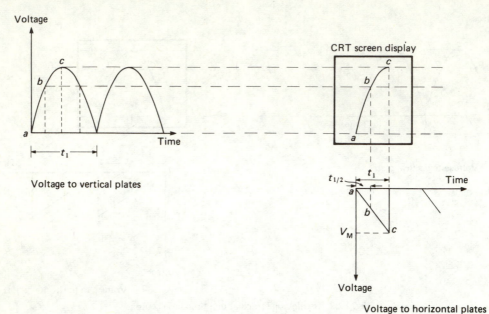

Fig. 12.4. Oscilloscope trace with time base half as long as that in Fig. 12.2.

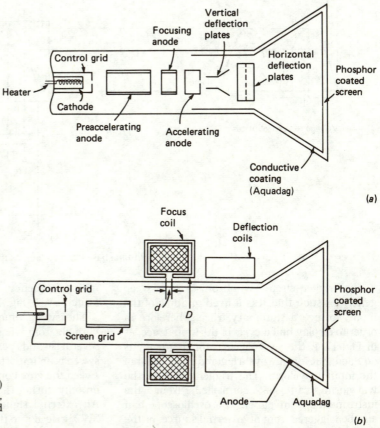

Fig. 12.5. CRT construction; (*a*) electrostatic focus and deflection, (*b*) electromagnetic focus and deflection.

used in applications such as phase and frequency measurement, as described in Section 12.11.

When triggering off the input waveform a small part of the beginning of the wave is lost, since it occurs before the start of the horizontal sweep. A delay line is usually incorporated into the oscilloscope, which delays the signal for about 0.2–0.3 μs so that the whole waveform can be viewed on the screen.

The remaining part of the basic oscilloscope, illustrated in Fig. 12.1 is the power supply. This provides the low voltage supply for the drive electronics, and the high voltage needed by the CRT, as described in Section 12.3.

12.3 The cathode ray tube

There are many different types of CRT, and they can be classified in several ways. (i) According to the number of electron beams used in a single glass envelope. This is described in Section 12.5. (ii) According to the method used to deflect the electron beam. (iii) According to whether or not any post acceleration is used, after the beam has been deflected. (iv) According to the characteristics of the screen, which includes the phosphor used and the graticule structure.

A CRT, of the type shown in Fig. 12.5, can generally be divided into five distinct regions; beam generation, focus, deflection, postacceleration, and the screen.

12.3.1 *CRT beam generation*

The electron beam is generated by a heated cathode. Indirect heating is usually used for the cathode, and the typical power required is 600 mA at 6.3 volts. High efficiency systems use 300 mA, at 6.3 volts and for special, low power designs, 140 mA at 1.5 V.

The electron flow is regulated by a control grid. This is usually a metal cup of low permeability steel, about 1.5 cm in diameter and 1.5 cm long. An aperture of about 0.25 mm is drilled in

the cap of the grid, and the electrons pass through this. The anode also has an aperture at one end, so a narrow beam of electrons emerge from it. A high potential difference, of several thousand volts, is needed to accelerate the electrons through the grid. This voltage also gives the electrons high energy as they leave the anode.

The intensity of the electron beam on the screen is controlled by a knob on the front panel of the oscilloscope, which regulates the voltage on the control grid, and therefore the energy of the electron beam. The control grid is also used to blank the beam, such as during Z blanking. This is known as grid blanking. An alternate method of blanking, called deflection blanking, consists of an anode with a double aperture and intermediate deflection plates. To blank the beam it is deflected so as to miss the second aperture. The advantage of deflection blanking is that the regulation on the power supply is not as critical as in grid blanking; the disadvantage is that the cathode is now continuously emitting electrons, so its life is shortened.

Many different designs are used for beam generation, or electron gun assembly. Sometimes the anode is used to accelerate the electron beam and also to focus it. In other structures a screen grid, held at about 200–400 V, is also used.

12.3.2 *Electron beam focus*

In an electrostatic focus system, shown in Fig. 12.5(*a*), the preaccelerating anode, focusing anode and the accelerating anode form an electron lens system. The preaccelerating anode and the accelerating anode are connected together to a high positive voltage, of about 1500 volts. The focusing anode is connected to a lower adjustable voltage, of about 500 volts. The three electrodes form equipotential surfaces at their interfaces. This causes the electron beam to

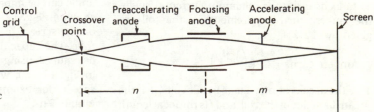

Fig. 12.6. Focus in an electrostatic CRT.

be refracted at these surfaces, and the combination therefore acts as an electron lens system, as illustrated in Fig. 12.6.

The electrons leaving the cathode are caused by the grid to reach a point called the crossover point. The beam then expands through the anode until it is caused by the electron lens system to converge to a spot on the screen. The ratio n/m is called the image to object ratio, and this, multiplied by the size of the spot at the crossover point, determines the minimum size of the electron beam spot on the screen. The spot is typically 0.2–0.5 mm in diameter.

An oscilloscope usually has two controls on its front panel, focus and astigmatism. The focus control adjusts the focal length of the lens system, by varying the voltage on the focusing anode, and shifts the focal point of the beam along the CRT axis. Focus is used to produce the smallest dot possible on the screen. Astigmatism makes the dot as round as possible, both in the centre of the screen and at its edges.

The electron beam tends to distort as it comes close to the deflection plates, and this leads to an increase in spot size at the edges of the CRT, compared to the centre. At high beam intensities the mutual repulsion of the electrons in a beam causes a rapid increase in spot size. This is called the space charge effect. Adjusting the control grid voltage, such as to adjust brightness, will also affect the position of the crossover point, resulting in defocusing on the screen, and the need to re-adjust the voltage on the focusing anode.

In electromagnetic focusing the focus coils consist of a large number of turns of fine wire wound on an insulating bobbin. The coil and bobbin are enclosed in a soft iron shell, of about 1.5 mm thickness, having a gap of about 2 cm at one end. In this system the magnetic field acts at right angles to the electron beam, and therefore does not give it any additional kinetic energy.

More than one value of coil current will give good focusing, and the system is usually designed to focus using the minimum current. Usually the ampere turn requirements of the coil are given by (12.2)

$$\text{Ampere turns} = 220\left(\frac{V_a d_c}{f}\right) \tag{12.2}$$

V_a is the accelerating voltage in kilovolts, d_c is the mean diameter of the coil, and f is the focal length

of the system. The coil should be well shielded to minimise ampere turns, and this is usually achieved when the gap in the soft iron shell is chosen such that $d = D/20$.

12.3.3 *Electron beam deflection*

There are many variations of the electron beam deflection system used in a CRT, each designed for a given cost/performance ratio. The main two types of deflection are electrostatic and electromagnetic, as illustrated in Fig. 12.5.

Electrostatic deflection uses an X and Y pair of deflection plates, to which potentials are applied to deflect the beam in a horizontal or vertical direction. The important parameters in an electrostatic system are sensitivity (or its inverse called the deflection factor), linearity, and scan area. Sensitivity is defined as the distance moved by the spot on the screen for a unit deflection voltage. Linearity indicates whether sensitivity is the same at the centre of the screen as at its periphery, and the scan area is the part of the screen which is scanned by the beam.

Fig. 12.7(a) illustrates a beam being deflected by plates of length l spaced a distance d apart. The sensitivity of the deflection is given by (12.3)

$$\frac{D}{V_d} = \frac{KlL}{V_a d} \tag{12.3}$$

V_d is the deflection voltage between the plates, V_a is the beam potential, between the anode and the cathode, and K is a constant which is primarily dependent on the postaccelerator field (Section 12.3.4). Therefore for maximum sensitivity the electrostatic deflection system should be designed with long plates spaced close together, a long tube, and a weak beam potential. Unfortunately reducing beam voltage also reduces the brightness of the display, unless postacceleration is used.

The deflection plates used in an electrostatic deflection system can be modified to meet special functions. Generally the scan angle is limited since the plates obstruct the electron beam. The angle can be improved by curving the plates, as shown in Fig. 12.7(b). This allows a larger scan on the screen, or the same scan with a shorter tube. The capacitive effect of the plates can also be reduced, using a segmented plate structure as in Fig. 12.7(c), which allows higher deflection speeds, for example full scan angle deflection within 5 ns.

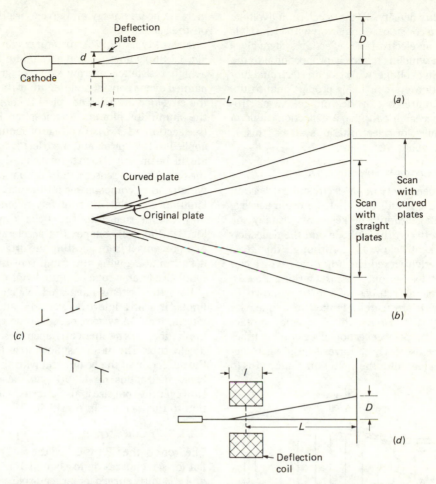

Fig. 12.7. Beam deflection in a CRT: (a) electrostatic deflection, (b) comparison of deflection using straight and curved plates, (c) segmented plates, (d) electromagnetic deflection.

The mechanism used in electrostatic deflection involves charging of capacitive plates, and requires lower amplification costs, but uses a more complex tube. It is capable of high speed operation, and is used in almost all CRTs with bandwidths above 1 MHz. Electromagnetic deflection involves changing a magnetic field by means of current in a coil. At high frequencies the number of turns on the coil needs to be reduced and the current increased, so that above about 25 kHz such a system has high amplification costs and excessive power dissipation.

All modern oscilloscopes operate with arbitrary waveforms at high frequencies, and use electrostatic deflection almost exclusively. Electromagnetic deflection is sometimes used for the X plates. The advantage of an electromagnetic deflection system is that it enables wider deflection angles, since there are no plates inside the tube, and is cheaper at bandwidths below about 25 kHz. It is mainly used on raster scan displays, like television. Electromagnetic deflection also enables a large diameter electron beam to be employed, as there are no obstructions in the CRT, resulting in a brighter image.

In an electromagnetic deflection system, shown in Fig. 12.7(d), the deflection is proportional to the flux, or coil current, and inversely proportional to the square root of the accelerating voltage. Deflection is given by (12.4)

$$D = K_1 LlHV_a^{-1/2} \qquad (12.4)$$

H is the flux density, V_a the accelerating voltage and K_1 is a constant. Comparing with (12.3) it is seen that an electrostatic deflection system gives deflection which is inversely proportional to the accelerating voltage, whilst for an electromagnetic deflection system this is proportional to the inverse square root of the voltage. An electromagnetic system can give a deflection angle of 110° while for electrostatic systems this is limited to about 40°.

12.3.4 *Electron beam postacceleration*

For good sensitivity in an electrostatic deflection system, (12.3) indicates that the beam accelerating voltage should be kept low. Usually the voltage between the cathode and the deflection plates is kept below 4 kV. Although this gives good sensitivity it reduces brightness, which can be especially serious at fast writing speeds. Monoaccelerator tubes, of the type shown in Fig. 12.5(*a*), are usually limited to frequencies below 10 MHz. Above this frequency postaccelerator, or post deflection acceleration tubes (PDA) are used. These have a large beam accelerating bias after the deflection plates, so as to increase beam energy and give a bright display on the screen.

Several types of PDA structures can be used. Fig. 12.8(*a*) shows a spiral accelerator in which a high resistance narrow spiral of graphite is painted over a considerable length of the inside of the envelope funnel. The spiral is connected to the aluminium film on the phosphor, if present (see Section 12.3.5.). A voltage of about 10 kV is applied to this spiral, and the effect is to accelerate the beam, but also to bend it slightly towards the axis. This decreases the deflection sensitivity, which can be compensated for by increasing the length of the tube. Alternatively spherical mesh may be inserted into the helix tube, as in Fig. 12.8(*b*). This shapes the accelerating field and prevents it from affecting the original beam deflection, so giving the same sensitivity as a monoaccelerator tube of equal length.

Both the systems illustrated in Fig. 12.8 are limited in scan angle to between 35° and 40°. An alternative PDA system, called a high expansion mesh tube, uses a spherical shaped mesh without a helix tube. The mesh is shaped so as to expand the deflection of the beam, and with this system beam deflections up to 90° can be obtained. However the spot size is also increased in proportion to the increase in sensitivity.

12.3.5 *The CRT screen*

The front of the CRT is called the face plate. It is flat for screen sizes up to about 10 cm × 10 cm, and is slightly curved for larger displays. The face plate is formed by pressing molten glass in a mould and then annealing it. Some CRTs have a face plate which is made entirely from fibre optics, which has special characteristics. It is also feasible to have a small fibre optic region which gives direct contact exposure of photographic film, if required, without needing a lens system.

The inside surface of the face plate is coated with phosphor. This consists of very pure inorganic crystalline phosphor crystals, about 2–3 microns in diameter, to which traces of other elements, called activators, have been added. Activators in current use are metals such as silver, manganese, copper and chromium, and they affect the characteristics of the phosphor, such as its luminous efficiency, spectral emission and persistence.

Many types of phosphors are used for CRTs, and the characteristics of a few of them are given

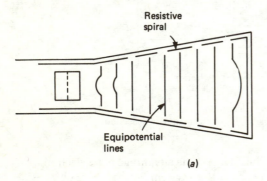

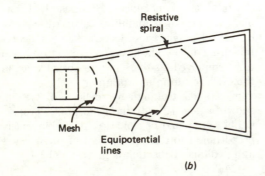

Fig. 12.8. CRT post accelerators; (*a*) spiral, (*b*) mesh.

Table 12.1 *Phosphor characteristics*

Type	Colour Fluorescent	Phosphorescence	Persistence (to 10% level)	Luminance*	Writing speed*	Intended use
P1	yellow–green	yellow–green	1–100 ms	4	7	oscilloscope, radar
P2	yellow–green	yellow–green	1–100 ms	2	4	oscilloscopes
P4	white	white	10 μs–1 ms	3	3	monochrome television
P7	blue	yellow	10 μs–1 ms(b) 100 ms–1 s(y)	4	2	radar, medical
P11	blue	blue	10 μs–1 ms	5	1	photographic
P15	ultra-violet	green	<1 μs(uv) 1–10 μs(g)	9	8	TV (flying spot scanner)
P16	ultra-violet	ultra-violet	<1 μs	10	8	TV (flying spot scanner), photographic.
P18	white	white	10 μs–100 ms	—	10	projection TV
P19	orange	orange	>1 s	5	10	radar
P22	white (red. blue, green)	white (red., blue, green)	10 μs–1 ms	—	—	tricolour TV
P26	orange	orange	>1 s	8	10	radar
P28	yellow–green	yellow–green	100 ms–1 s	3	5	radar, medical
P31	green	green	10 μs–1 s	1	3	oscilloscopes, bright TV
P33	orange	orange	>1 s	6	9	radar
P39	yellow–green	yellow–green	100 ms–1 s	3	6	radar, computer graphics
P40	blue	yellow–green	10 μs–1 ms(b) 100 ms–1 s(yg)	—	—	low repetition rate P16
P44	yellow–green	yellow–green	1–100 ms	3	—	bistable storage
P45	white	white	1–100 ms	4	—	monochrome TV display

*1 = Highest luminance or fastest writing speed

in Table 12.1. A phosphor converts electrical energy to light energy. When an electron beam strikes phosphor crystals it raises their energy level. This is known as cathodoluminescence. Light is emitted during phosphor excitation and this is called fluorescence. When the electron beam is switched off the phosphor crystals return to their initial state, and release a quantum of light energy. This is called phosphorescence or persistence.

The writing speed of a phosphor is measured by its fluorescence rise time, which is the time from the beginning of excitation to reach 90% of the maximum emission state, and by the decay time, which is the time to fall from the maximum state to the 10% level. Writing speed is determined by the phosphor type, crystal size, impurity content, and the manufacturing process. A phosphor must be refreshed by electrons, before the end of its decay time, in order to give a flicker free display. Short persistence phosphors require more frequent refreshes, whereas long persistence phosphors may result in characters fading slowly, giving a trace afterglow or trails. Long persistence phosphors are also more prone to permanent discolourations, and loss of luminous efficiency, at high beam current or fixed data use. Most modern oscilloscopes use short persistence phosphors (e.g. P1, P2, P11, P31) because of their fast refresh capability, whilst the slower speed signals seen in medical applications need longer persistence phosphors (P7 and P39). For radar use very long persistence is needed (P19, P26, P33) or even storage oscilloscopes (see Section 12.7).

The electron beam on striking the phosphor gives off light and heat. Excessive beam current on one spot of the screen, over a long time, can degrade the light output from the phosphor, and can burn a hole in it. Phosphors can be classified as having low burn resistance (P19, P26, P33), medium burn resistance (P1, P2, P4, P7, P11) and high burn resistance (P15, P31).

The luminance of the phosphor is a measure of its brightness. It is determined by the luminance efficiency of the phosphor, and by the beam energy, which is a product of the beam current density, the accelerating potential and the writing time. Therefore one can have high light output, at low beam accelerating potential, provided the sweep speed is low, but for fast single-shot transient recording a high beam current and accelerating voltage, and phosphor with high luminance, are required.

Each phosphor has a unique colour. The human eye response peaks at 555 nm, in the yellow–green region, so phosphors which give light in these regions (P2, P31, P39) would appear to be brighter than others such as P11, P16, P19. A camera film, however, is sensitive to ultraviolet or blue, so for high speed photography phosphors P11 or P16 may be better. The spectral emission from a phosphor is broadband, i.e. a range of wavelengths go to determine its colour. The gradation of colour (hue) and the chromatic purity with respect to white (saturation) can be precisely specified by X and Y coordinates on a chromaticity diagram.

A thin film of metal, such as aluminium (1000–1500 Å) is usually deposited on the non viewing side of the phosphor. This has three effects. (i) The metal layer acts as a heatsink and reduces the danger of phosphor burn. (ii) The light scatter from the phosphor is reduced, and the aluminium reflects light going back into the tube towards the viewer, so increasing the brightness. (iii) Electrons which strike the phosphor release secondary electrons. These secondary electrons are usually collected by the aquadag. Under certain conditions the electrons may remain on the screen, and cause a negative voltage build up. This would decrease the accelerating voltage of the beam. This charge build up is prevented by the aluminium layer, which is connected electrically to the aquadag layer.

The electron beam needs more energy to penetrate the aluminium layer in front of the phosphor. Therefore at low acceleration potentials the aluminised phosphor screen has lower efficiency than a standard screen, but the efficiency increases rapidly at higher potentials. This is illustrated in Fig. 12.9.

The phosphor does not convert beam energy (beam current times accelerating voltage) linearly into radiated flux. This is illustrated in Fig. 12.10, which shows the offset energy component (IV_T) followed by the linear working region. Voltage V_T is due mainly to the electron beam loss as it goes through the aluminizing layer, and is about $1 - 3$ kV.

The ideal phosphor for a laboratory oscilloscope CRT would be one which gave off a colour to which the eye had maximum response; had a high burn resistance, to prevent accidental

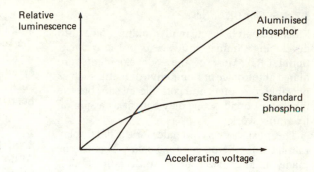

Fig. 12.9. Luminescence comparison between standard and aluminised phosphors.

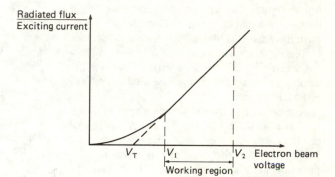

Fig. 12.10. Illustration of phosphor non linearity.

damage; had a high luminance and writing speed; and had short persistence to avoid multiple images when recording fast displays. A typical choice of phosphor is P31. Phosphors have a persistence which follows a mix of exponential law (e^{-kt}) and a power law (t^{-m}). The decay characteristics of the phosphor are determined by variables such as the anode voltage, beam current density, pulse repetitive rate, and excitation duration. The persistence values given in Table 12.1 are usually obtained empirically.

Graticules are scale markings on the face plate of the CRT, which are used to make measurements of the trace in the X and Y directions. There are three types of graticules. (i) External graticule. This is scribed on a Plexiglas acrylic plastic and fitted to the screen. The graticule can be easily changed, to make different types of measurements, and its position can be adjusted to align it with the trace of the CRT. External graticules suffer from parallax since they are not in the plane of the phosphor. Therefore the alignment of the trace and the graticule varies with the viewing angle. (ii) Internal graticule.

This is deposited on the internal surface of the CRT face plate, and is therefore on the same surface as the phosphor. Although there is now no parallax, an internal graticule cannot be changed, and also needs some method of electrical trace alignment. The graticule is also difficult to illuminate for photography, unless special illumination is provided in the oscilloscope. (iii) Projected graticule. These are available with some cameras. It provides flexibility and can include additional features such as legends on the graticule.

The face plate is sometimes tinted neutral-grey to reduce ambient light interference, or external filters can be used to reduce glare. Black wire mesh filters increase the contrast ratio and restrict the viewing angle, so reflections from oblique light sources are minimised. Polarised filters can also be used in high glare applications. Light filters are sometimes utilised to affect selective colour components of the phosphor. For example an amber filter on a P7 phosphor would eliminate the short persistence blue component, and increase the longer persistence yellow component; and a blue filter would do the reverse.

12.3.6 *Flat CRT displays*

Much research is currently taking place into developing slimmer equivalents of the conventional CRT. Most of these developments are aimed at the office or home environment, such as portable computers and portable TV sets, but the laboratory oscilloscope may benefit as a spin off from this work.

The most serious contender for the flat display market is the liquid crystal display. This can be made in large sizes and multicolour, and it dissipates very low power. The disadvantages of liquid crystal displays are poor contrast, high cost of associated drive electronics, and low switching speed. This last is the most critical for oscilloscope applications since a typical switching speed of liquid crystals is 20 ms. However the display provides acceptably fast media for computer displays.

A more conventional development of a flat CRT is that shown in Fig. 12.11. A traditional electron gun is used, and the electrons are projected from the side of the screen, in a direction parallel to the screen. Horizontal and vertical deflection plates are used as in conventional CRTs. In addition a third set of deflection plates is formed by the phosphor screen and the front faceplate. This bends the electron beam through 90° onto the phosphor screen. The third set of deflection plates provides the focusing field and ensures that the angle of beam incidence does not vary across the screen, so a good round spot is always obtained.

The phosphor coated screen is at the back of the tube and not on the face plate. Since the tube is thin the phosphor can be seen through the transparent front electrode and the image can be magnified by the lens. By placing the phosphor at the rear of the tube, and viewing the side on which the electrons strike, a brighter image is obtained, so the power requirements are reduced. It is also now much easier to adequately heatsink the back of the phosphor, so reducing the risk of phosphor heat damage.

12.3.7 *Colour CRT displays*

Although colour CRTs, such as shadow mask tubes, are common in television and computer displays, they are not used for oscilloscopes. One of the reasons is that they are relatively expensive, and colour does not give many advantages when viewing single or multiple traces which are separated from each other. However, limited colour can have advantages, such as in medical oscilloscopes, and an instrument is available for this application. It uses a standard CRT, and a liquid crystal shutter to produce a three colour display, as shown in Fig. 12.12.

The CRT has a multicomponent phosphor,

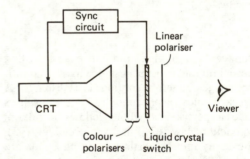

Fig. 12.12. Exploded view of a liquid crystal switch used with a CRT.

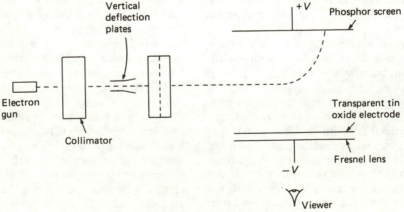

Fig. 12.11. A flat CRT construction.

which gives peak outputs of orange and cyan light. Two colour polarisers are used after the CRT and these polarise the orange light vertically, the cyan light horizontally, and absorb white light. The liquid crystal switch will not affect the light passing through it when on, but it will rotate the light through it by 90° when off. The linear polariser will pass horizontally polarised light only.

The liquid crystal switch is synchronised to the CRT trace. In operation the CRT first sweeps out the cyan colour display field with the liquid crystal switch on. Only cyan light will be displayed since this will be passed by the linear polariser. The CRT then sweeps out the orange coloured portion of the display with the liquid crystal switch off. Now both the cyan and orange lights from the colour polarisers are rotated, so that the orange is horizontally polarised and is passed by the linear polariser. If any

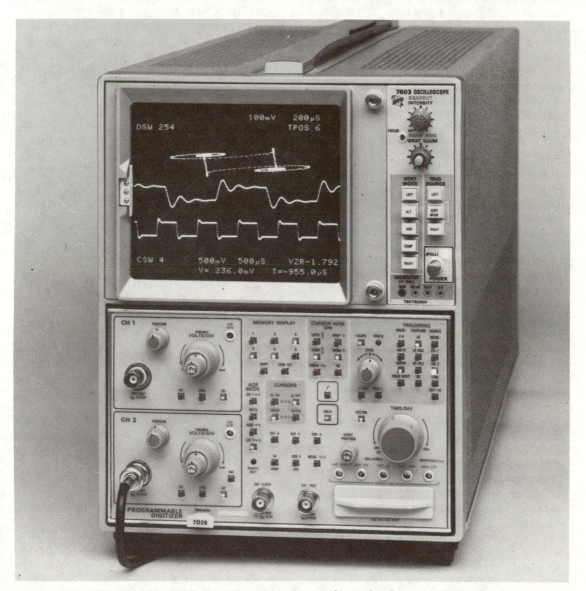

Fig. 12.13. An oscilloscope front panel controls. (Photograph reproduced courtesy of Tektronix, Inc.)

part of the display is to be yellow then it is traced by the CRT, when the liquid crystal switch is both on and off, so the cyan and orange colours combine to give yellow.

The liquid crystal shutter oscilloscope has a good contrast ratio since the shutter attenuates the non lighted areas giving a black background. However the brightness is low since the light is emitted 50% of the time for any one colour, due to the duty cycle, and the polarisers also attenuate about half the light going through them. Overall the brightness is less than 20% of that of a CRT without a liquid crystal shutter. The most serious problem with the liquid crystal shutter system is that it is limited in switching speed. By making the liquid crystal switch narrow, and the rotational distance of the molecules small, the switching time can be reduced to about 0.5 ms. Even then the bandwidth of the oscilloscope is limited to under 100 kHz.

12.4 Control circuits

The basic functions of an oscilloscope were introduced in Section 12.2, and illustrated in Fig. 12.1. In this section the different control circuits are described in greater detail under the sub-sections, input selector, attenuator, amplifier, delay line and time base.

Fig. 12.13 shows the front panel of a typical oscilloscope. Several types use fibre optic projections for presenting alphanumeric data on the screen, or use an LED display for warnings, or to give information. The internal controls are often microprocessor based, and this enables calculations to be performed, such as conversion between period and frequency, and the results are then shown on the LED display.

12.4.1 *Input selector*

The input selector switch, which passes the vertical signal to the attenuator, usually consists of three positions, d.c., a.c. and ground. In the d.c. position the signal is connected directly to the attenuator. The a.c. position connects a capacitor in series with the signal path, to remove any d.c. components of the input, and so measure a.c. only. The ground connection earths the input. It is used for setting a zero voltage level on the oscilloscope display. The ground connection is located between the a.c. and d.c. positions on the input selector switch. This ensures that the input line is earthed when

switching between a.c. and d.c., which removes any stored charge from the coupling capacitor.

12.4.2 *Attenuators*

The attenuator is used to change the magnitude of the signal fed into the amplifier. It usually has an attenuation factor of 1000:1 in about 10 steps. This, combined with the amplifier gain, which is in the region of 2000:1, gives a typical vertical deflection sensitivity variation from 10 mV to 50 volts per division.

The attenuator must give a constant signal loading at all settings of the attenuator, and it should attenuate all frequencies, in the input signal, equally. A compensated $R–C$ attenuator, as shown in Fig. 12.14, is commonly used. If Z_o

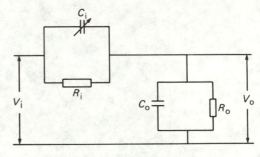

Fig. 12.14. A compensated $R–C$ attenuator.

and Z_i are the impedances of the shunt and series arms respectively, then the ratio of output voltage V_o to input voltage V_i is given by (12.5)

$$\frac{V_o}{V_i} = \frac{Z_o}{Z_o + Z_i}$$

$$= \frac{R_o(1 + j\omega R_i C_i)}{R_o(1 + j\omega R_i C_i) + R_i(1 + j\omega R_o C_o)} \quad (12.5)$$

If capacitor C_i is adjusted to make $R_i C_i = R_o C_o$ then (12.5) reduces to (12.6)

$$\frac{V_o}{V_i} = \frac{R_o}{R_o + R_i} \quad (12.6)$$

This equation does not contain any frequency dependent components therefore all frequencies in the input signal are attenuated equally. If frequency compensation was not used then at high frequencies, above about 10 kHz, the measurement leakage capacitances would need to be considered, and these would give time constants which changed at different frequencies. Capacitors C_1 and C_0 swamp the effects of these leakage capacitors.

The input impedance of most oscilloscopes is about 1 Mohm $\pm$ 1%, shunted by a capacitance of $10 - 100$ pF. Some very high frequency oscilloscopes have attenuators with 50 ohm terminators, for use with transmission lines.

Attenuators have external knobs for compensation adjustment, which changes the value of C_1. To compensate an attenuator a square wave test signal is usually applied to it, as in Fig. 12.15(a).

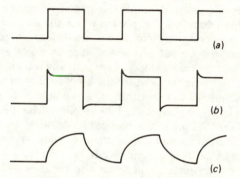

Fig. 12.15. Effect of attenuator compensation on the displayed waveform; (a) input waveform, (b) display with over compensation, (c) display with under compensation.

The value of capacitor C_1 is then altered to show a square wave display on the screen. If C_1 is too large there is over compensation, and the display shows spikes as in Fig. 12.15(b). If C_1 is too small then it under compensates, and the oscilloscope display gives a rounding of the edges of the square wave, as in Fig. 12.15(c).

12.4.3 *Amplifiers*

Amplifiers are used to drive the horizontal and vertical deflection systems in an oscilloscope, and for Z control. Their function is to provide gain for low level signals, and they should have a high input impedance, to minimise loading on the signal source.

Typical gain of an amplifier is 2000:1, and this is fixed, making it easier to design the amplifier for stability and bandwidth. Usually a pre amplifier is followed by a main amplifier, and the first stage of the pre amplifier is an FET source follower, for high input impedance.

The vertical deflection plates of an oscilloscope are driven in push–pull by the amplifier. This can therefore have a differential construction throughout, or it can have a single ended input with differential output. Differential amplifiers as shown in Fig. 12.16, are preferred since the drift and stability are better, and they are also less affected by noise, whether this is due to radiation or conduction.

An important parameter of differential amplifiers is the common mode rejection ratio (CMRR). If the same value of differential signal is applied to both inputs of the amplifier, then there should be no difference between the differential outputs. Therefore if $+ V_1$ is applied to inputs I_1 and I_2 in Fig. 12.16 the difference between outputs O_1 and O_2 should be zero. In practice the

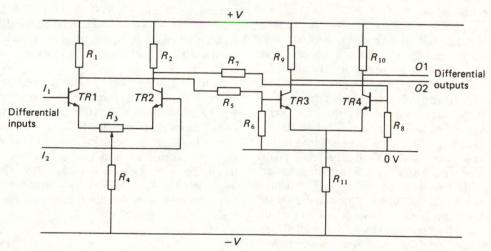

Fig. 12.16. A two stage differential amplifier circuit.

output will have a small differential voltage, equal to ΔV_1. If a signal of $+ V_2$ is applied to input I_1 and $- V_2$ to I_2 then the difference between O_1 and O_2 should exist, due to the differential gain of the amplifier; let it be ΔV_2. Then the common mode rejection ratio is given by (12.7)

$$\text{CMRR} = \frac{\Delta V_1}{\Delta V_2}$$

$$= \frac{\text{Gain with common mode signal}}{\text{Gain with differential signal}}$$

$$(12.7)$$

The common mode rejection ratio is usually specified in decibels. So a CMRR of $- 100\,\text{dB}$ will mean that a common mode signal of $1\,\text{V}$ will have as much effect as a differential signal of $10\,\mu\text{V}$. CMRR is important when measuring low level signals, such as the output from transducers, in the presence of electrical noise arising from sources like transformers, power supplies and motors. CMRR is also lower at high frequencies, and is degraded by input circuits such as attenuators used before the amplifier.

There is a limit to the maximum value of differential and common mode voltage which can be applied to an amplifier. For large differential voltages the amplifier output will run into saturation, so that it will not respond to any further change in the input. This is called clipping. Large common mode inputs make the output signal progressively more non-linear, so that the peak value of common mode voltage is usually determined by the distortion allowable by the application.

The gain of an amplifier falls with frequency, and a phase shift is introduced between the output and input signals. At low frequencies the gain of the amplifier is equal to its d.c. value. At high frequencies the gain falls off ('rolls' off) at the rate of 20 dB per decade, which is the same as 6 dB per octave. This means that for every doubling in frequency the gain falls by 6 dB, or every time the frequency is increased by a factor of ten the gain falls by 20 dB. The break point frequency is the frequency at which the gain has fallen 3 dB below its d.c. value, and at which there is a phase shift of $45°$. One can design an amplifier to have several break points. At each of these the roll off increases by a further 20 dB per decade. However the phase shift also increases by $45°$ at each point, so that it will eventually

approach $180°$ and result in positive feedback. For stability the gain at this point must be less than one.

Another important consideration in the design of an amplifier for oscilloscope deflection is its response to a step input voltage, as this affects deflection speed. After a short recovery or dead time the output voltage commences to change, and rises relatively linearly towards its final voltage. The time to change from 10% to 90% of this final value is called the slew time, and the rate at which it achieves this is measured in volts per microsecond, and is the amplifier slew rate. This determines the speed at which the amplifier can operate without undue distortion. Thereafter the output of the amplifier will oscillate about the final value, before gradually attaining it. The time required for the amplitude of these oscillations to fall within an acceptable band is called the settling time of the amplifier. The effect of the CRT plate capacitance on operating speed is also important; a segmented plate system should be used for high speeds, as in Fig. 12.7(c). A higher plate capacitance also requires more drive current to attain the same rise time, and this results in increased dissipation within the system.

12.4.4 *Delay lines*

In all oscilloscopes there is a delay in the start of the horizontal sweep, due to delays in the trigger circuit, amplifier, and the start of the sweep generator. This delay is of the order of 100 ns. To be able to view the whole of the waveform, at the vertical signal input, it is necessary to delay this by about 200–300 ns. This is done by the delay line. To be effective the trigger circuit pick up point must be before the delay line, as shown in Fig. 12.1.

Early delay lines were of the lumped parameter design having some 50 L–C segments. Fig. 12.17(a) shows a T section arrangement. If each T section is terminated by its characteristic impedance Z_0 then the impedance looking into the input terminals is also Z_0. The T section now has a characteristic of a low pass filter whose attenuation and phase shift vary with frequency. The upper limit of the filter, called its cut off frequency f_0, is given by (12.8)

$$f_0 = \frac{1}{\pi \sqrt{(LC)}} \qquad (12.8)$$

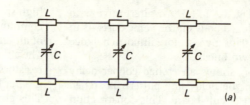

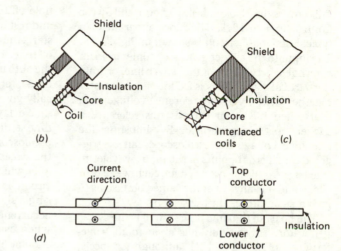

Fig. 12.17. Delay lines; (*a*) lumped parameter, (*b*) distributed parameter, twin core,
(*c*) distributed parameter, single core, (*d*)
etched.

If the input signal has frequencies which are all
much below the value of the cut off frequency,
given by (12.8), then the output signal will be a
reproduction of the input, but delayed by a time
t_d, given approximately by (12.9)

$$t_d \simeq \frac{1}{\pi f_o}$$

$$\simeq [LC]^{1/2} \qquad (12.9)$$

t_d is the delay for a single T section, and if there
are *n* sections in the lumped delay line then the
total delay is given by $n t_d$.

The lumped parameter delay line has a sharp
cut off frequency, which results in amplification
and phase distortion problems, at high signal
frequencies. The response at high frequencies
can be improved by using *m* derived sections,
which have mutual coupling between adjacent
halves of the inductors in the T sections.

Each segment of the lumped parameter delay
line has to be adjusted carefully, so that the two
halves are matched when feeding into a differential amplifier. Although this is tedious, the

delay line has a very stable performance when
adjustment has been completed.

A distributed parameter delay line has a
helical coil wound on a core. Core and coil are
insulated from the outer covering, called the
shield (Fig. 12.17). The shield is usually made of
braided insulated wire, connected at the end of
the cable to reduce eddy currents. This arrangement behaves like an *L–C* delay line, in which
the inductance is determined by the helical
winding, and the capacitance by the two coaxial
cylinders separated by dielectric. The value of
inductance can be increased by using a ferromagnetic core, and the capacitance by using
thinner spacing between the outer and inner
conductors. Typical values of characteristic impedance and delay time, for a distributed parameter delay line, are $1\,k\Omega$ and $200\,ns/m$
respectively.

The distributed parameter delay line occupies
less space than the lumped parameter delay line,
and it does not need careful setting up. Two
separate lines are required for differential circuits, or two cores can be placed inside the same

shield, as in Fig. 12.17(*b*). An alternative is to use two interlaced coils on the same core, as in Fig. 12.17(*c*). This gives a more uniform balance between the two lines.

Fig. 12.17(*d*) shows an etched copper delay line, in which the delay line pattern is etched onto a printed circuit board. This construction gives low cost, good reliability and uniformity.

12.4.5 *Time bases*

This covers the blocks labelled Trigger and Sweep Control in Fig. 12.1. The block diagram of a typical time base circuit is shown in Fig. 12.18. The input to the trigger generator comes from an external source, or from the power line, or from the vertical signal. This is fed into a comparator together with a variable reference voltage such that when the input exceeds this reference a trigger pulse is generated. By adjusting the reference voltage the oscilloscope can be triggered on different points of the input waveform. One trigger pulse is provided for each waveform cycle, and it is always at the same point, so as to give a steady trace on the screen.

The heart of the time base system is the sweep generator. This usually has a sawtooth waveform, as in Fig. 12.2–12.4, although for special purpose oscilloscopes it could be sinusoidal, exponential or parabolic. The sweep times are variable, from about 5 ns per division to 5 s per division, with an accuracy better than 2% and a linearity below 1%. A times ten expander is sometimes included for the horizontal trace, and this gives a sweep time of 0.5 ns per division for

high speed work. A Miller integrator is the most commonly used time base generator. It consists of an amplifier with a Miller feedback capacitor. The circuit allows a wide range of capacitors and resistors to be used, of values from 10 pF to greater than 1 μF, and from 100 kΩ to 100 MΩ. This gives a good range of sweep times.

After the sweep has been completed the hold off circuit allows flyback on the sawtooth, whilst the Miller integrator recovers. Following the hold off time the reset circuit operates the gate and restarts the integrator. The free run control starts the integrator automatically after the hold off time. The oscilloscope can also be operated in the auto mode. In this the auto circuit checks the time after a sweep has occurred, and automatically gives a trigger after a period of time. This time delay can be set by a control on the sweep time setting. Sometimes it is preferable to run the oscilloscope in a single shot mode. In this mode the sweep will run once from the trigger input, after which no more sweeps occur until the reset circuit is cleared by an external trigger. This single shot mode is useful for showing random occurrences of single shot signals, especially when using photography.

Some oscilloscopes have a second time base generator, and this can be used with the main sweep generator in three operating modes: delayed sweep, mixed mode sweep and switched sweep.

Delayed sweep. In this mode the display is of the second time base generator only, as illustrated in

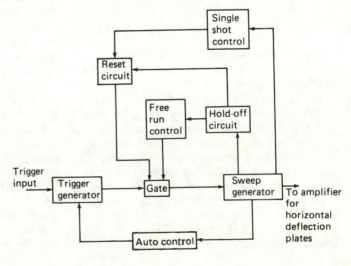

Fig. 12.18. Block diagram of a time base circuit.

Fig. 12.19. The main sweep is started by a trigger pulse at t_1. When this reaches a value set by the variable control of the trigger level, at time

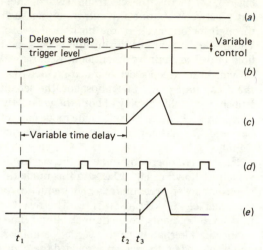

Fig. 12.19. Illustration of delayed sweep; (a) main sweep trigger, (b) main sweep, (c) delayed sweep with no delayed sweep trigger, (d) delayed sweep trigger, (e) delayed sweep with delayed sweep trigger.

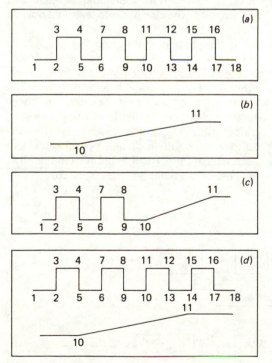

Fig. 12.20. Oscilloscope displays in various sweep modes; (a) single sweep, (b) delayed sweep, (c) mixed sweep, (d) switched sweep.

t_2, the delayed sweep is started. This is a fast sweep and shows an expanded portion of the trace. For example if the input signal is a square wave, as in Fig. 12.20(a) and the delayed sweep is started at point 10 on the waveform, and the sweep time is set to the rise time of the square wave, then the oscilloscope display will show the trace given in Fig. 12.20(b). This enables the rise time of the square wave to be precisely measured.

Once the delayed sweep generator has started it can run in two modes. If it is set on auto it runs from time t_2, as in Fig. 12.19(c). The main sweep generator is now used as a time delay generator. If the delayed sweep is operated in a trigger mode by internal or external pulses, then it will start not at t_2 but at t_3 (Fig. 12.19(d) and (e)) when the next delayed sweep trigger occurs. The main sweep is now used as a trigger delay generator, and it is said to arm the delayed sweep, since it is readied for operation at the next trigger pulse.

Mixed mode sweep. This is illustrated in Fig. 12.21. It is similar to the delayed sweep

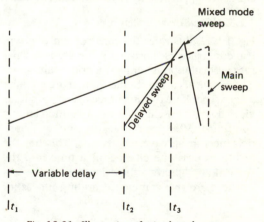

Fig. 12.21. Illustration of mixed mode sweep.

mode except that the main sweep is also displayed. The main sweep starts at t_1 and this shows the first part of the display (Fig. 12.20(c)). At t_2 an external trigger starts the delayed sweep. At t_3 the delayed sweep exceeds the main sweep, and this is now displayed for the rest of the cycle. Assuming that t_3 corresponds to point 10 on the trace shown in Fig. 12.20(a), the trace in mixed mode will be as shown in Fig. 12.20(c). The reference signal is obtained up to t_3, and then the expanded portion is shown. The point at which t_3 occurs can be altered by varying t_2, so as to expand any desired part of the input signal.

Switched mode. This shows the main and delayed sweeps as two traces, by electronically switching between them. The switching is similar to the chop and alternate modes, described in Section 12.5, and problems can occur regarding beam brightness if the two sweeps are far apart in sweep times. Fig. 12.20(*d*) illustrates the oscilloscope display in switched mode.

12.5. Multi input oscilloscopes

Oscilloscopes can have multiple input and display facilities. Two inputs is the most common, although four and eight inputs are available for special applications. There are two primary types; single beam which can then be converted into several traces, and dual beam, which may also subsequently be converted into a further number of traces. Two input oscilloscopes are described in this section, under the headings of dual trace and dual beam, although the principles are applicable to any number of inputs.

12.5.1 *Dual trace oscilloscopes*

Fig. 12.22 illustrates the construction of a typical dual trace oscilloscope. There are two separate vertical input channels, *A* and *B*, and these go through separate attenuator and preamplifier stages. Therefore the amplitude of each input, as viewed on the oscilloscope, can be individually controlled. After preamplification the two channels meet at an electronic switch. This has the ability to pass one channel at a time into the vertical amplifier, via the delay line.

There are two common operating modes for the electronic switch, called alternate and chop, and these are selected from the instrument's front panel. The *alternate* mode is illustrated in Fig. 12.23. In this the electronic switch alternates between channels *A* and *B*, letting each through for one cycle of the horizontal sweep. The display is blanked during the flyback and hold-off periods, as in a conventional oscilloscope. Provided the sweep speed is much greater than the decay time of the CRT phosphor, the screen will show a stable display of both the waveform at channels *A* and *B*. The alternate mode cannot be used for displaying very low frequency signals.

The *chopped* operating mode of the electronic switch is shown in Fig. 12.24. In this mode the electronic switch free runs at a high frequency, of the order of 100 kHz to 500 kHz. The result is that small segments from channels *A* and *B* are connected alternately to the vertical amplifier, and displayed on the screen. Provided the chopping rate is much faster than the horizontal sweep rate, the display will show a continuous line for each channel. If the sweep rate approaches the chopping rate then the individual segments will be visible, and the alternate mode should now be used.

The time base circuit shown in Fig. 12.22 is similar to that of a single input oscilloscope. Switch *SW2* allows the circuit to be triggered on either the *A* or *B* channel waveforms, or on line frequency, or on an external signal. The horizontal amplifier can be fed from the sweep generator, or the *B* channel via switch *SW1*. This is the *X–Y* mode and the oscilloscope operates from channel *A* as the vertical signal and channel *B* as the horizontal signal, giving very accurate *X–Y*

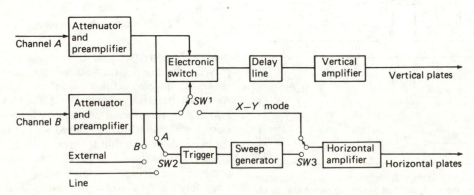

Fig. 12.22. A dual trace oscilloscope block diagram.

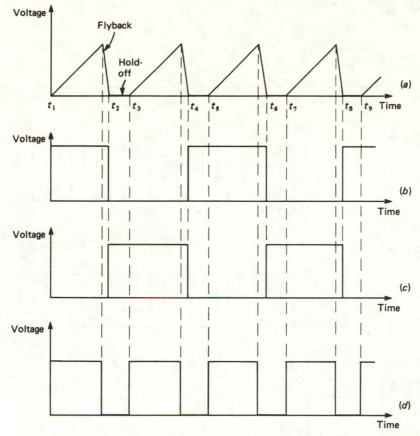

Fig. 12.23. Waveforms for a dual channel oscilloscope operating in alternate mode: (a) horizontal sweep voltage, (b) voltage to channel A, (c) voltage to channel B, (d) grid control voltage.

measurements. Several operating modes can be selected from the front panel for display, such as channel A only, channel B only, channels A and B as two traces, and signals $A + B$, $A - B$, $B - A$ or $-(A + B)$ as a single trace.

12.5.2 *Dual beam oscilloscope*

The dual trace oscilloscope cannot capture two fast transient events, as it cannot switch quickly enough between traces. The dual beam oscilloscope has two separate electron beams, and therefore two completely separate vertical channels, as in Fig. 12.25. The two channels may have a common time base system, as in Fig. 12.22, or they may have independent time base circuits, as in Fig. 12.25. An independent time base allows different sweep rates for the two channels but increases the size and weight of the oscilloscope.

Two methods are used for generating the two electron beams within the CRT. The first method uses a double gun tube. This allows the brightness and focus of each beam to be controlled separately but it is bulkier than a split beam tube.

In the second method, known as split beam, a single electron gun is used. A horizontal splitter plate is placed between the last anode and the Y deflection plates. This plate is held at the same potential as the anode, and it goes along the length of the tube, between the two vertical deflection plates. It therefore isolates the two channels. The split beam arrangement has half the brightness of a single beam, which has disadvantages at high frequency operation. An alternative method of splitting the beam, which improves its brightness, is to have two apertures in the last anode, instead of one, so that two beams emerge from it.

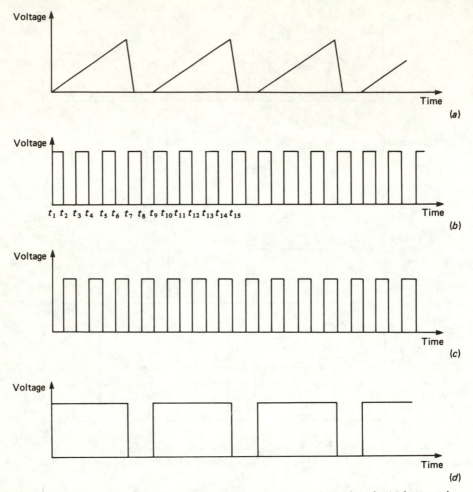

Fig. 12.24. Waveforms for a dual channel oscilloscope operating in chopped mode: (*a*) horizontal sweep voltage, (*b*) voltage to channel *A*, (*c*) voltage to channel *B*, (*d*) grid control voltage.

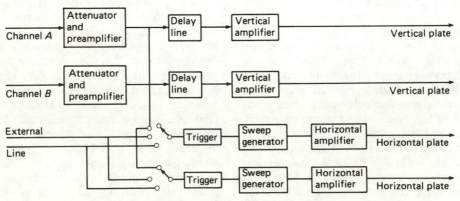

Fig. 12.25. Dual beam oscilloscope with independent time bases.

The disadvantage of the split beam construction is that the two displays may have noticeably different brightness, if operated at widely spaced sweep speeds. The brightness and focus controls also affect the two traces at the same time.

12.6 Sampling oscilloscopes

A sampling oscilloscope is used to examine very fast signals. It is similar in principle to the use of stroboscopic light to look at fast mechanical motion. Samples are taken at different portions of the waveform, over successive cycles, and then the total picture is stretched, amplified by relatively low bandwidth amplifiers, and displayed as a continuous wave on the screen. The advantage of a sampling oscilloscope is that it can measure very high speed events, which require sweep speeds of the order of 10 ps per division, and amplifier bandwidths of 15 GHz, using instruments having bandwidths several orders lower. The disadvantage of a sampling oscilloscope is that it can only make measurements on repetitive waveform signals.

Fig. 12.26 shows a block schematic for a typical sampling oscilloscope. The input signal is delayed, and then sampled by a diode gate. The sampled signal is saved on a capacitor store, then amplified and fed to the vertical plates. Unity feedback is used from the amplifier to the sampling diode gate. This ensures that the voltage on the capacitor store is only increased by the incremental value of the input voltage change, between each sample.

The time base waveforms are shown in Fig. 12.27. The trigger circuit increments the staircase generator by one step, and also starts

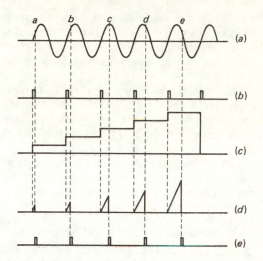

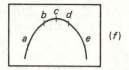

Fig. 12.27. Sampling oscilloscope waveforms; (a) input signal, (b) trigger pulses, (c) staircase generator waveforms, (d) ramp generator waveforms, (e) Z control, (f) oscilloscope screen display.

the ramp voltage. When this equals the staircase value, as detected by the comparator, a sampling pulse is produced to open the diode gate for about 400 ps. At the same time the screen is brightened by a Z control signal of about 2 μs duration. Because of the staircase waveform, sampling of the input waveform occurs later and later in

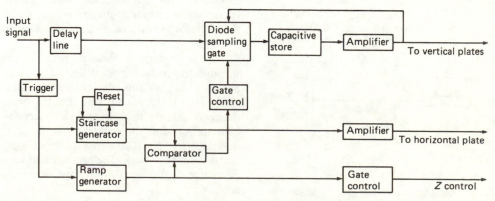

Fig. 12.26. Sampling oscilloscope block diagram.

each cycle. The staircase is reset after a certain number of steps, typically 100 to 1000, and it then starts again. Therefore up to about 1000 points are used to create the waveform on the screen.

The staircase waveform also feeds the horizontal plates of the CRT, and it is used to move the spot across the screen in a series of rapid movements. Fig. 12.27(f) shows the screen display.

The sample frequency used in sampling oscilloscopes can be as low as one hundreth of the signal frequency, so a signal frequency of 1 GHz needs an amplifier bandwidth of only 10 MHz.

12.7 Analogue storage oscilloscopes

Storage oscilloscopes, capable of retaining the image on the screen for longer than that possible with conventional high persistence phosphors, have many applications. Examples of these are the capture and storage of transients, and the steady display of a very low frequency signal. Two techniques are used to store signals in an oscilloscope, and these are called analogue and digital storage. Analogue storage is capable of higher speeds, but is less versatile than digital storage. This section describes the principles of analogue storage, and the two methods most frequently used, called variable persistence and bistable storage. Digital storage oscilloscopes are covered in Section 12.8.

12.7.1 *Principle of secondary emission*

Both the variable persistence and bistable storage oscilloscopes depend, for their operation, on the principle of secondary emission. Fig. 12.28 shows a simple electrode arrange-

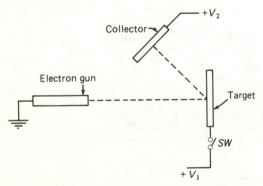

Fig. 12.28. Simple electrode arrangement to illustrate secondary emission.

ment to illustrate this principle. When a beam of electrons from the electron gun strikes the target it emits secondary electrons, which are gathered by the collector. The collector is at a positive potential of V_2, the target is at a potential of V_1 which can be varied, and the electron gun is at ground potential.

If I_p is the value of the current in the primary electron beam, coming from the electron gun, and I_s is the electron current emitted from the target and collected by the collector, then the ratio I_s/I_p is called the secondary emission ratio. The value of this ratio depends on the primary electron velocity and intensity, and on the chemical composition of the target.

Fig. 12.29 shows the variation of the secondary emission ratio with target voltage, for the arrangement of Fig. 12.28. When the target voltage is much greater than the collector voltage, all secondary electrons emitted from the target are attracted back to it. Therefore the collector current, or secondary current, is zero, and so also is the secondary emission ratio. The operating point is now well to the right of V_2 in Fig. 12.29. If the target voltage is slightly negative, as at point F, then all the electrons from the gun are deflected onto the collector, before they reach the target. Therefore although there is no secondary emission the collector current equals the beam current, and the secondary emission ratio is unity. This point is known as the lower stable point. As the target voltage increases from this point electrons are attracted away from the collector, but they do not have enough energy to release secondary electrons from the target. Therefore the secondary emission ratio falls to a minimum at A. Beyond the minimum point secondary emission from the target starts to occur, and these electrons are accumulated by the collector, so increasing the secondary emission ratio.

The secondary emission ratio increases through the crossover point E until it reaches a peak at C. Beyond this point secondary electrons emitted from the target are attracted back, in numbers greater than those which reach the collector, so that the secondary emission ratio decreases sharply. The curve reaches the upper stable point at G, where the primary and secondary currents are equal, and then decreases to zero.

The lower and upper stable points represent

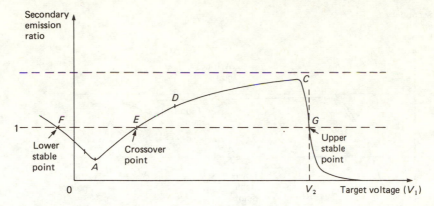

Fig. 12.29. Variation of secondary emission ratio for the electrode arrangement shown in Fig. 12.28.

the erased and written conditions of the CRT screen, and in the absence of a target voltage the target can remain in one of these two stages only. Suppose for instance that the target is in a state to the left of the crossover point E and switch SW is opened to remove the target supply voltage. Because the secondary emission ratio is less than one, fewer electrons leave the target than arrive at its surface, so the target is driven progressively negative, until point F is reached. Now the number of electrons reaching the target equals the number which are emitted from it. This represents a stable erased state. If the target was to the right of the crossover point when switch SW is opened, then, since the secondary emission ratio is greater than one, more electrons leave the target than arrive at its surface. The target is therefore driven positive until it reaches the upper stable point G, which represents the written condition of the screen. The target will stay at the upper or lower stable point so long as the electron gun is on. At the crossover point the target is unstable, and it will move up or down on the curve depending on the direction in which it is sent by noise.

Fig. 12.30 shows an arrangement using a segmented multiple target, with a collector mesh in front through which the electrons can pass. Two guns are used, a high energy writing gun, which emits a narrow beam of electrons, and a flood gun which covers the target area with a continuous stream of low energy electrons. The flood gun remains on all the time, and maintains the target in either of the two stable states. The writing gun is used to move the target from the lower to the upper stable state.

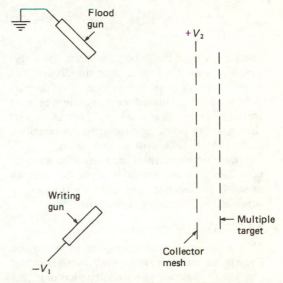

Fig. 12.30. Multiple target and flood gun arrangement.

Fig. 12.31 shows the secondary emission ratio curves for the electron arrangement of Fig. 12.30. Suppose that the target is at point A, the erased state. The writing gun is turned on and its high energy electrons result in a large amount of secondary emission, and drive the target towards the written state. Up to point C the flood gun opposes the writing gun, but thereafter it aids it; so the writing gun can be switched off and the target will reach the written state at D. To erase the target the collector mesh is momentarily taken to a negative voltage. This repels secondary emissions back into the target, moving its potential negative due to the electrons

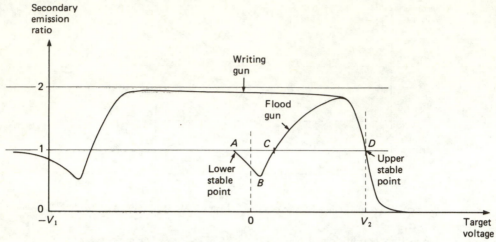

Fig. 12.31. Secondary emission ratio curves for the arrangement of Fig. 12.30.

reaching it from the flood gun. When the target reaches point A no more charging occurs. The collector mesh is now returned to its original voltage of $+ V_2$. This is done gradually to prevent the target from being accidentally driven beyond the crossover point, due to capacitive coupling, and so reaching a written state again.

The principle of secondary emission storage, described in this section, is applicable to both variable persistence and bistable storage, as described in the next two sections.

12.7.2 *Variable persistence storage*

The variable persistence storage technique is also known as halftone storage or mesh storage. Fig. 12.32 illustrates the construction of a CRT using this storage technique. There are two

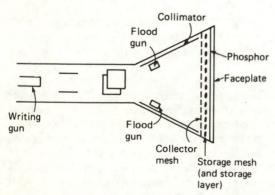

Fig. 12.32. Construction of a variable persistence storage tube.

screens, a storage mesh which retains the image traced on it by the writing gun, and the phosphor screen, which is very similar to that used in a conventional CRT.

Dielectric material, consisting of a thin layer of material such as magnesium fluoride, is deposited on the storage mesh, and this acts as the storage target. The writing gun is at a high negative voltage, the flood gun at a few volts negative, the collector mesh at about 100 V positive, and the storage mesh at ground potential or a few volts negative. The collimator consists of a conductive coating on the inside surface of the CRT. It is biased so as to distribute the flood gun electrons evenly over the surface of the target, and causes the electrons to be perpendicular to the storage mesh.

When the writing gun is aimed at the storage target it causes areas where it strikes to be charged to a positive potential, due to secondary emission, as in Fig. 12.31. These areas are maintained at their upper stable point, even after the writing gun is switched off, due to the action of the flood gun. Electrons from the flood gun also pass through those areas which are positively charged, causing the phosphor beyond to glow, and displaying the original signal traced by the writing gun.

The pattern stored on the storage target lasts for about one hour, but it can display a bright image for about one minute. The stored pattern fades due to electrons from the flood gun charging other parts of the storage surface, giving an impression that the whole screen has been written. This is known as fading positive. To erase an image which has been stored the storage mesh is momentarily raised to the same positive potential as the collector mesh.

The variable persistence storage tube is capable of producing halftone or gray scale images. As the charge on the storage surface is increased, beyond a critical threshold value, more electrons pass through it onto the display phosphor. The charge on different parts of the storage screen can be varied either by control of the time spent by the beam on different areas, or by varying the beam current over different areas. Usually four levels can be readily obtained, for example a mesh voltage of $-10\,\text{V}$ represents an off state, $-5\,\text{V}$ and $-2.5\,\text{V}$ represent intermediate states, and $0\,\text{V}$ represents full brightness.

Variable persistence tubes enable the persistence or storage time of the image to be varied from a few milliseconds to several hours. It can be done in the CRT shown in Figure 12.32 by applying a repetitive negative pulse, of $-4\,\text{V}$ to $-11\,\text{V}$, to the storage mesh, so as to discharge the positively charged areas. The control over the discharge time is usually obtained by operating the pulses at a constant frequency of about $1\,\text{kHz}$, but varying the pulse width; the narrowest pulse gives the largest persistence as it takes longer to discharge the storage layer.

Variable persistence storage finds many applications, such as the storage of an entire waveform of a slow moving signal, which then fades before the next trace is written. It can also be used to store several traces before the first one fades, so as to see how the signal changes with time.

12.7.3 *Bistable storage*

The bistable storage tube is between two and ten times slower than a comparable variable persistence tube. However it is capable of much longer storage times, measured in hours rather than in minutes as for variable persistence. The bistable tube is also capable of operating in a split screen mode, where half the screen has storage capability and the other half is a conventional phosphor tube.

Fig. 12.33 shows the construction of a bistable storage tube. Unlike the variable persistence tube the same phosphor screen is used for both storage and display. The screen consists

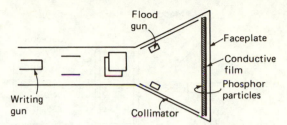

Fig. 12.33. Construction of a bistable storage tube.

of P1 phosphor, doped to have good secondary emission characteristics and deposited on a conductive backplane made from a transparent metal film. The phosphor layer consists of a thin coating of scattered particles, so as to give a discontinuous surface. This stops the boundary migration of stored charge. The thin phosphor coating also has a short life since it suffers from light output reduction with time.

The conductive film is held at a low positive potential, so as to attract a cloud of low energy electrons from the flood gun. These electrons have insufficient energy to penetrate the phosphor, and are gathered by the collimator. When the write gun is switched on, its high energy electrons result in secondary emission from areas traced on the screen, moving these to the upper stable point, in Fig. 12.31. The trace is therefore at a high positive potential, and this is maintained due to the low leakage of the phosphor. The low energy electrons from the flood gun are now attracted to the positive areas of the screen, and go through the phosphor to reach the metal film at the back. In passing through the phosphor they cause it to glow, displaying the area traced out by the writing gun. The screen can be erased by setting the metal film to a negative voltage, repelling the electrons back into the storage area, and returning the phosphor to the low stable point.

12.7.4 *Fast storage oscilloscopes*

Two modifications are made to the conventional storage CRTs, to enhance the speed with which they can capture transient information. These methods are known as transfer storage and expansion storage.

Transfer storage uses an intermediate mesh target, called the fast mesh, which has been optimised for speed. The waveform is initially written onto this mesh and it is then transferred,

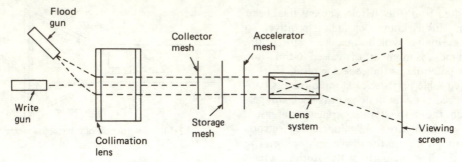

Fig. 12.34. Expansion storage tube construction.

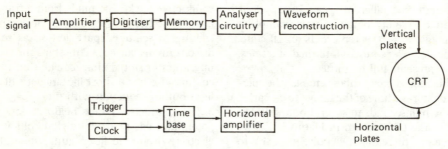

Fig. 12.35. Block diagram of a basic digital storage oscilloscope.

at lower speed, onto a second mesh which has been optimised for storage time. The second mesh can be bistable or variable persistence.

Expansion storage, shown in Fig. 12.34, uses an intermediate mesh which is about one fifth the size of the viewing screen. This mesh can be written on very quickly, and it is placed between the collector and accelerator meshes. The image on the storage mesh is magnified by a static electron lens system, to give a full size display on the viewing screen. This system is capable of writing speeds of 2000 cm/μs, and can capture signals at frequencies of 300 MHz.

12.8 Digital storage oscilloscopes

12.8.1 *Principle of operation*

The availability of electronic circuitry at low cost has enabled many digital features to be added to analogue oscilloscopes. Examples of these are, generation of a trigger after an elapsed time or after a count of a number of pulses; digital display of the parameters; integral digital voltmeter and counter; remote control. However the basic oscilloscope still remains analogue, and uses an analogue storage CRT, as described in Section 12.7.

A digital oscilloscope digitises the input signal, so that all subsequent signals are digital. A conventional CRT is used, and storage occurs in electronic digital memory. Fig. 12.35 shows a block diagram of a basic digital storage oscilloscope. The input signal is digitised and stored in memory in digital form. In this state it is capable of being analysed to produce a variety of different information. To view the display on the CRT the data from memory is reconstructed in analogue form.

Digitising occurs by taking a sample of the input waveform at periodic intervals. In order to ensure that no information is lost, sampling theory states that the sampling rate must be at least twice as fast as the highest frequency in the input signal. If this is not done then aliasing will result, as shown in Fig. 12.36. This requirement for a high sampling rate means that the digitiser, which is an analogue to digital converter, must have a fast conversion rate. This usually requires expensive flash analogue to digital converters, whose resolution decreases as the sampling rate is increased. It is for this reason that the bandwidth and resolution of a digital oscilloscope is usually limited by its analogue to digital converter.

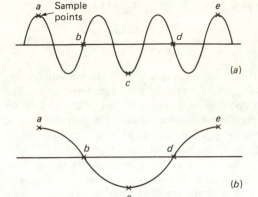

Fig. 12.36. Effect of a low sampling frequency; (*a*) input signal, (*b*) aliased signal.

One method of overcoming the need for a high performance converter is to use an analogue store, as in Fig. 12.37. The input signals are sampled, and these are stored in an analogue shift register. They can then be read out at a much slower rate to the analogue to digital converter, and the results stored in a digital store. This method allows operation at up to 100 megasamples per second, and has the advantage that a low cost analogue to digital converter can be used, whose resolution does not decrease as the sampling rate is changed. The disadvantage is that the oscilloscope cannot accept data during the digitising period, so it has a blind spot. At low sweep speed operations it is usual to switch out the analogue memory, feeding the analogue to digital converter in real time.

Many different input channels are used with digital storage oscilloscopes. However if all these channels share a common store, through a multiplexer, then the memory available to each channel is reduced. Oscilloscopes with up to 40 channels are commercially obtainable, with a storage capability of 25 000 dots. Several oscilloscopes also have floppy disc storage capability to allow non volatile storage of waveforms, which can later be recalled into the oscilloscope and manipulated.

12.8.2 *Waveform reconstruction*

Although the input signal may be sampled at greater than twice the highest signal frequency, aliasing can still result when the output is present as a series of dots, corresponding to the sampled values. This is illustrated in Fig. 12.38(*a*), where the user's mind connects

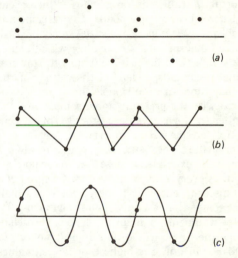

Fig. 12.38. Illustration of interpolation methods; (*a*) without interpolation, (*b*) linear interpolation, (*c*) sinusoidal interpolation.

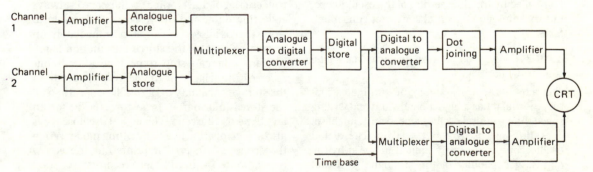

Fig. 12.37. A digital oscilloscope which uses analogue storage to eliminate the need for a very fast analogue to digital converter.

together the dots which are physically closest to each other, rather than those which are closest on the time scale.

In the illustration of Fig. 12.38(*a*) it is difficult to visualise the final waveform, and oscilloscopes generally have the facility to interpolate between the dots, if required by the user. Two techniques are used, linear interpolation and sinusoidal interpolation. In *linear interpolation*, shown in Fig. 12.38(*b*) a straight line is used to connect the dots together. This works well on a pulsed or square waveform, but not on a sinusoidal wave. Fig. 12.38(*c*) shows that *sinusoidal interpolation* gives a much better fit for sine waves, although it is not suitable for pulse or square waves.

Another problem with the sampling technique used in digital oscilloscopes is that it can miss short term transients, or 'glitches', which occur in between the sample points. To overcome this problem *envelope mode oscilloscopes* may be used. These have special logic circuitry which causes the sample and digitising circuitry to run at a high speed, independent of the setting of the display time. At each sample the value is compared with the previous stored sample, and the higher (or lower) value is stored. This is continued for the screen interval, so that for that interval the highest and lowest points are always stored. For example, suppose that an oscilloscope digitises every 2 ms, at a given sweep speed. If a 0.1 ms transient were to occur there is a high probability that a conventional digital oscilloscope would miss it. In an envelope mode oscilloscope the input would be sampled, say, every 200 ns, but only the highest, or lowest, values that occur within a 2 ms window would be stored in memory. Therefore the transient would be recorded. The sample rate of the oscilloscope is controlled by the time setting of the oscilloscope, but the analogue to digital converter runs much faster.

12.8.3 *Comparison between analogue and digital storage oscilloscopes*

The advantage of the analogue storage oscilloscope is that it has a higher bandwidth and writing speed than a digital oscilloscope, being capable of operating speeds of about 15 GHz. The digital oscilloscope is primarily limited in speed by the digitising capability of the analogue to digital converter. Aliasing effects also limit the useful storage bandwidth (usb) of the oscilloscope to a value given by the ratio in (12.10).

$$usb = \frac{\text{maximum digitising rate}}{\text{constant } C} \tag{12.10}$$

The value of constant C is dependent on the interpolation method used between the dots. For a dot display C should be about 25, to give an eligible display; for straight line interpolation it should be about 10; and for sinusoidal interpolation C should be about 2.5.

The digital storage oscilloscope has a CRT which is much cheaper than an analogue storage oscilloscope, making replacements more economical. The digital oscilloscope is also capable of an infinite storage time, using its digital memory. Furthermore it can operate with a constant CRT refresh time, so giving a bright image even at very fast signal speeds. The digital storage oscilloscope is not, however, capable of functioning in a variable persistence storage mode.

The time base in a digital oscilloscope is generated by a crystal clock, so that it is more accurate and stable than an analogue oscilloscope, where the time base is generated by a ramp circuit. The analogue to digital converter used in a digital oscilloscope also gives it higher resolution than an analogue oscilloscope. For example a twelve bit digitiser can resolve one part in 4096. A conventional analogue oscilloscope typically resolves to about 1 mm on the screen, that is to about one part in 50, equivalent to 6 bit resolution.

Digital storage oscilloscopes are also capable of operating in a look back mode, as described for waveform recorders in Section 11.5. An analogue oscilloscope collects data after it has been triggered. A digital storage oscilloscope is always collecting data, and the trigger tells it when to stop. The oscilloscope can stop immediately on trigger, so that all the stored information is pre trigger, or it can stop some time after it has received the trigger. If the delay is longer than the storage capability of the oscilloscope, then all the stored information is post trigger, as for an analogue oscilloscope. The digital oscilloscope is also able to operate in a baby sitting mode. When the scope is triggered it prints out the stored results onto a hard copy recorder (or disc storage), and then re-arms itself ready for another reading.

The digital oscilloscope is capable of local processing. This can be used in many ways, such as to expand selected parts of the waveform. It is also possible to simultaneously display a previously stored trace along with an updated real time trace, so that changes can be compared.

12.9 Oscilloscope performance

Many different parameters are used to specify the performance of an oscilloscope, and therefore to select from a large number of different types of oscilloscopes. An obvious area of comparison is the *features* provided, such as flexibility of the trigger system (e.g. single sweep, TV trigger, delayed time base), ease of use (e.g. beam finder, auto focus and auto intensity controls) and number of channels. In every case the performance and features should match the application requirements.

An important measure of an oscilloscope's performance is its *sensitivity*. Vertical sensitivity or deflection factor measures deflection for a specific input signal, and measures its ability to see the smallest signal on the screen. Vertical sensitivity is usually measured in millivolts per centimetre of millivolts per screen division, which can vary from 0.75 cm to 1.3 cm. It is the smallest deflection factor on the attenuator, and can be read off from the control knob. Although sensitivity usually refers to the vertical system of an oscilloscope, it can also apply to the horizontal and trigger circuits.

A trade off is usually required between sensitivity and *bandwidth* in an oscilloscope. Bandwidth is the frequency range over which the vertical amplifier gain is within ± 3 dB of the specified gain at midband frequency. Since most oscilloscopes perform well at d.c. this is usually taken as the reference frequency, and bandwidth is the highest frequency which has less than 3 dB loss of gain from its d.c. value. A high bandwidth oscilloscope picks up more noise, and therefore requires a lower sensitivity than a low bandwidth oscilloscope. Alternatively for high sensitivity stages the bandwidth may be deliberately suppressed.

Bandwidth is important when measuring amplitudes of a signal, and *rise time* when measuring signal time. The rise time of the vertical amplifiers measures how well the oscilloscope will reproduce pulse waveforms. The oscilloscope rise time must be better than the rise time of the signal being measured, or errors will result.

If the true rise time of a signal is T_s, the rise time of the oscilloscope is T_o, and the rise time of the signal displayed on the oscilloscope is T_D, then the true rise time can be found from (12.11)

$$T_s = [T_D{}^2 - T_o{}^2]^{1/2} \qquad (12.11)$$

This equation can be used to plot the error in the measured rise time, equal to $100(T_D - T_s)/T_s$, as a function of the ratio T_o/T_s, and is shown in Fig. 12.39. The error is seen to increase rapidly

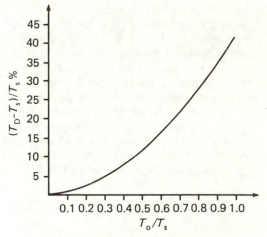

Fig. 12.39. Rise time error in an oscilloscope. (T_o = rise time of the oscilloscope. T_D = rise time of the display. T_s = true rise time of the signal.)

as the signal rise time approaches that of the oscilloscope. As an example of the use of (12.11) suppose that an oscilloscope has a rise time of 10 ns and it displays a waveform as having a rise time of 15 ns. Then the true rise time of the signal is $[15^2 - 10^2]^{1/2}$ or 11.2 ns.

Another error in oscilloscope measurements is known as *aperture error*, caused by the input signal changing during the time taken to digitise it. If f_m is the maximum frequency present in the signal, and t_a is the aperture time of the oscilloscope, then in the worst case the aperture error e_a is given by (12.12)

$$e_a = 2\pi f_m t_a \times 100\% \qquad (12.12)$$

The phosphor used has an impact on parameters such as the screen colour, persistence, luminance and writing speed. The *writing speed* is usually specified as the maximum speed at which a beam can move, and still give a visible trace on a screen or photographic plate.

The *accuracy* of the oscilloscope is an indication of how closely the measured value will equal the actual value. Accuracy in vertical measurements depends to a large extent on the linearity of the input amplifiers, and is usually 1–3%. An analogue oscilloscope has a time base accuracy of 1–2%, although this is better than 0.01% for a digital oscilloscope. The *stability* of the digital oscilloscope is also very good since it is based on counting circuits, rather than on ramps as in an analogue oscilloscope.

Several parameters are specified when comparing the performance of oscilloscopes. The useful storage bandwidth, given by (12.10), is determined by the digitising rate of the input, and by the method used to re-create the waveform from the measured points. The *digitising rate* must be adequate for the application. For example rise time measurements require about ten samples during the rise period to re-create the waveform. Therefore the maximum rise time which can be measured is given by (12.13).

$$\text{Maximum rise time} = \frac{10}{\text{Sample rate}} \quad (12.13)$$

This means that to measure a waveform with a 100 ns rise time would require an oscilloscope with a sample rate of 100 MHz. The sample rate is usually stated in samples per second (or Hz) or in bits per second. The sample rate must of course be at least twice the highest frequency to be measured, as required by sampling theory and explained in Section 12.8.

The *vertical resolution* of an oscilloscope is its ability to discriminate between signals which are close together. In a digital oscilloscope this is determined by the number of bits used in the analogue to digital converter. For example an 8 bit converter can give a vertical resolution of one part in 256, or 0.391%. Many factors affect the actual achievable resolution, such as noise and aperture uncertainty, and screen resolution is a function of spot size and shape. These usually limit the resolution to about 2% of full scale. *Horizontal resolution* is a measure of the amount of digital memory used to store the waveform. If the signal is stored in 512 data words then the horizontal resolution is 1/512 or 0.195%.

12.10 Oscilloscope accessories

Two main accessories, used with oscilloscopes are described in this section, probes and cameras.

12.10.1 *Oscilloscope probes*

An oscilloscope can be connected to the circuit being measured by a piece of wire. Although simple, this connection suffers from two problems. First it is prone to pick up stray signals, especially 50 Hz power signals. Secondly the loading on the circuit is now equal to that of the oscilloscope, about 1 MΩ and 10–100 pF, which can be relatively large. Because of these disadvantages a simple wire connection is useful for low impedance signal source measurements only.

A coaxial cable may be used to connect the oscilloscope to the circuit being measured. The cable would protect against stray signal pick up, but a typical 50 ohm coaxial cable has a distributed capacitance of about 100 pF per metre, which would increase the loading on the signal source. An unterminated coaxial cable can also resonate, at a frequency dependent on the cable length, and this would give a false signal indication on the oscilloscope.

The functions of an oscilloscope probe are (i) to transmit an accurate representation of the signal from probe tip to the oscilloscope (ii) to attenuate or amplify the signal before it reaches the oscilloscope (iii) to prevent the oscilloscope from unduly loading the signal source, and so giving a false reading, and (iv) to convert special signals, such as current, high voltage, and amplitude modulated signals, into a form which can be measured on the oscilloscope.

There are several types of probes, such as passive probes, active probes, current probes, high voltage probes, and demodulator probes. All these consist of three sections, a probe head, the interconnecting cable, and a cable termination. The length of the probe cable is important. Special low capacitance cables are often used, having a resistive centre conductor of several hundred ohms, which provides an approximation to a terminating characteristic impedance, and damps ringing.

Passive probes can only attenuate, and attenuation ratios of 10:1, 50:1 and 100:1 are typical. Some instruments incorporate a system which automatically changes the gain of the

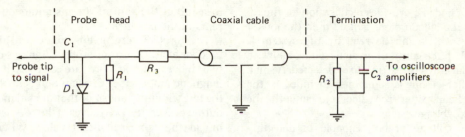

Fig. 12.40. A typical passive compensated probe.

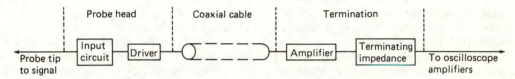

Fig. 12.41. Block diagram of an active probe.

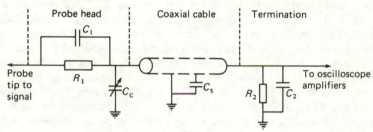

Fig. 12.42. A typical demodulator probe.

oscilloscope amplifiers to compensate for the probe attenuation, or whch indicate the change in sensitivity on a display when a probe is used. This minimises operational errors when changing between probes of different attenuation ratios.

To reduce the loading effect of the cable and oscilloscope a compensated probe is used, as shown in Fig. 12.40. R_2 and C_2 form the cable termination, and this is usually the input impedance of the oscilloscope i.e. typically 1 MΩ and 10–100 pF. The cable has a relatively high distributed capacitance C_s of value dependent on the cable length. R_1 and C_1 are typically 9 MΩ and 10–15 pF and together with C_c they form the probe head. The value of C_c is adjusted on a square wave, as illustrated in Fig. 12.15, to compensate for the capacitance of the cable. This ensures that the system attenuates all frequencies equally, as in the compensated attenuator of Section 12.4.1., the attenuation ratio being $(R_1 + R_2)/R_2$.

Active probes are used when the input impedance of the probe and oscilloscope system must be very high; or where very long cables are to be used, which would require drivers in the probe head; or where the loss of signal sensitivity is critical. Fig. 12.41 shows the block diagram of a typical active probe. The input circuit is usually an amplifying device, such as field effect transistor (FET). This gives a high input impedance, of about 10 MΩ shunted by 0.5 pF. However the FET limits the dynamic range which can be handled to between 0.5 V and 5 V. The driver in the probe head provides low impedance drive capability for long cables. The terminating impedance can also consist of active devices.

Demodulator probes are used to convert an r.f. signal into a d.c. voltage, which is proportional to the peak value of the r.f. signal. The d.c. voltage can then be measured on the oscilloscope. Demodulator probes are used in applications such as aligning receivers, and for signal tracing. Fig. 12.42 shows a typical demodulator probe circuit. Diode D_1 provides rectification for the r.f. signal. If a silicon diode is used then the lowest voltage which can be read by the system is 600 mV, but a germanium diode would allow r.f. potential down to 200 mV to be detected. A small R–C filter circuit is sometimes included in the probe head, after the diode, to smooth the rectified voltage.

High voltage probes, capable of detecting voltages in the 50 kV region, consist of a case made from thick insulating plastic to prevent operator shock. A high valued series resistor is included in the probe, and the instrument's shunt resistance is often used to give attenuation ratios of 1000:1 or more.

Current probes measure the value of current flowing in a conductor. A.C. current probes consist of a toroidal core, which can be opened to fit around the current-carrying conductor (Fig. 12.43). The core is wound with a multi turn secondary coil, so that it acts as a transformer having a single turn primary. The output from the secondary is amplified and displayed on the screen of the oscilloscope. A typical scale factor for the current probe is 1 mV per milliampere. The frequency response of the system is good over the range 100 Hz–100 MHz, and it provides a loading effect to the circuit of about 50 milliohms and 0.05 μH in series with the conductor, and a shunt capacitance to ground of 2 pF.

The sensitivity of the probe can be improved by looping extra primary turns through the toroidal core. This also increases the inductive loading effect, and increases the inductance and loop capacitance, which results in ringing in the MHz frequency region. Several separate current-carrying conductors can be passed through the same core, so that the output from the probe is a measure of the sum of the instantaneous currents in the conductors.

The toroidal core arrangement can also be used for a direct current probe. Windings on the two halves of the core are connected as a magnetic amplifier. A.C. excitation is applied to the two windings, but since they are balanced no output occurs, the two halves of the core saturating at the same excitation value. When d.c. current flows in the central conductor half the core saturates first, giving an output voltage at twice the excitation frequency. This is amplified, demodulated, and then displayed on the oscilloscope. The probe adds negligible resistance into the circuit being measured, but introduces an inductance of typically 0.5 μH in series, and a capacitance to ground of about 1 pF. The typical range of the system is 1 mA–10 A. Good shielding is needed to minimise the effect of the earth's magnetic field. A Hall effect device (F.F. Mazda, *Discrete Electronic Components*, C.U.P., 1981) is used in most modern d.c. current probes to sense the magnetic effect of the current. This gives a probe with good response from d.c. to 50 MHz.

Several parameters are used in characterising oscilloscope probes. The bandwidth is the maximum -3 dB frequency which can be expected with the probe and oscilloscope. The maximum voltage rating of the probe is specified as the d.c. and peak a.c. value. This voltage needs to be derated with frequency, usually for frequencies above 100 kHz. The maximum allowable voltage at any frequency is determined by the terminating elements, or the resistive centre conductor in the probe. The compensation range is the range of the oscilloscope input capacitance over which the probe can compensate to give a flat attenuation over a band of frequencies. The maximum probe length is also important. Extra length

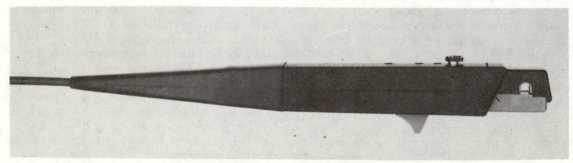

Fig. 12.43. A split toroidal core current probe. (Photograph reproduced by courtesy of Tektronix, Inc.)

reduces bandwidth, increases capacitance loading, and introduces extra propagation at the rate of about 5 ns per metre.

12.10.2 *Oscilloscope cameras*

The most effective and widely used method of obtaining a hard copy output of the waveform display on the oscilloscope is to photograph it. A hand held camera may be used, but the best are those which have been designed to bolt directly onto the oscilloscope. The camera mounting should block out all leakage light, although an operator viewing window is usually available, as shown in Fig. 12.44. It is also convenient to be able to swing the camera out so as to get a direct view of the CRT.

The camera lens is usually wide aperture, wide angle, and flat field. It has a short focal length, and the lens is coated to reduce reflections to below 1%. A mechanical or electrical shutter system may be used, an electrical shutter being preferred for high speed photography and for remote control. Shutter speed is usually adjustable from 1/60 s to several seconds. The camera may be used to record a continuous trace or a transient waveform. For single shot photography the shutter may be opened before the start of the trace, and can be closed some time after the end of the trace, as long as there is low light leakage.

Films of 1000 ASA to 10 000 ASA are common with 3000 ASA being standard. Both polaroid and negative film may be used, although polaroid film is more popular. Some cameras have a built in ultra violet light which gives a low level excitation of the background phosphor. This displays the graticule in black against a grey background, with the trace in white.

An important parameter in specifying oscilloscope cameras is writing speed, usually measured in cm per ns. This is the maximum spot speed which can be photographed, and depends on the film exposure threshold and oscilloscope characteristics. The writing speed can be increased at high speed low light levels by first exposing the film to a uniform low level of light. This is called

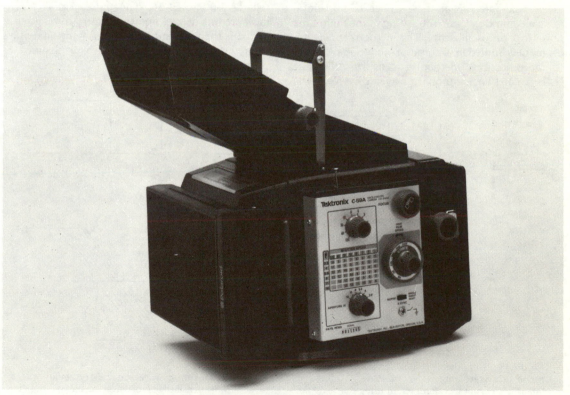

Fig. 12.44. Oscilloscope camera. (Photograph reproduced courtesy of Tektronix, Inc.)

fogging and it gives a development centre in the film emulsion. Special equipment can be bolted onto the camera body which performs film fogging. This can be triggered remotely, or from the oscilloscope, to give pre-fogging, post-fogging or simultaneous fogging i.e. the fogging occurring before, after or during the trace sweep. Simultaneous fogging gives the greatest gain in writing speed.

12.11 Measurements using oscilloscopes

The oscilloscope is a very versatile instrument, and its use to observe and measure waveforms has already been introduced in earlier sections. It has also been explained that modern day oscilloscopes often incorporate other instruments, such as counters, so that composite measurements can be made on a signal. The present section will look at further ways in which an oscilloscope may be used, to measure time, frequency and phase shift.

12.11.1 *Time measurements*

The time between two points on a waveform can be calculated from the number of horizontal scale markings, and the setting of the main time base of the oscilloscope. This method of measurement is limited in accuracy to about 5%. A much more accurate method of measuring the time between two points on the waveform is to use the delay time position control. This has a ten turn calibrated dial, so the delay can be set precisely.

Fig. 12.45 illustrates how the delay time setting can be used to measure the period of a square wave. The main time here is set at a convenient value, say $S\,\mu$s per division, to display the main waveform as in Fig. 12.45(*a*). The time delay control is then adjusted to intensify the leading edge A of the waveform. With the oscilloscope on alternate mode the intensified part of the waveform is displayed, and the delayed time base is adjusted so that its half amplitude point is on a convenient vertical graticule, say the central one, as in Fig. 12.45(*b*). The reading T_1 of the delay dial is noted. The delay time control is now adjusted to intensify edge B of the waveform, and this is again displayed on alternate mode with its half amplitude point on the same graticule line as before. The reading T_2 of the delay dial is noted. The period of the waveform T is then given by (12.14)

$$T = (T_2 - T_1) \times S \tag{12.14}$$

The pulse jitter time can similarly be measured, as in Fig. 12.46. The pulse period T is measured as before, and the uncertain or jitter period is then measured by intensifying edge A, as shown in Fig. 12.46(*b*), and measuring time t.

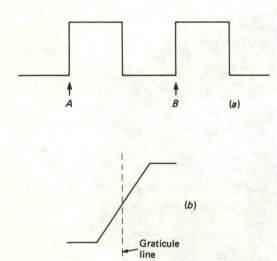

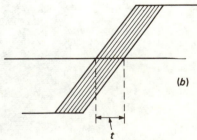

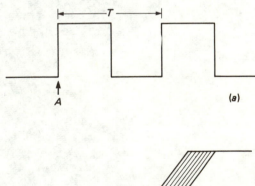

Fig. 12.45. Using the delay time setting to measure the period of a waveform; (*a*) square wave, (*b*) display of intensified part of waveform on (*a*) using the dealy time base.

Fig. 12.46. Measuring waveform jitter; (*a*) square wave, (*b*) display of intensified part of waveform on (*a*) using the delay time base.

The percentage jitter J is given by (12.15)

$$J = \frac{100t}{T} \qquad (12.15)$$

Pulse rise time measurements can be made by noting the time between the 10% and 90% amplitude points using the delay time base. Correction for the finite rise time of the oscilloscope must be made using (12.11). If the rise time of a circuit is being measured, then an arrangement such as in Fig. 12.47 is used. The pulse

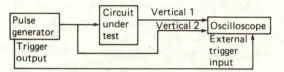

Fig. 12.47. Measuring pulse rise time of the circuit under test.

generator output goes directly to one vertical input of the oscilloscope, and it also goes, via the circuit under test, to the other vertical input. The rise time of both traces on the oscilloscope can therefore be noted. However a correction is now required for the finite rise time of the pulse generator and the oscilloscope. If t_p is the rise time of the pulse generator, t_o is the rise time of the oscilloscope, t_m is the rise time of the circuit under test, as measured by the oscilloscope, then its actual or true rise time t_t is given by (12.16)

$$t_t = [t_m{}^2 - t_o{}^2 - t_p{}^2]^{1/2} \qquad (12.16)$$

12.11.2 *Frequency measurement*

The period of a waveform can be measured, as in Section 12.11.1, and the frequency found as a reciprocal of this value. Alternatively Lissajous figures may be used to measure the frequency. In this method the unknown frequency is compared with an accurately known frequency, and

Lissajous figures are used to determine the ratio between the two frequencies.

To obtain a Lissajous figure the unknown frequency is fed into the vertical input of the oscilloscope. The internal time base of the oscilloscope is switched off and the output from an accurate signal generator is fed into the horizontal input. The sensitivity of the two inputs is adjusted so that the display fills the screen as much as possible. The signal generator frequency is then adjusted to give a stationary display on the screen of the oscilloscope. There is now a fixed ratio between the frequencies of the two inputs. This ratio is found as the number of crossing points, in the display, over the horizontal and vertical lines. Stated another way it is the ratio of the number of loops, in the display, which touch the horizontal and vertical tangents.

Fig. 12.48 shows some examples of Lissajous figures. If the two frequencies are equal then the trace is a circle, as in Fig. 12.48(a). Now the unknown frequency equals that of the signal generator setting, which can be known with an accuracy of 0.001%. The display obtained depends on the frequency ratio and the phase relationship, of the two input frequencies, but in every case (12.17) holds true.

$$\frac{\text{Vertical input frequency}}{\text{Horizontal input frequency}}$$

$$= \frac{\text{Number of loops which touch horizontal tangent}}{\text{Number of loops which touch vertical tangent}} \qquad (12.17)$$

The waveform shown in Fig. 12.48(d) is called a double image, and is obtained when the high frequency signal is 90° leading the low frequency signal. The electron beam, after travelling to the

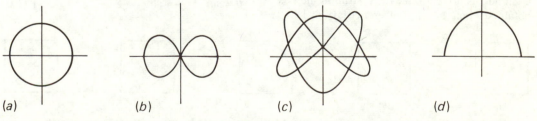

Fig. 12.48. Examples of Lissajous figures for different frequency ratios; (a) 1:1, (b) 2:1, (c) 3:2, (d) 2:1.

end of the trace, reverses its direction and returns to the start. To apply (12.17) to this trace the horizontal tangent must be across the open end.

12.11.3 *Phase measurement*

The phase shift between two waveforms can be obtained by displaying them both on the oscilloscope, in alternate mode, and then setting the sweep speed to provide a convenient display to measure T and t, as in Fig. 12.49. The phase

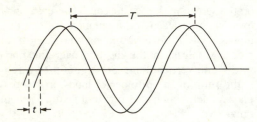

Fig. 12.49. Measuring phase shift between two waveforms.

difference P between the two waveforms is then given in degrees by (12.18)

$$P = 360\frac{t}{T} \qquad (12.18)$$

An alternative method for measuring phase shift is by the use of Lissajous figures as in Fig. 12.50.

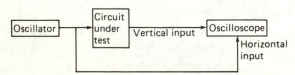

Fig. 12.50. Arrangement for measuring the phase shift through the circuit under test using Lissajous figures.

For equal amplitudes and frequencies, of the two inputs, the effect of the phase shift is to vary the figures from a straight line, through an ellipse to a circle, as in Fig. 12.51. A straight line is obtained when the phase difference is 0° or 180°.

The angle to the horizon is 45° when the two amplitudes are equal. The line leans toward the side having the greater voltage, for example if the voltage on the horizontal input is higher than that on the vertical input, then the angle to the horizontal will be less than 45°. The tilt angle θ is given by (12.19)

$$\theta = \tan^{-1}\left(\frac{\text{Vertical voltage}}{\text{Horizontal voltage}}\right) \qquad (12.19)$$

A circle is obtained for 90° or 270° phase shift when the amplitudes of the two inputs are equal. If they are not equal then an ellipse is obtained, with the main axis along the vertical or horizontal depending on whether the vertical or horizontal signal, respectively, has the greater amplitude.

The phase angle between two signals can be determined from the ellipse, as shown in Fig. 12.52. The gains of the vertical and horiz-

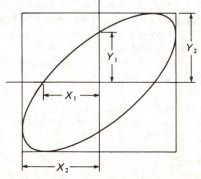

Fig. 12.52. Measuring phase shift using Lissajous figures.

ontal amplifiers are usually adjusted such that the ellipse fits into the box shown. The magnitude of the phase angle is found as the ratio of the ellipse dimensions as in (12.20). These should be interpreted together with Fig. 12.51, depending on which way the ellipse leans.

$$\theta = \sin^{-1}\frac{Y_1}{Y_2} = \sin^{-1}\frac{X_1}{X_2} \qquad (12.20)$$

| (a) | (b) | (c) | (d) | (e) |

Fig. 12.51. Effect of phase shift on Lissajous figures; (a) 0°, (b) 30° or 330°, (c) 90° or 270°, (d) 150° or 210°, (e) 180°.

Part 3: Application areas and special instruments

13. Electronic component tests

13.1 Introduction

Electronic component testers cover a wide spectrum of instruments. The present chapter will concentrate on laboratory measurement instruments, and will only briefly introduce large semiconductor testers, used by manufacturers for semiconductor characterisation or for goods incoming testing. Also not covered in this chapter is equipment intended for environmental testing of components, such as temperature cycling and vibration.

The first half of this chapter is concerned with measurements on passive components, and the second half on semiconductor measurements. Resistors, capacitors and inductors are complex devices, having elements of each in their make up. For example an inductor, which is operated at a frequency greater than its self resonance frequency, will behave like a capacitor. Therefore the concept of impedance is first introduced.

There are three main methods for measuring passive components: bridges, voltage–current, and Q. Bridges, described in Chapter 7, balance the unknown component against known standard components, and calculate the unknown from the values of the standard. The voltage–current method applies a constant voltage or constant current to the component being measured, and measures the other parameter i.e. current or voltage. Ohm's law is then used to

determine the value of the unknown component. The Q method uses the series resonant properties of the unknown component to find its Q factor, and then calculates the component from this.

Bridge based instruments were very widely used, and are still employed where high accuracy is required. For most applications the accuracy obtained from modern voltage–current instruments, using automatic digital control, is adequate.

In this chapter the principles of instruments which measure resistance (R), inductance (L), capacitance (C), Q factor, and overall impedance (Z), are described in separate sections, although it should be realised that most modern instruments combine the functions of measuring R, L, C, Q and Z in a single instrument.

13.2 Impedance concepts

The impedance Z of a circuit is a measure of its opposition to the flow of current through it when an a.c. voltage is applied across it. The impedance is made up of a real or resistive part R, and an imaginary or reactive part X, and its value may be stated by (13.1) in Cartesian form, or by (13.2) in polar form.

$$Z = R \pm jX \tag{13.1}$$

or

$$Z = |Z| \underline{/\theta} \tag{13.2}$$

where

$$|Z| = (R^2 + X^2)^{1/2} \tag{13.3}$$

and

$$\theta = \tan^{-1}(X/R) \tag{13.4}$$

The reciprocal of impedance is called admittance $(Y = Z^{-1})$, that of resistance is called conductance $(G = R^{-1})$, of reactance is called susceptance $(B = X^{-1})$, and they are connected by (13.5)

$$Y = G \pm jB \tag{13.5}$$

Fig. 13.1 shows a vector diagram of impedance.

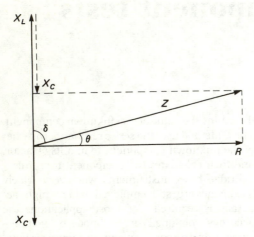

Fig. 13.1. Vector diagram of impedance.

Reactance can be inductive (X_L) or capacitive (X_c), and the resultant value of X is obtained from (13.6)

$$X = X_L - X_c \tag{13.6}$$

The values of X_L and X_c for an inductance L, capacitance C and frequency f are given by (13.7) and (13.8)

$$X_L = 2\pi f L \tag{13.7}$$

$$X_c = (2\pi f c)^{-1} \tag{13.8}$$

True reactance is a measure of a component's ability to store energy, and resistance is a measure of its power loss. For a component having reactance and resistance its quality factor, or Q factor, is a measure of the component's stored energy to lost energy. It is numerically equal to 2π times the ratio of the maximum instantaneous energy stored during one cycle, to the energy dissipated over that cycle.

Usually Q factor is applied to inductive circuits, and a similar term, the dissipation factor (D) is applied to capacitive circuits. These are connected by (13.9)

$$Q = D^{-1} \tag{13.9}$$

The dissipation factor is also connected to the angle δ in Fig. 13.1 by (13.10), so that δ is often called the loss angle.

$$D = \tan \delta \tag{13.10}$$

For large valued capacitors having a high loss (e.g. electrolytic capacitors) it is more usual to specify the power factor, $\cos \theta$. For values of Q greater than or equal to 10 the power factor is numerically very close to the value of the dissipation factor (D).

All components, resistive, capacitive or inductive, have parasitics. The true value of the component is its theoretical value, which excludes all parasitic effects, and is therefore only of academic interest. The effective value, or a.c. value, of the component is its value including all parasitic effects. The measured value of the

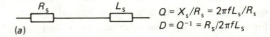

$$Q = X_s/R_s = 2\pi f L_s/R_s$$
$$D = Q^{-1} = R_s/2\pi f L_s$$

(a)

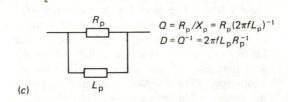

$$Q = X_c/R_s = (2\pi f C_s R_s)^{-1}$$
$$D = Q^{-1} = 2\pi f C_s R_s$$

(b)

$$Q = R_p/X_p = R_p(2\pi f L_p)^{-1}$$
$$D = Q^{-1} = 2\pi f L_p R_p^{-1}$$

(c)

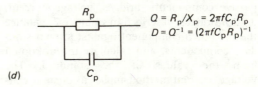

$$Q = R_p/X_p = 2\pi f C_p R_p$$
$$D = Q^{-1} = (2\pi f C_p R_p)^{-1}$$

(d)

Fig. 13.2. Series and parallel lumped circuits: (a) series R–L, (b) series R–C, (c) parallel R–L, (d) parallel R–C.

component includes parasitic effects as well as the errors inherent in the measuring instrument.

Component parasitics can be treated as lumped parameters, and placed in series or parallel, as in Fig. 13.2. Conversion between the series and parallel modes can readily be done, as illustrated in Fig. 7.2.

A circuit consisting of inductance and capacitance can resonate at a frequency f_r given by (13.11)

$$f_r = [2\pi\sqrt{LC}]^{-1} \tag{13.11}$$

A parallel circuit has a high impedance at resonance and a series circuit has low impedance.

13.3   Resistance measurement

13.3.1   *Equivalent circuit*

A simplified equivalent circuit of a resistor is shown in Fig. 13.3. The inductance of the leads,

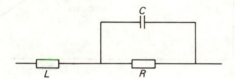

Fig. 13.3. Simplified equivalent circuit of a resistor.

or of the windings, is denoted by lumped inductor L, and the parasitic capacitance across the resistor is shown as a lumped capacitor C. If this resistor is measured using d.c. conditions then the true value of resistance R will be indicated. However, many resistors are used in a.c. applications, and in these instances the a.c. or effective resistance value is more relevant. Even at d.c. the indicated resistance could, of course, by different from the true value, due to instrument errors.

The impedance of the equivalent circuit shown in Fig. 13.3 is given by (13.12)

$$Z = \frac{R}{1 + \omega^2 C^2 R^2}$$
$$+ j\frac{\omega L + \omega^2 L C^2 R^2 - \omega C R^2}{1 + \omega^2 C^2 R^2} \tag{13.12}$$

The first term of this equation is the effective resistance. It is also called the equivalent series resistance (ESR), which is the magnitude of the

real value of a component when it is modelled using a series equivalent circuit. Note that although the effective resistance is real, and has no imaginary part, its magnitude is dependent on frequency. The greater the parasitics the higher the frequency dependance of the resistance.

13.3.2   *Measurement methods*

Several methods may be used to measure resistance. The bridge technique is described in Section 7.3, and the method using voltmeter, ammeter, and ohmmeter, is explained in Section 6.2.4. The problem of measuring effective resistance is that, for any current, the voltage across the resistor is due to both the ESR and the imaginary part of (13.12), so the total impedance is measured. This limitation is overcome by using a synchronous detection technique, as in Fig. 13.4. The signal source is

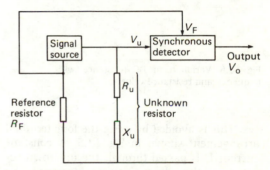

Fig. 13.4. A synchronous detector circuit.

applied to the unknown resistor, and to a reference resistor which has negligible parasitics. The voltages V_u and V_F across the two devices are then compared in the synchronous detector, which measures only those elements of V_u which are in phase with V_F. Therefore the output V_o is an indication of the ESR of the unknown resistor.

Several precautions need to be taken when measuring low valued resistors. The voltage leads should be twisted together, and so should the current leads, to prevent errors due to mutually induced emfs. Thermal emfs, caused by dissimilar metal junctions, and contact resistance in the voltage sense leads, must also be minimised. The greatest problem encountered in measuring low valued resistors is due to the voltage drop across connecting lead impedan-

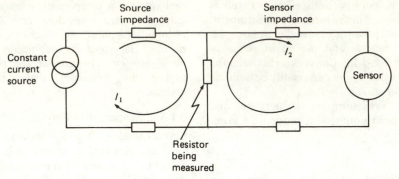

Fig. 13.5. Four terminal resistor measurement.

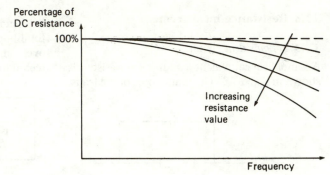

Fig. 13.6. Variation of DC resistance with frequency and resistance value.

ces. This is avoided by using the four terminal arrangement shown in Fig. 13.5. A constant current I_1 is passed through the unknown resistor, and the voltage across it is measured. The impedance of the source and the leads may be much larger than that of the resistor being measured, without affecting the results. The sense current I_2 is small so the voltage loss across the sensor impedance does not result in appreciable error.

The parasitics present in high valued resistors have a much larger influence on the measurements than those in low valued resistors. Fig. 13.6, for example, shows typical curves illustrating the frequency performance of high valued resistors. To make measurements, a constant voltage is applied across the resistor, and the value of current sensed, since even a small current in a constant current mode would generate a high voltage across the resistor. The surface of the high valued resistor must be kept clean, since leakage across the surface insulation would result in a parallel resistive path.

13.4 Capacitance measurement

13.4.1 *Equivalent circuits*

The formula for the capacitance formed by two parallel plates of area A and separated by a distance d is given in (13.13)

$$C = \frac{K\varepsilon_o A}{d} \qquad (13.13)$$

In this equation ε_o is the permittivity of free space, and K is the dielectric constant of the material between the plates.

The equivalent circuit for a capacitor is shown in Fig. 13.7. The dielectric forms the single most

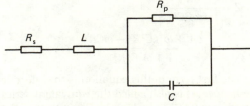

Fig. 13.7. Simplified equivalent circuit of a capacitor.

important part of the capacitor. It suffers from loss due to dielectric absorption, insulation resistance, and leakage current when operating with a d.c. bias. The loss in the dielectric is represented by the parallel resistor R_p.

The resistance of the capacitor leads, the contact resistance, and the plate resistance, are represented by resistor R_s. This resistance remains relatively constant as the frequency across the capacitor is changed. The inductance of the capacitance leads and plates is small, at low frequencies, and is represented by L. The true value of the capacitor is C.

Fig. 13.7 can be analysed to yield the value of impedance given in (13.14)

$$Z = R_s + \frac{R_p}{1 + \omega^2 R_p^2 C^2}$$

$$+ j\frac{\omega L - \omega R_p^2 C + \omega^3 R_p^2 C^2 L}{1 + \omega^2 R_p^2 C^2} \qquad (13.14)$$

The real part of (13.14) is the equivalent series resistance (ESR). Its value is greater than the series resistance R_s due to a value contributed by R_p. This second term is frequency dependent.

Fig. 13.7 can be reduced to a series or parallel circuit, as in Fig. 13.8. Resistor R_{SER} is the ESR of

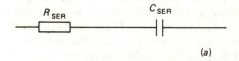

(a)

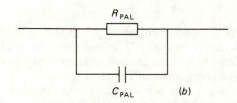

(b)

Fig. 13.8. Alternative capacitor equivalent circuits; (a) series, (b) parallel.

the circuit, and there is an equivalent parallel resistor R_{PAL}. The conversion between the series and parallel circuits can be made using the equations given in Fig. 7.2(b). Since D is frequency dependent this conversion can only occur at the same frequency. The smaller the value of D, that is the closer the capacitor is to a perfect

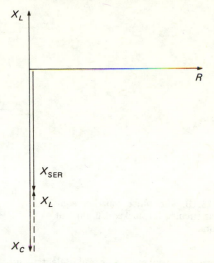

Fig. 13.9. Vector diagram of a capacitor equivalent circuit.

device, the closer the series and parallel equivalent circuits will be to each other.

The vector diagram of the capacitor equivalent circuit is shown in Fig. 13.9. From this it is seen that the equivalent series capacitance C_{SER} is the resultant of L and C, and is given by (13.15)

$$C_{SER} = \frac{C}{1 - \omega^2 LC} \qquad (13.15)$$

Therefore the equivalent capacitance value increases when an inductor L is added in series with capacitor C.

If the resonant frequency of the circuit is denoted by (13.16), then (13.15) can be rewritten in terms of the resonant frequency as shown in (13.17)

$$f_r = [2\pi\sqrt{(LC)}]^{-1} \qquad (13.16)$$

$$C_{SER} = \frac{C}{1 - (f/f_r)^2} \qquad (13.17)$$

From (13.17) it is evident that as the applied frequency approaches the resonant frequency of the circuit, the equivalent series capacitor increases sharply.

Many different capacitor constructions have been used to increase the performance under various circuit conditions, and these all give variations to the capacitor equivalent circuit. For example, Fig. 13.10(a) shows the equivalent circuit of a stacked foil capacitor, consisting of

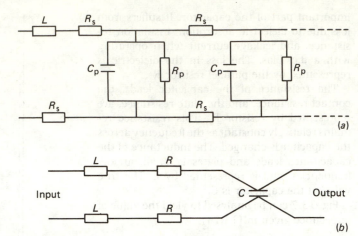

Fig. 13.10. Alternative capacitor equivalent circuits; (*a*) stacked foil, (*b*) four terminal.

several stacks of electrode and dielectric, which is intended to give a very low ESR. Four terminal electrolytic capacitors, having an equivalent circuit as shown in Fig. 13.10(*b*), have separate input and output terminals, and also present a low ESR to the load.

Fig. 13.11 shows a typical curve of

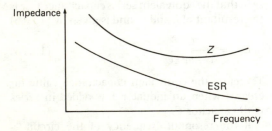

Fig. 13.11. Typical impedance vs. frequency curves for a capacitor.

capacitance variation with frequency. Capacitors are much more susceptible than resistors to changing stimuli, and these need to be considered when making capacitor measurements. The prime cause of capacitance variation is the characteristic of the dielectric, and this is affected by temperature, d.c. bias, test voltage levels, frequency and ageing. The way in which the capacitance changes with these parameters varies from one capacitor type to another.

13.4.2 *Measurement methods*

Several methods may be used to measure capacitance. The capacitance bridge was described in Section 7.5, and it is the most accurate

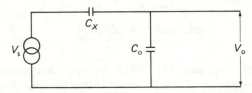

Fig. 13.12. Capacitor ratio method of measuring capacitance.

method of measuring capacitance. Fig. 13.12 shows an alternative method in which voltage V_s is applied across the unknown capacitor C_x and a standard capacitor C_o, and the voltage across C_o is measured. The value of C_x can be found from (13.18), assuming perfect capacitors.

$$\frac{V_o}{V_s} = \frac{C_x}{C_x + C_o} \tag{13.18}$$

In a capacitance meter the output voltmeter, which measures V_o, is calibrated to read C_x directly, and range changing is accomplished by switching different values of C_o.

Fig. 13.13 shows a similar method to Fig. 13.12 but a resistor (*R*) is now used instead of a capacitor (C_o) to measure current. A phase or

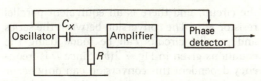

Fig. 13.13. Phase or synchronous detector method of measuring capacitance.

synchronous detector is also used, to measure the voltage which is in quadrature to the input, thus avoiding errors due to capacitor losses or the inductance of the resistor.

The measurement method shown in Fig. 13.14 uses the unknown capacitor C_x in the

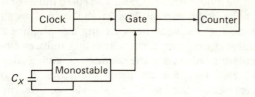

Fig. 13.14. Electronic counting method of measuring capacitance.

timing circuit of a monostable. This monostable operates the gate which lets through pulses from a high frequency clock onto a counter. The greater the value of C_x the longer the monostable period, keeping the gate open for a longer time and hence giving a high count. The counter can be calibrated in terms of C_x. Circuits which control and reset the counter between readings are not shown in Fig. 13.14.

The series inductance shown in Fig. 13.7 becomes important when measuring large values of capacitance i.e. a low reactance. Four terminal connections must now be used, and the mutual inductance between leads minimised by twisting them together. When measuring low values of capacitance the system needs to be shielded to minimise RFI effects. The influences of test fixtures and leads are very important since shunt capacitances as low as 0.5 pF can be significant when measuring a capacitor of 1 pF. The amount of fixtures and leads used should be minimised, and the capacitor attached directly to the instrument terminals. Even then, at high frequencies, the lead length can be critical. The terminals of the instrument should be shielded to minimise their capacitive effect, using the three terminal construction described in Section 2.4.3.

13.5 Inductance measurement

13.5.1 *Equivalent circuit development*

The inductance of a coil of N turns, wound on a core of permeability μ_c, cross sectional area A and mean magnetic length L, is given by (13.19)

$$L = \mu_c \frac{AN^2}{L} \tag{13.19}$$

An air gap is usually introduced into the magnetic path, to increase the linearity of the inductor. This reduces the permeability to an effective value μ_e, where ratios of μ_e/μ_c between 0.05 and 0.2 are typical.

The permeability is equal to the ratio of flux density (B) to magnetic intensity (H), and since the relationship between B and H is not linear the permeability varies, as shown in Fig. 13.15. This figure assumes zero initial magnetisation, which is not usually the case. Most magnetic materials exhibit residual magnetisation, as shown by their hysteresis curve, as in Fig. 13.16. The area enclosed by this hysteresis curve is a measure of hysteresis loss in the magnetic material. When a small a.c. magnetisation is superimposed onto the d.c. magnetisation force, an incremental permeability curve is obtained, as in Fig. 13.16.

The value of inductance varies considerably, depending on the conditions of measurement, particularly for iron cored inductors. This presents many problems in repeatability of measure-

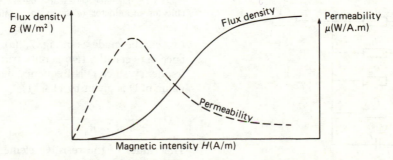

Fig. 13.15. Permeability and magnetisation curves for a typical iron cored inductor.

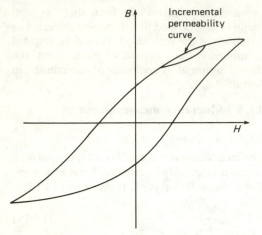

Fig. 13.16. Hysteresis curve for a typical iron cored inductor.

ment. The main reason for inductance variation, in iron cored inductors, is the variation of permeability. The permeability changes with the level of the test signal and the d.c. bias. The previous history is also important since the core may have been left in a residual magnetisation state. Prior to any measurement it should be demagnetised by applying a large (saturating) a.c. signal, and then slowly reducing this to zero. The permeability also usually decreases with frequency, and for materials with low μ_c the permeability increases with temperature, whilst for materials with high μ_c it may increase or decrease with temperature. Because of these variations it is important to simulate, as closely as possible, the actual usage conditions when making inductance measurements.

The equivalent circuit for an inductor is shown in Fig. 13.17. L is the true inductance and R_s is the actual (d.c.) winding resistance. At

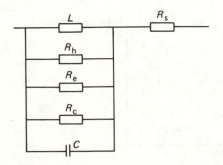

Fig. 13.17. Equivalent circuit of an inductor.

high frequencies this resistance increases due to skin and proximity effects in the conductors. R_h represents the hysteresis loss, and is only applicable for iron cored inductors. Its value depends on the core material, the frequency, and the flux density. The eddy current loss is represented by resistor R_c. It applies to both iron cored and air cored inductors, and for an iron cored inductor it is the eddy current loss in the core. For all inductors the windings carrying a.c. produce a varying magnetic field, whose flux induces circulating currents in the conductors, giving eddy current loss. For air cored inductors R_e also represents the loss in shields or nearby conductors.

Capacitor C, in the equivalent circuit of Fig. 13.17 is the distributed capacitance of the inductor. The dielectric loss associated with this capacitor is denoted by R_c. It is given by (13.20).

$$R_c = K\omega^3 L^2 C \tag{13.20}$$

K is the power factor of the distributed capacitor and ω is equal to $2\pi f$, where f is the frequency.

The equivalent circuit of the inductor can be reduced to an equivalent series or parallel circuit, as in Fig. 13.18. The conversion between these

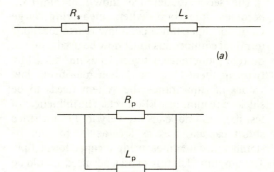

(a)

(b)

Fig. 13.18. Effective series and parallel equivalent circuits for an inductor; (a) series, (b) parallel.

two can be made as in Fig. 7.2(a). The equivalence between the two is only true at the same frequency since Q is frequency dependent. The value of Q is given by (13.21)

$$Q = \frac{\omega L}{R_s + R_e + R_h + R_c} \tag{13.21}$$

The values of the resistive elements vary in a complex fashion with frequency, giving a typical

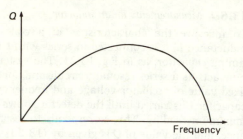

Fig. 13.19. Effect of frequency on the Q of an inductor.

Q versus frequency plot as in Fig. 13.19.

Using the correct series or parallel equivalent circuit mode, shown in Fig. 13.18, gives an indicated inductance which is closest to the effective inductance. When inductance is large the value of its reactance is high, and the effect of the parallel resistor is more significant; the parallel mode should now be used. At low values of inductance the reactance is small, so that the effect of the series resistance is more significant, and the series mode of measurement should be used. The value of inductance at which the choice between the two modes is made is dependent on reactance, and therefore on frequency.

13.5.2 *Measurement methods*

Several methods may be used to measure inductance. The most accurate method involves inductive bridges, as described in Section 7.4. Fig. 13.20 shows a phase detection method, which is very similar to that used for capacitors, as in Fig. 13.13. In fact the two systems are often housed in the same instrument, and a switch used to select between capacitance and inductance measurement. The position of the standard resistor R and unknown inductor L_x shown in Fig. 13.20 can be interchanged, but now the output will be proportional to $1/L_x$ and not L_x.

Iron cored inductors use materials with permeabilities in the region of several hundred thousand, so they have a high volumetric efficiency. However the core also results in higher losses, and less predictable behaviour, unless an air gap is introduced in the flux path to improve the stability. An air cored inductor is much more stable, and shows little change in inductance with frequency or current. Temperature changes are as a result of dimensional changes in the coil, or a change in the resistance of the wire. The distributed capacitance causes most of the inductance change with frequency, and the measurement frequency (f_m) should be kept well below the resonant frequency (f_r). Under these conditions the incremental inductance δL over the low frequency inductance L, is given by (13.22).

$$\frac{\delta L}{L} = \omega^2 LC$$

$$= (f_m/f_r)^2 \qquad (13.22)$$

Air cored inductors are affected by adjacent conducting materials, and this is specially noticeable in r.f. coils, which should be shielded.

Several precautions are required when measuring high values of inductance. Because of the higher reactances involved measurements should be made in an RFI-free environment. The test leads should be shielded and kept short. The inductor would probably be made from high permeability material, and detailed information would be required on the core characteristics. Large inductances also mean many turns of fine wire, which gives high values of distributed capacitance and series resistance. The capacitance results in readings which are frequency dependent. High resistance gives lower Q, and a greater difference between values obtained by the equivalent series and parallel measurement modes. Large values of inductance also result in long response times, resulting in apparent instabilities when changing signals, or the d.c. bias levels, on automatic test instruments.

Low valued inductors have low impedances, so four terminal measurement techniques need

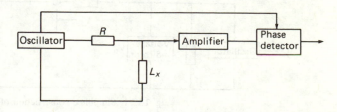

Fig. 13.20. Measurement of inductance using phase detection.

to be used. Test leads should be avoided if possible, and the inductor fixed directly to the equipment. Precaution should also be taken of not placing the inductor close to objects which can affect its value, such as instrument posts and even the operator's hands.

13.6 Measurements using a *Q* meter

13.6.1 *Q meter construction*

The construction of a *Q* meter is shown in Fig. 13.21. It works on the principle of resonating an *LC* circuit and then using (13.16). The *Q* meter can measure the *Q* factor of a circuit, at the operating frequency, as well as the inductance, capacitance and effective resistance.

In Fig. 13.21 the frequency of the oscillator can be varied, usually with range select switches and continuous adjustment dials. The selected frequency is usually indicated on a meter. Resistor R_i is a low valued, about 0.02 ohm, internal resistor. The oscillator voltage is developed across this resistor, and is measured using thermocouple systems. Because of the low value of R_i it does not significantly affect the circuit being measured.

Capacitor *C* is also internal to the *Q* meter, and its value can be adjusted by dials and range select buttons. The dial usually has a plus–minus indication, to show the variation from the selected range setting. A high frequency high impedance detector, connected across the capacitor, is calibrated directly in *Q*, and the value of the *Q* being measured is equal to this meter reading multiplied by that on meter V_o.

The component under test can be connected in series or in parallel with *C*, by means of several terminals on the body of the instrument. Commercial *Q* meters are also supplied with a range of calibrated inductors and capacitors, for extending the measurement range. *Q* meters typically operate over a range of 10 kHz–100 MHz, and can measure *Q* values between 5 and 1000.

13.6.2 *Measurements on an inductor*

To measure the characteristics of a coil of inductance L_x it is connected in series with the tuning capacitor, as in Fig. 13.21. The system now acts as a series resonant circuit and, for a fixed value of oscillator voltage and frequency, capacitor *C* is varied until the detector shows a maximum reading. This corresponds to resonance, and the value of *Q* is given by (13.23)

$$Q = \frac{V_D}{V_o} \tag{13.23}$$

Therefore the *Q* factor is given by the detector V_D reading, with the indicator V_o acting as a multiplying factor. Also at resonance the values of the inductance L_x and series resistance R_x, of the unknown coil, can be found from (13.24) and (13.25), knowing the values of the resonant oscillator frequency f_r, capacitor *C* and the *Q* reading.

$$L_x = [(2\pi f_r^2)C]^{-1} \tag{13.24}$$
$$R_x = [2\pi f_r CQ]^{-1} \tag{13.25}$$

The parasitic or self resonance capacitance of an inductor can also be measured by the *Q* meter, using the connections of Fig. 13.21. *C* is initially set to a known high value, say C_H, and the oscillator frequency is adjusted for resonance, as indicated by the maximum reading on the detector. The value of the oscillator frequency f_r is noted. The oscillator is then set to twice this frequency, that is $2f_r$, and the capacitor is adjusted to (say) C_L, to again give resonance. The value of the self resonance capacitance C_s of the inductor is given by solving the two resonant equations (13.26) and (13.27) to yield (13.28).

$$f_r = [2\pi(L\{C_H + C_s\})^{1/2}]^{-1} \tag{13.26}$$
$$2f_r = [2\pi(L\{C_L + C_s\})^{1/2}]^{-1} \tag{13.27}$$
$$C_s = \frac{C_H - 4C_L}{3} \tag{13.28}$$

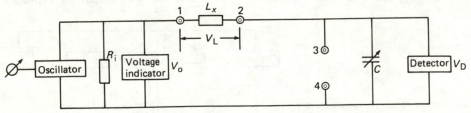

Fig. 13.21. Simplified construction of a *Q* meter.

13.6.3 *Measurements on a capacitor*

The Q meter can be used to determine the dissipation factor and effective capacitance of a capacitor, based for example on the equivalent parallel circuit of Fig. 13.8(b). The same instrument configuration as Fig. 13.21 is used initially in which the unknown inductor L_x, connected between terminals 1 and 2, is replaced by one of the standard inductors L supplied with the instrument. The capacitor is again set close to its maximum value, say C_H, and the oscillator frequency adjusted for resonance. The oscillator frequency f_r and the Q of the circuit, Q_H, are noted.

The unknown capacitor is now connected in parallel with capacitor C using terminals 3 and 4. The frequency of the oscillator is kept at f_r but C is adjusted to C_L to again give resonance. The value of Q, equal to Q_L, is noted. The parallel capacitance C_{PAL}, dissipation factor D and shunt resistor R_{PAL} can be found from (13.29), (13.30) and (13.31)

$$C_{PAL} = C_H - C_L \tag{13.29}$$

$$D = \frac{Q_H - Q_L}{Q_H Q_L} \cdot \frac{C_H}{C_H - C_L} \tag{13.30}$$

$$R_{PAL} = \frac{Q_H Q_L}{Q_H - Q_L} \cdot \frac{1}{2\pi f_r C_H} \tag{13.31}$$

13.6.4 *Measurements on a resistor*

High frequency components of a resistor can be determined with a Q meter, and using the equivalent parallel circuit of Fig. 13.22, which is derived from the general equivalent circuit of Fig. 13.3. The measurement circuit of Fig. 13.21 is again used, with L_x replaced by a standard inductor L. Initially the system is resonated to find f_r, C_H and Q_H, as was done in Section 13.6.3. The unknown resistor is then connected in parallel with C, whose value is adjusted to C_L to again give resonance, and the value of Q_L noted. The components of the resistor are then given by (13.32), (13.33) and (13.34)

$$R_p = \frac{Q_H Q_L}{Q_H - Q_L} \cdot \frac{1}{2\pi f_r C_L} \tag{13.32}$$

$$C_p = C_H - C_L \tag{13.33}$$

$$L_p = [(2\pi f_r)(C_H - C_L)]^{-1} \tag{13.34}$$

13.6.5 *Measurement of low impedances*

Low impedances, that is low values of resistance and inductance, or large value of capacitance, are best measured using the series circuit of Fig. 13.23. The unknown impedance is connected in series with a standard inductor L, also called a work inductor. The unknown impedance Z_x is first short circuited, using the shorting strap S_W, and the frequency and capacitance adjusted for resonance. The values of C_H and Q_H are noted. The shorting strap S_W is then removed and, without changing the frequency f_r, the capacitor is adjusted for resonance. The values of C_L and Q_L are noted. The unknown reactance X_x, resistance R_x and Q factor Q_x are found from (13.35), (13.36) and (13.37).

$$X_x = \frac{C_H - C_L}{2\pi f_r C_H C_L} \tag{13.35}$$

$$R_x = \frac{C_H Q_H - C_L Q_L}{2\pi f_r C_H C_L Q_H Q_L} \tag{13.36}$$

$$Q_x = \frac{(C_H - C_L) Q_H Q_L}{C_H Q_H - C_L Q_L} \tag{13.37}$$

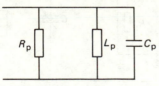

Fig. 13.22. Equivalent parallel circuit of a resistor operating at high frequency.

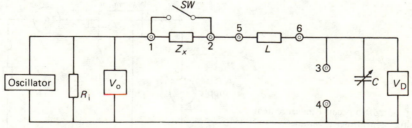

Fig. 13.23. Q meter connection for measuring low impedance.

If the impedance is inductive then C_H is greater than C_L, and the value of inductance L_x is given by (13.38). If the impedance is capacitive then C_H is less than C_L and the unknown capacitance C_x is given by (13.39)

$$L_x = \frac{C_H - C_L}{(2\pi f_r)^2 C_H C_L} \tag{13.38}$$

$$C_x = \frac{C_H C_L}{C_L - C_H} \tag{13.39}$$

If the impedance is purely resistive then no adjustment will be needed to C from its initial value of C_H, and the value of resistance is derived from (13.36) for $C_H = C_L$ as in (13.40)

$$R_x = \frac{Q_H - Q_L}{2\pi f_r C_H Q_H Q_L} \tag{13.40}$$

13.6.6 *Measurement of high impedances*

To measure high values of impedance, that is large resistance, inductance above about 100 mH or capacitance below about 400 pF, a parallel circuit connection should be used, as in Fig. 13.24. A standard inductor L is first connected in series with C and the system adjusted for resonance. The values of f_r, C_H and Q_H are noted. The unknown impedance Z_x, shown in Fig. 13.24 by its parallel components X_p and R_p, is then connected in parallel with C, which is

again adjusted for resonance. The values of C_L and Q_L are noted. Equations (13.41), (13.42) and (13.43) enable X_p, R_p and Q_p to be calculated.

$$X_p = [2\pi f_r(C_H - C_L)]^{-1} \tag{13.41}$$

$$R_p = \frac{Q_H Q_L}{2\pi f_r C_H (Q_H - Q_L)} \tag{13.42}$$

$$Q_p = \frac{(C_H - C_L)Q_H Q_L}{C_H(Q_H - Q_L)} \tag{13.43}$$

If Z_x is inductive then $X_p = 2\pi f_r L_p$ and the value of L_p is found from (13.41), as in (13.44). If Z_x is capacitive then $X_p = (2\pi f_r C_p)^-$ and the value of C_p is given by (13.45)

$$L_p = [(2\pi f_r)^2 (C_H - C_L)]^{-1} \tag{13.44}$$

$$C_p = C_H - C_L \tag{13.45}$$

13.7 Impedance measurement

Impedance measurements can be made with considerable accuracy using a wattmeter, voltmeter and ammeter, connected as in Fig. 13.25. If P is the reading of the wattmeter and I and V the current and voltage, then the value of the impedance Z_x is given by (13.46).

$$Z_x = \frac{V}{I} \tag{13.46}$$

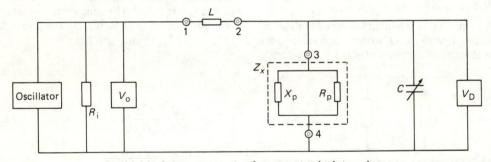

Fig. 13.24. *Q* meter connection for measuring high impedance.

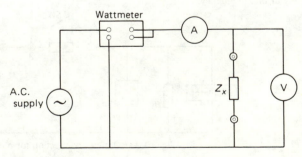

Fig. 13.25. Impedance measurement using a wattmeter, ammeter and voltmeter.

The value of the a.c. resistance in this impedance is found from (13.47), and the reactance from (13.48)

$$R_{ac} = \frac{P}{I^2} \tag{13.47}$$

$$X = (Z_x^2 - R_{ac}^2)^{1/2} \tag{13.48}$$

For good accuracy the voltage loss across the ammeter and the current coil of the wattmeter must be low, and the voltmeter should have a high impedance compared to the impedance being measured (Z_x).

Modern *LCR* meters use a variety of techni-

ques, mainly digital, to measure impedance. A single instrument can usually measure impedance, *D*, *Q* and phase angle. The system can operate at several frequency and bias levels, to make measurements on the impedance under conditions which simulate its application.

Another instrument, used to measure the impedance of a device at many different frequencies, is the vector impedance meter shown in Fig. 13.26. The oscillator frequency can be varied over a wide range, by a select switch and a continuous control knob as in the instrument of Fig. 13.27. The device amplifier can operate as a constant voltage or a constant current source.

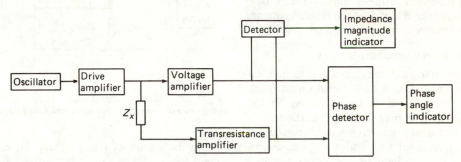

Fig. 13.26. Simplified diagram of a vector impedance meter.

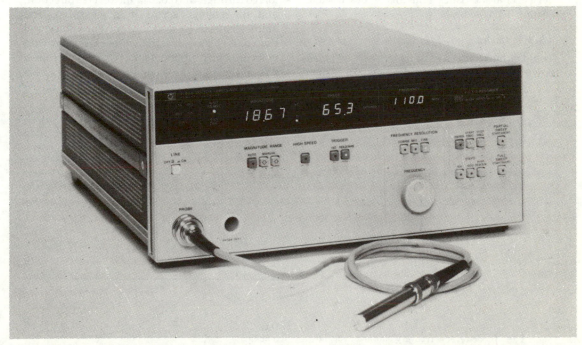

Fig. 13.27. A vector impedance meter (reproduced by courtesy of Hewlett Packard Inc).

The constant voltage operating mode is selected for impedances greater than one kilohm, and the constant current mode for impedances below this value. The detector measures the voltage and current through the unknown impedance Z_x and calculates its numerical value, which is displayed on the magnitude indicator. The phase difference between the voltage and current is measured by the phase detector, and this is shown on the phase angle indicator. The transresistance amplifier converts current to a proportional voltage. The instrument shown in Fig. 13.27 is also capable of operating in a sweep mode, in which the frequency is swept linearly over any portion of the frequency range, or is swept logarithmically over the entire frequency range of the instrument.

13.8 Discrete semiconductors

There are many different types of discrete semiconductors, and each has several unique parameters which need to be measured. In this section only the most important parameters of diodes, transistors, and thyristors are considered. The measurement techniques will use general purpose test equipment; the curve tracer is extensively used for measuring semiconductor device characteristics and this is described in Section 13.10.

13.8.1 *Diodes*

The d.c. parameters of a diode, such as breakdown voltage and voltage drop, can be measured

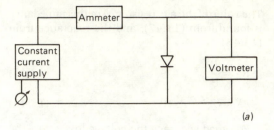

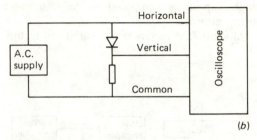

Fig. 13.28. Measurement of diode d.c. parameters; (*a*) static measurement, (*b*) dynamic measurement.

using the circuit of Fig. 13.28(*a*). To measure breakdown voltage a constant current, equal to the maximum permissible leakage current, is forced through the diode in the reverse direction, and the voltage across it is measured. The ammeter needs to read low current values and the voltmeter reads large voltages. For forward volts drop measurements the required current is passed through the diode in the forward direc-

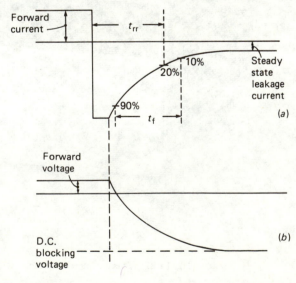

Fig. 13.29. Diode reverse recovery characteristic: (*a*) current, (*b*) voltage.

tion, and the volts drop across it is measured. The ammeter must now be capable of measuring large values of current, whereas the voltmeter reading will be typically 0.5 V.

Fig. 13.28(*b*) shows an arrangement for measuring the diode's d.c. characteristic dynamically, by displaying it on an oscilloscope. A full four quadrant trace will be obtained, since the voltage across the diode drives the horizontal plates of the CRT, whilst the current through it drives the vertical plates.

An important switching parameter of a diode is its reverse recovery time, as shown in Fig. 13.29. If the voltage across a forward conducting diode is suddenly reversed then it will continue to conduct in the reverse direction, the current gradually dying away to the steady state leakage value. The peak reverse current can be very large, usually limited by impedances in the system. Fig. 13.30 shows a test arrangement for displaying the reverse recovery current and voltage curves on a dual trace oscilloscope, and from these traces the recovery time can be measured. The recovery time is dependent on the magnitude of the forward conducting current, and the peak reverse current, so these can be adjusted as part of the test parameters.

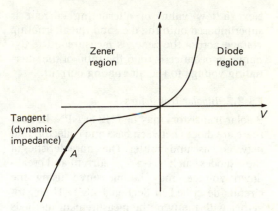

Fig. 13.31. Zener diode characteristic.

Zener diodes have characteristics similar to conventional diodes except that their reverse breakdown voltage is lower and less sharp, as shown in Fig. 13.31. The value of the zener voltage, and the dynamic impedance, are important parameters. The zener voltage can be measured at any zener current, as for a diode, using either of the circuits of Fig. 13.28. The dynamic impedance is measured with the circuit of Fig. 13.32. The direct current is first adjusted to bias the diode to the required zener current.

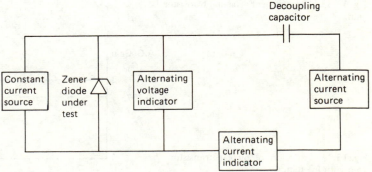

Fig. 13.30. Measurement of diode reverse recovery time.

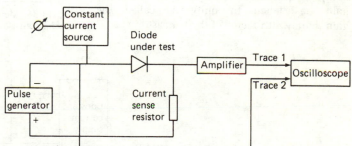

Fig. 13.32. Measurement of Zener diode dynamic impedance.

Next a low value of alternating current is superimposed onto the d.c., and the alternating voltage across the zener is measured. The dynamic impedance is then the ratio of the alternating voltage to the alternating current.

13.8.2 *Bipolar transistors*

Bipolar transistors may be PNP or NPN. In both there are diodes between base and collector, and between base and emitter. The characteristics of these diodes, such as leakage current and breakdown voltage, may be measured using the circuits described in Section 13.8.1. These, together with many of the measurement methods covered in the present section, are available in a variety of transistor testers.

An important parameter for transistors is the d.c. gain H_{fe} or β_{dc}. Its value varies with temperature, the collector current, and the collector voltage. It can be measured using the arrangement of Fig. 13.33. The transistor base current I_B is adjusted to give the required value of collector current I_c, and the d.c. gain is then given by the ratio I_c/I_B.

At low frequencies the d.c. gain β_{dc} and the a.c. gain G_{ac} are almost equal. However, as shown in Fig. 13.34, the gain of the transistor falls sharply at high frequencies. The frequency, at which the gain has fallen to unity, is called the gain–bandwidth factor of the transistor.

The transistor a.c. gain at frequency f is given by (13.49)

$$G_{ac} = \frac{\beta_{dc}}{[1 + (f/f_\beta)^2]^{1/2}} \tag{13.49}$$

For a frequency f which is much smaller than f_β the a.c. gain equals the d.c. gain β_{dc}. At frequency f_β the a.c. gain has fallen to 0.707 of the d.c. value. The frequency f_T, at which the a.c. gain falls to unity, is given by (13.50), since the d.c. gain is usually much greater than one.

$$f_T = f_\beta \cdot \beta_{dc} \tag{13.50}$$

Usually f_T is very high, but its value can be determined by measuring f_β, using the circuit shown in Fig. 13.35, and then calculating f_T from (13.50). The constant current source is first set to the desired value of base current I_B, the frequency generator is set to a low frequency,

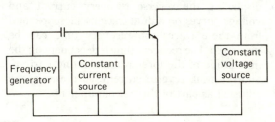

Fig. 13.35. Measurement of gain–bandwidth factor of a bipolar transistor.

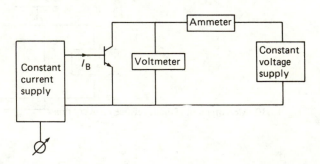

Fig. 13.33. Measurement of bipolar transistor d.c. gain.

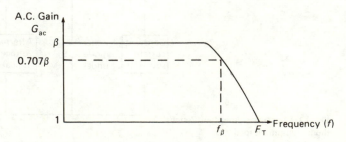

Fig. 13.34. Variation of bipolar transistor gain with frequency.

and the collector current I_c is measured. The d.c. gain β_{dc} is then calculated as the ratio I_c/I_B. Next, with the same value of I_B, the frequency of the generator is increased until the collector current has fallen to $0.707\, I_c$. The value of the frequency at this point is f_β, and the gain–bandwidth factor f_T can be found from (13.50).

The transistor switching characteristics are primarily defined by the four times illustrated in Fig. 13.36. They can be measured by observing the base drive and collector output waveforms

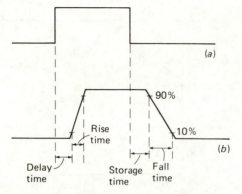

Fig. 13.36. Transistor switching characteristics; (*a*) base drive waveform, (*b*) collector output waveform.

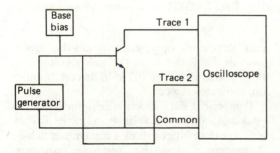

Fig. 13.37. Measurement of transistor switching times.

on a dual trace oscilloscope, as shown in Fig. 13.37. It is important when making these measurements to correct for the rise times of the pulse generator and the oscilloscope, using (12.16).

Several instruments exist for measurements on transistors when they are located within a circuit. These in-circuit testers are convenient to use, since the device does not need to be removed from the circuit prior to test, but their accuracy is limited, mainly by the values of the components

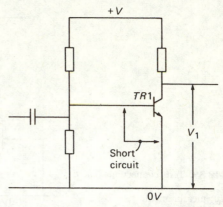

Fig. 13.38. Example of a transistor in-circuit test.

which surround the transistor. Fig. 13.38 illustrates a basic system to test transistor $TR1$. The voltage V_1 is first measured, and a short circuit is then applied across the base and emitter of the transistor. If the device is operational the value of V_1 should now rise from a low value to close to the supply voltage.

13.8.3 *Field effect transistors*

There are two types of field effect transistors, junction field effect (JFET) and metal oxide semiconductor field effect (MOSFET). Field effect transistors can be P channel or N channel, and these can be compared by analogy to bipolar NPN and PNP transistors.

A JFET operates in the depletion mode, that is it is conducting with no gate bias. A MOSFET can operate in depletion, enhancement, or enhancement–depletion modes. The saturation drain current I_{DSS} is the current which flows with zero gate to source voltage V_{GS}. It is a function of the drain to source voltage V_{DS} when V_{DS} is below the pinch off voltage V_p. For V_{DS} greater than V_p the value of I_{DSS} is largely independent of V_{DS} and varies with V_{GS}.

An FET has many parameters which need to be measured. Only a few are mentioned here with reference to a JFET, although the same circuits may be used for MOSFETs. The value of I_{DSS} for a JFET can be found using the circuit of Fig. 13.39(*a*). The gate and source are connected together to give $V_{GS} = 0$. The value of V_{DD} is such that V_{DS} is greater than V_p, and the current I_{DSS} is read off the meter.

Fig. 13.39(*b*) is used to measure the pinch off

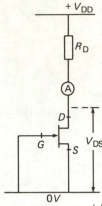

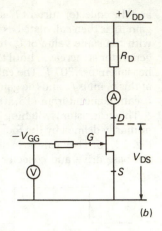

Fig. 13.39. Test connections for a JFET.
(a) I_{DSS} measurement.
(b) V_p measurement.

voltage V_p. The value of V_{GG} is made progressively more negative until I_D falls below a specified low value. V_{GG} is now equal to the pinch off voltage.

Another parameter of importance is the transconductance of a JFET. It is the change in output current I_D, induced by a change in the input gate to source voltage V_{GS}, and is given by (13.51)

$$g_m = \frac{\Delta I_D}{\Delta V_{GS}} \tag{13.51}$$

The transconductance is the slope of the I_D/V_{GS} characteristic.

The same circuit as Fig. 13.39(b) may be used to measure g_m. The values of V_{GG} and R_D are adjusted such that V_{DS} is greater than V_p and a typical value of I_D flows through the meter. Voltage V_{GG} is now altered slightly, and the change in I_D is noted to give values for ΔV_{GG} and ΔI_D. The transconductance is then found from (13.51).

The switching characteristics of an FET are important in pulse circuits. They are measured with a pulse generator, for gate drive, and a dual trace oscilloscope, using a circuit and method very similar to that shown in Fig. 13.37.

13.8.4 *Unijunction transistor*

The unijunction transistor has two layers, similar to a diode, and is primarily used in oscillator and trigger circuits. Fig. 13.40 shows its static characteristics. As the emitter current increases it reaches a peak value I_p at a peak voltage V_p. The device now goes into a negative resistance region, and it is this region in which

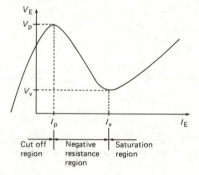

Fig. 13.40. UJT static emitter characteristic.

oscillator circuits operate. Eventually the current reaches the valley I_v, at the valley voltage V_v. Beyond this point the UJT is in its continuously on, saturated region.

Component data sheets specify the values of peak and valley point voltage and current. They also give the emitter to base 1 and emitter to base 2 resistance, and the interbase resistance. Although the emitter–base 2 resistance is fairly constant the emitter–base 1 resistance varies with emitter current. Another parameter of interest in UJT circuits is the intrinsic stand off ratio (η). This is given by (13.52)

$$V_p = \eta V_B + V_D \tag{13.52}$$

V_B is the interbase supply voltage and V_D is the forward voltage drop across the emitter–base 1 diode, which is in the region of 0.7 volt. η determines the firing point of UJT oscillator circuits. It varies only slightly with changes in V_B and temperature. V_D is usually measured at the peak emitter current I_p, which is the current at which it is required for (13.52).

The static emitter characteristics of the UJT can be measured using the circuit of Fig. 13.41.

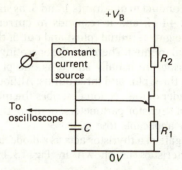

Fig. 13.41. Circuit arrangement to measure the static emitter characteristics of a UJT.

With the UJT off the constant current is gradually increased, whilst the voltage across the capacitor is monitored by the oscilloscope. Just beyond the peak point the oscilloscope will record a sawtooth waveform, indicating that the UJT is oscillating and has reached its negative resistance region. The current is now decreased slowly until the oscillation stops. The value of the current at this point equals I_p, and the voltage across the UJT, which can be measured by a high impedance voltmeter, is V_p. The current from the constant current source is then increased, well into the negative resistance region, until the valley point is reached. Beyond this point the oscillations will once again stop, and the UJT will remain permanently on. The current now equals I_v and the voltage across the UJT is V_v.

Fig. 13.41 can also be used to measure the rise time of the UJT. This parameter is required for pulse circuit applications. With the UJT biased in the negative resistance region the pulses across

resistor R_1 are recorded on an oscilloscope, and the rise time measured.

The value of V_D can be measured by the circuit shown in Fig. 13.42. The current through the

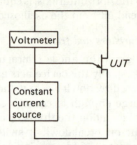

Fig. 13.42. Simple arrangement to measure V_D in a UJT.

emitter is set to I_p by the current source, and the emitter–base 2 voltage is measured by a high impedance voltmeter. The value of the intrinsic stand off ratio can also be found by noting the resistance between emitter–base 1 (r_{B1}) and base 1–base 2 (r_{BB}) and using (13.53)

$$\eta = \frac{r_{B1}}{r_{BB}} \qquad (13.53)$$

A better method for measuring η and V_D is shown in Fig. 13.43. The signal generator is set to a low frequency of about 10 Hz to provide a swept interbase voltage. The values of R_2 and C are chosen such that the UJT oscillates at a much higher frequency of about 5 kHz. The swept interbase voltage is connected to the horizontal or X input of an oscilloscope. One end of the Y input is connected to measure the emitter voltage, the other is connected to a fraction (n) of the interbase voltage (V_B) via potentiometer R_1. The oscilloscope will therefore show the trace of V_D $+ (-n)V_B$ on its vertical axis against V_B. This is similar to the form of (13.52). The value of n is adjusted by R_1 until the upper envelope of the display is horizontal. At this point n is equal to η.

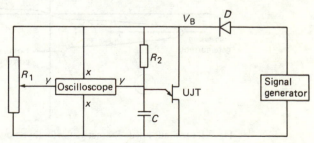

Fig. 13.43. Circuit arrangement for measuring η and V_D of a UJT.

If R_1 is a precision multiturn potentiometer the value of η can be determined with an accuracy better than 0.05%. Also, with this setting of R_1, the displacement of the upper envelope of the trace, from the horizontal axis, is equal to V_D, and it can be read off from the oscilloscope.

13.8.5 *Thyristors and triacs*

Fig. 13.44 gives the anode characteristics of a thyristor. Basically the device acts as a rectifier. In the reverse direction it blocks voltage until this becomes large enough to cause breakdown. In the forward direction the thyristor again blocks voltage, but on breakdown it switches into a conducting mode, and remains in this mode provided the anode current exceeds a minimum

value, called the holding current. The point of breakdown in the forward direction can be controlled by the gate current, which must always flow into the gate terminal.

A triac operates like a bi-directional thyristor. It can conduct in quadrants 1 and 3, as shown in Fig. 13.44(*b*), and it responds to current flow into the gate terminal (plus) and out of the gate terminal (minus). Therefore the operating modes for a triac are usually stated as one plus, one minus, three plus and three minus. Although the remainder of this section describes the measurement of thyristor parameters, many of these are equally applicable to a triac.

The gate of a thyristor acts as a diode, and has characteristics as shown in Fig. 13.45. The

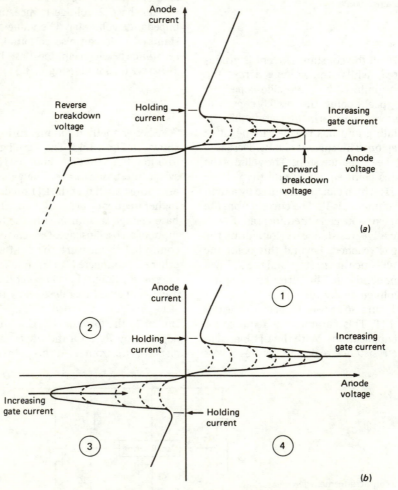

Fig. 13.44. Static anode characteristics for thyristors and triacs: (*a*) thyristor, (*b*) triac.

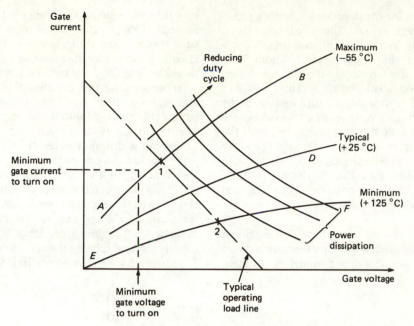

Fig. 13.45. Thyristor gate characteristic.

gate–cathode diode curves have a spread between AB and EF, so that for any load line, as shown, the operating point can be situated between 1 and 2. These must be outside the box bounded by the minimum turn on voltage and current, whilst at the same time being less than the relevant power dissipation curve. The shorter the gate duty cycle the larger the permitted gate drive, and high power pulse firing is often used for thyristors to ensure rapid turn on.

Fig. 13.46 illustrates a circuit arrangement for measuring the static characteristics of a thyristor. To measure blocking characteristics the gate supply is switched off, so as to short circuit the gate to the cathode, or to connect a fixed resistor between them. For forward breakdown the anode voltage is increased until the thyristor breaks over into forward conduction. This is indicated by a rapid increase in anode current, which is limited by the impedance of the supply. The voltage at this point is the forward breakover voltage of the thyristor. For reverse characteristics the reverse voltage is increased, to a specified value, and the leakage current noted.

The gate characteristics can also be measured using the circuit of Fig. 13.46 by setting the anode supply at a fixed value, and gradually

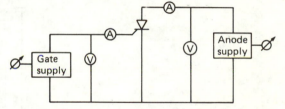

Fig. 13.46. Circuit arrangement for measuring thyristor static anode and gate characteristics.

increasing the gate current and voltage until the thyristor turns on. The gate current and voltage vary with temperature and the value of the anode voltage. Pulse firing may be used, the pulse amplitude and width being adjusted by the gate supply. The holding current is found by setting the anode supply to a constant current mode, and triggering the thyristor. The anode current is then reduced, with gate drive removed, until the thyristor goes off. The current at this point is the holding current. Another parameter, which is usually higher than the holding current, is the thyristor latching current. This is the minimum current which must flow through the thyristor to enable it to latch on. It is found by setting the anode supply at a low constant current, and then pulsing the gate. The thyristor will come on for the duration of the gate pulse but will then go off. The anode current is progressively increased until the thyristor latches on, even after the gate pulse has been removed; this is the latching current of the device.

The arrangement shown in Fig. 13.46 can also be used to measure the turn on time of the thyristor. The gate voltage and anode voltage are observed on a dual trace oscilloscope, with the anode supply set at a fixed value V_B. The delay time t_d is the time for the anode voltage to fall to 90% of its blocking value, the rise time t_r is the time for the voltage to go from 90% to 10%, and the turn on time t_o is the sum of the delay and rise times.

Fig. 13.47 shows an alternative arrangement which can be used to measure thyristor turn on time, and other dynamic parameters. To mea-

sure turn on time the anode supply is fixed at a constant voltage. The gate is triggered, under supervision of the control circuit. The control circuit also monitors the anode voltage and can therefore determine the turn on time.

Another parameter of interest in thyristors and triacs is their capability to withstand a steeply rising voltage without turning on. This is referred to as dV/dt performance. Usually a device can withstand greater dV/dt if it has not just come out of a conduction mode. The circuit shown in Fig. 13.48 can be used to measure the dV/dt capability of a thyristor. The anode supply is set to a constant current mode, so as to give a ramp anode voltage due to the charging of capacitor C. The discharge circuit is turned on periodically to discharge the capacitor, resulting in a sawtooth waveform across the thyristor. The

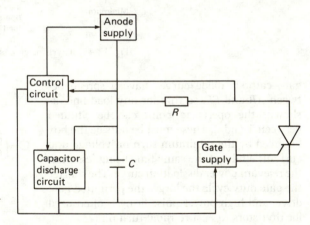

Fig. 13.47. Circuit arrangement for measuring thyristor dynamic characteristics.

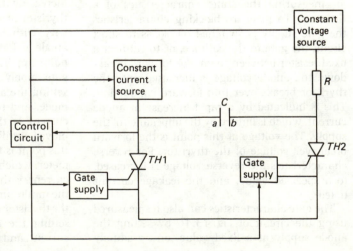

Fig. 13.48. Circuit arrangement for measuring thyristor dV/dt and turn off characteristics.

control circuit gradually increases the capacitor charging current, to give a progressive increase in dV/dt until eventually the thyristor will turn on. Knowing the value of C and the charging current I, the dV/dt rating of the thyristor can be calculated from (13.54).

$$dV/dt = \frac{I}{C} \qquad (13.54)$$

During the dV/dt test the gate of the thyristor is short circuited to its cathode, or connected to it via a resistor of specified value.

The turn off time of a thyristor is the time between the anode of a conducting thyristor going negative, and when the anode can again be made positive without the thyristor turning on. The value of the turn off time depends on the thyristor construction, the forward current prior to turn off, the reverse recovery current and the reapplied dV/dt. Fig. 13.48 shows an arrangement for measuring the turn off time of a thyristor $TH1$. Initially $TH1$ is turned on and the current through it is adjusted by the constant current source. Capacitor C will charge via resistor R, with plate b positive, to a voltage V_R set by the constant voltage source. To attempt to turn off $TH1$ thyristor $TH2$ is fired. This applies V_R across $TH1$, as shown in Fig. 13.49. The capacitor now starts to charge, at a rate determined by the constant current source, to give the dV/dt across $TH1$. Thyristor $TH1$ will be reverse biased for time t_R, and if this time is greater than the turn off time of the thyristor it will remain off. After $TH2$ has fired the constant voltage supply is

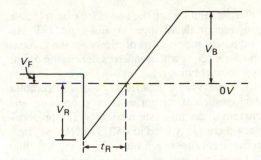

Fig. 13.49. Thyristor waveforms during turn off time testing.

switched off, so that $TH2$ will turn off once C has charged to the peak voltage of the constant current source (V_B).

13.9 Integrated circuits

13.9.1 Types of integrated circuits

Integrated circuits can be broadly classified as being digital or linear, as shown in Fig. 13.50. Digital circuits can be differentiated into small and medium scale integration devices (SSI and MSI), or large and very large scale integration circuits (LSI and VLSI). SSI and MSI consist mainly of standard functions such as gates, counters and shift registers. They are available in a variety of technologies such as low power Schottky transistor–transistor logic (LSTTL), complementary metal oxide semiconductor (CMOS) and emitter coupled logic (ECL). The parameters which need to be measured for the

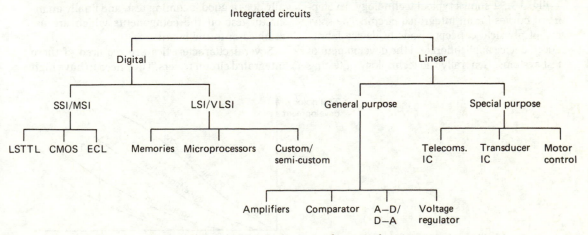

Fig. 13.50. Example of a few types of integrated circuits.

different families are the same, for example gate propagation delay, logic '0' and logic '1' levels, and power supply current. However the value of the parameters will be different depending on the type of technology.

LSI and VLSI components consist of complex devices such as memories, microprocessors, and custom and semi-custom circuits. The predominant technology used in CMOS or NMOS. These components are much more difficult to test than SSI and MSI devices. However they represent a higher proportion of usage, by value, and this is increasing. The testing of memories and microprocessors is described in Section 13.9.4.

Until relatively recently linear integrated circuits consisted mainly of operational amplifiers, and linear testers were designed primarily for these. Not only has the range of general purpose linear components expanded to include devices such as comparators, analogue-to-digital and digital-to-analogue converters, and voltage regulators, but special purpose devices with unrelated characteristics, such as telecommunication components, now dominate the linear integrated circuit market. These are considered in Section 13.9.5.

13.9.2 *Testing integrated circuits*

As the manufacturing yield of a device increases so its fabrication cost falls. The cost of testing the component also decreases, but at a slower rate, so that the ratio of testing cost to device cost increases, as shown in Fig. 13.51. This illustrates the importance of optimal testing for a component.

Fig. 13.52 shows typical technology development curves for an integrated circuit. The solving of physical concepts leads to device fabrication, device application and the development of test systems. Generally the technology of testing

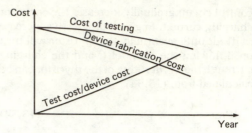

Fig. 13.51. Curves illustrating the increasing contribution of testing costs to the overall cost of an IC.

lags fabrication, and this is known as the test gap. However it is important that, as the speed and complexity of integrated circuits increase, so test systems keep pace, since an integrated circuit cannot be developed successfully if its performance cannot be verified.

Three types of tests are carried out on integrated circuits; d.c., functional and dynamic. D.C. tests measure the basic parameters such as leakage current, power consumption, breakdown voltage, and logic zero and one levels. Functional tests check that the integrated circuit performs as expected. It is therefore a logic test, and is verified using patterns and truth tables. Dynamic tests involve time measurements such as delay time, memory access and hold times, and rise and fall times.

Testing occurs in several places during the life of an integrated circuit. Initially it is required during product development, in engineering. The wafer is probe tested after diffusion, and the packaged devices are tested prior to shipment to the customer. The equipment manufacturer usually does a good incoming test, and finally an in-circuit test on the components which are assembled onto the board.

Several characteristics are required of these integrated circuit testers. They need to have high

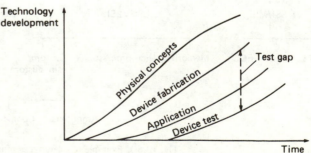

Fig. 13.52. Illustration of the 'Test Gap'.

accuracy and reliability, and their measurement capabilities, such as speed, must exceed those of the circuit being tested. Instruments used for manufacturing tests should have a high throughput, with automatic calibration facilities and ease of use. Software development should be economically achievable, and the tester should be supported by a selection of utility programmes. Many testers need to have network capabilities, with standard interface to peripheral controllers and device handlers.

The large number of different types of integrated circuit testers can usually be grouped into four categories, as shown in Fig. 13.53. As expected, a direct relationship exists between the cost and the versatility of an instrument. On one end of the spectrum there are the very flexible testers, with excellent programme development capabilities, which are primarily used in engineering applications. These range in price to over one million pounds sterling. At the other end of the scale are the dedicated or focused testers, which are designed to test a limited range of components. These can cost less than fifty thousand pounds, and are generally used for applications such as goods incoming tests.

Fig. 13.54 shows the construction of a typical digital integrated circuit tester. The component is mounted in the test head, which can accommodate up to 256 pins, and each pin is individually

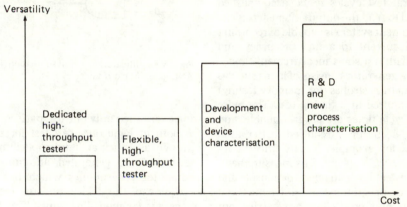

Fig. 13.53. Versatility vs. cost trade off in IC testers.

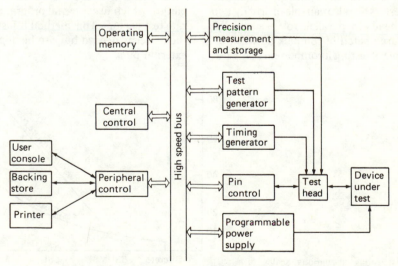

Fig. 13.54. Simplified block diagram of a digital integrated circuit tester.

controlled by a high speed driver and detector. Often the test head contains the drivers and detectors so that they are as close as possible to the device being tested, thus avoiding stray signal pick up. Several programmable power sources, at constant voltage and constant current, are needed for parametric testing.

The patterns required for testing the device can be generated on line, for example by using counters and random number generators. Complex test patterns, which are usually as a result of fault simulation, need to be generated off line and stored in operating memory, prior to use. A complex integrated circuit can need between 50 000 and 60 000 test patterns.

A response store is also sometimes used, which allows several test cycles to be accomplished without need to read the output after each cycle. The measurement system is capable of recording voltage and current to a high accuracy and resolution, within a short measurement time.

The timing generator consists of items such as clocks and address strobes (for memory testing) which are required in a timing test. The pulse widths and cycle times are programmable, and good timing accuracy is required; better than 0.5 ns since, for example, a 1 ns uncertainty would give a 20% error in a 5 ns measurement.

All large testers have an operator console and the facility for programme development. Data manipulation and processing are carried out within the central control to present the measurements in a form more meaningful to the user. Many testers are also capable of driving two to four test heads in parallel, to speed up the testing of a large batch of complex devices.

The principle of testing a component using test patterns is to apply a series of patterns to its input pins, such that there is a different output signal between a good and a bad component. Usually it is not possible to cover all possible combinations, since this would give an excessively long test time. In these instances the device is simulated for the case of no fault and of modelled faults, to see which cases give different results between good and bad devices. This is known as fault simulations, and indicates the amount of fault test coverage with any set of test patterns.

In the example of the two input AND gate shown in Fig. 13.55, if the output C is stuck at

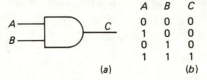

	A	B	C
	0	0	0
	1	0	0
	0	1	0
	1	1	1
(a)		(b)	

Fig. 13.55. Illustration a fault simulation: (a) TWO input AND gate, (b) truth table.

logic zero then only one test pattern, with A and B both at logic one, can test for this. This is known as 'stuck-at' fault modelling. Test patterns can be generated automatically using computer programmes which use logical description of the device, and the expected fault coverage for modelling faults. This is especially useful for combinational logic. Complex components often have self test or built in test capabilities, which use internal programmes to generate test vectors. This method is faster than if all the patterns needed have to be applied through external pins.

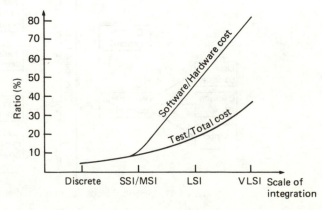

Fig. 13.56. Cost ratios at various scales of integration.

13.9.3 *The importance of software*

As the complexity of the integrated circuit which is to be tested increases, so also does the testing cost of the device relative to its unit cost, and the software costs of the test system relative to the hardware cost. These ratios are illustrated by the curves shown in Fig. 13.56. Therefore for a complex VLSI device 40% of the total cost of the component may be due to its testing cost, and 80% of the test system's cost is due to the software development.

Software performs three main functions within a test system. First the software is used to generate the function test patterns. Secondly data analysis software is used to reduce the masses of test data into meaningful results for the operator. Many standard software packages are available for data analysis, such as SHMOO plots, histograms, bit maps, pattern simulation, and data logging. Finally software is also necessary for the host controller, if the test system is networked with others.

Fig. 13.57 shows the software structure of a

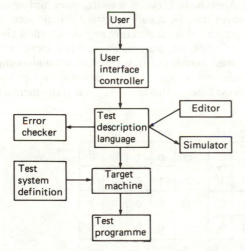

Fig. 13.57. The software structure of a typical tester.

typical tester. The test description language is usually high level, such as Pascal, to allow easier programme development. The operating system contains the support functions such as user interface, editor, error checker and simulator. Usually a large tester has facilities for off line programme development, using a secondary processor, so that programmes can be written

whilst the machine is still testing. The programmes are then checked for errors and simulated before being loaded onto the target machine, based on the test system definition, to obtain the test programme.

13.9.4 *Memory and microprocessor testing*

SSI and MSI devices can be tested by the truth table approach i.e. the inputs of the components are driven according to the device truth table, and the output response is monitored. For complex devices this method is not suitable since often no easily defined truth table exists.

One method of testing a microprocessor is to try it in the application to see if it works. This method is cheap but the application software may not exercise all the internal storage elements, or all the input–output lines, so that problems can show up later under exceptional conditions. Also, without control over the voltage levels and timing parameters, it is not possible to test for marginal devices, which may fail later in the field when device parameters begin to drift.

Fig. 13.58 shows a comparison method of

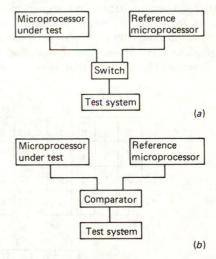

Fig. 13.58. Comparison testing of a microprocessor; (*a*) switching method, (*b*) comparison method.

testing complex devices, such as microprocessors. In the switching arrangement the test system is operated first with the reference microprocessor and then with the microprocessor being tested. The performance of the test system should be the same in both cases. In the comparison method both the microprocessors are run at the same time by the test system, and the differences in performance are measured by the comparator. The software for the tests can be written based on the expected device characteristics, or a pseudorandom sequence generator can be used to provide a set of instructions and data to exercise the two microprocessors.

When testing integrated circuits it is necessary to exercise the whole device during the test; for example exercising dynamic decoders for a static memory. A random access memory is fully functional only if: (a) A dynamic memory cell maintains its store for the required time period. (b) The addressing circuitry can access the required cell. (c) Memory cells can store both a logic zero and logic one state without interfering with adjacent cells. (d) The memory sense amplifiers function correctly. (e) The logic zero and logic one voltage levels are within the specified tolerance range.

Memories are prone to data dependent errors, so 'stuck at' fault models are not suitable. Special test patterns need to be put on the rows and columns, and multiple read-out operations done on these. Many patterns and sequences are available for testing memories of different sizes and configurations, only a few of these are described in this section. The shifted diagonal test, shown in Fig. 13.59, is mainly designed to

test memory sense amplifier recovery time. The initial pattern consists of a diagonal of logic ones in a field of logic zeros, as in Fig. 13.59(a). The memory is read out by column, so the sense amplifiers see a single one in a long string of zeros. The memory is then re-written with the diagonal shifted, as in Fig. 13.59(b). The sequence is continued until the diagonal reaches the original position as in Fig. 13.59(a) through to Fig. 13.59(e). The whole test is then repeated with a single logic zero diagonal in a field of logic ones.

Another test pattern, which checks for decoder problems and is a functional test, is known as marching ones and zeros. Initially the whole memory is filled with logic zeros. Then the zero from the first cell is read and a logic one is written into it. This process is continued with adjacent cells until the whole memory is filled with logic ones. Then working backwards the logic ones are read out in sequence and a logic zero is written back into the cell. This is continued until the first cell is reached. Finally the entire test is repeated with the data reversed.

Another test, called walking ones and zeros, proves that each cell can store a logic zero or logic one without affecting any other cell. It also checks the sense amplifier recovery speed, and address decoding. However since all addressing in the test is sequential it does not show up slow access time. In this test every cell in the memory

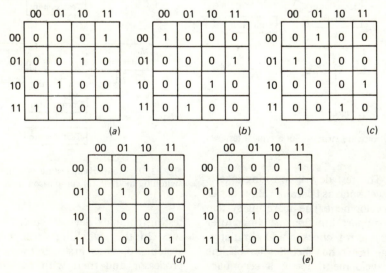

Fig. 13.59. Shifted diagonal test sequence for a 4 × 4 array.

is initially set to logic zero. A logic one is then written into the first cell and the content of every cell is checked. Next the first cell is restored to logic zero and a logic one is written into the second cell. All the cells are again tested to verify their content. This sequence is repeated for all the memory cells. The entire test is then repeated starting with all the memory cells set at logic one, and walking a logic zero through it.

The checkerboard pattern, shown in Fig. 13.60(a), can be used as a starting pattern,

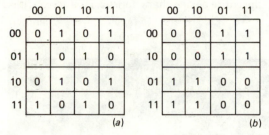

Fig. 13.60. Checkerboard patterns.

instead of all ones or all zeros, for some of the tests. It should however be noted that the pattern is dependent on the layout of the chip. For example if the chip is constructed such that address columns 00 and 01 are next to each other, and rows 00 and 10 are next to each other, then loading the pattern as in Fig. 13.60(a) will actually give the pattern shown in Fig. 13.60(b). It is therefore essential for the test engineer to have knowledge of the device construction.

The checkerboard pattern is useful for testing the refresh time in dynamic memory cells. The pattern in loaded into the cell and, after a specific time delay, the contents of the cells are read out and verified by column. This automatically refreshes all the cells as well. The test is then repeated with a reverse checkerboard pattern.

The time needed to test a memory increases rapidly with size. For example it would take about 2 seconds to test a 16 kbit memory using a shifted diagonal test pattern. This test time would increase to about 3 minutes for a 256 kbit memory and 10 minutes for a 1 Mbit memory. The test time also varies with the type of test pattern used. For example the test time, measured in number of cycles, for a B bit square memory array is equal to $2B^2$ for the shifted

diagonal test, 8B for a marching ones and zeros test, and $2B(B + 2)$ for a walking ones and zeros test.

The sequencing of the memory cells during a test is also important. When addressing the cells in order, as in Fig. 13.61(a), very few bit changes

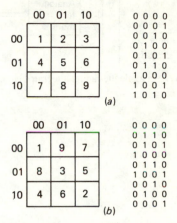

Fig. 13.61. Illustration of different memory addressing sequences; (a) sequential, (b) non sequential.

occur between steps, and so the test does not stress the address decoder. The sequencing arrangement of Fig. 13.61(b) gives many more bit changes between steps. The addressed cells are also spaced away from each other (which is more effectively illustrated in a large memory).

13.9.5 *Analogue testing*

Early analogue integrated circuits consisted mainly of operational amplifiers, and the testers were primarily used to measure their characteristics, such as loop gain and slew rate. Modern analogue integrated circuits cover a wide spectrum of devices such as A to D and D to A converters, voltage controlled amplifiers, phase locked loops, transducers, voltage regulators, motor speed controllers, and telecommunication integrated circuits. Many of these components have a mixture of analogue and digital circuits.

Analogue signals can be converted to a digital number by sampling in time, and then representing the sample by a digital value. Fast Fourier transforms are used to analyse the digital data. Dedicated test systems are available to test a family of analogue circuits, such as telecommun-

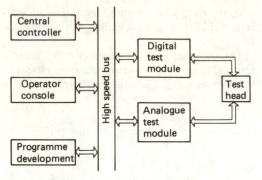

Fig. 13.62. An analogue/digital integrated circuit tester.

ication devices. Alternatively testers are available which can test general purpose analogue and digital components, as in Fig. 13.62.

13.10 Curve tracers

Automatic component test equipments (ATE), of the types described in earlier sections, are widely used. However they suffer either from the fact that they are not very flexible, or if flexible then they require programme development. For quick tests, or for testing a few devices, a curve tracer is more convenient. Since the curve tracer shows the entire device characteristic, on a screen, it is

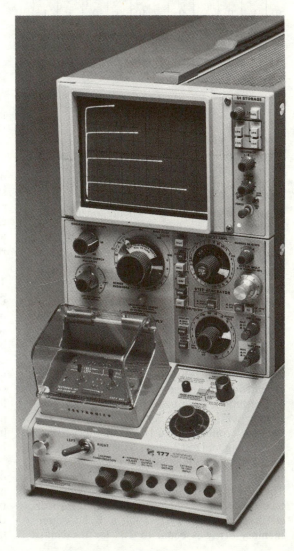

Fig. 13.63. The Tektronix Type 577 curve tracer (reproduced courtesy of Tektronix Inc).

also able to show peculiarities which would have been missed on an ATE.

Curve tracers are primarily used to make measurements on discrete semiconductors, and on a wide range of linear integrated circuits such as operational amplifiers, comparators and voltage regulators; they are not suitable for testing the performance of digital integrated circuits. In this section the use of a curve tracer is illustrated primarily in relation to discrete semiconductors.

Fig. 13.63 shows a typical curve tracer. It consists of a CRT display on which the characteristics of the component are plotted. Various plug-in modules are available, with special power supplies, sensors and drivers, for measuring different types of components.

Fig. 13.64 shows a simplified block diagram of a curve tracer. The sweep and step generators are capable of operating in a constant current or voltage mode, and the measurements made, through the device under test, feed the horizontal and vertical controls of the integral CRT.

Fig. 13.65 illustrates the curve tracer display when it is connected to make measurements on an NPN transistor. The sweep generator is operated in a voltage mode, and applied across the collector of the transistor. This value of V_{CE} is fed into the horizontal input of the CRT. The step generator is operated in constant current mode, and is fed into the base of the NPN transistor so as to give a series of traces I_{B1}, I_{B2}, I_{B3}, etc. on the CRT. The collector current I_C through the transistor is detected and applied to the vertical input of the CRT. The sweep and step generators are

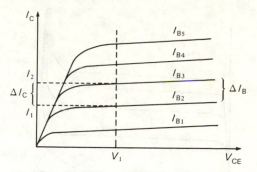

Fig. 13.65. Collector current characteristics of an NPN transistor as displayed on a curve tracer.

synchronised so that the collector voltage is swept from zero to its maximum value for each base current. This gives the series of curves shown in Fig. 13.65.

Several parameters for the transistor can be deduced from the characteristics shown in Fig. 13.65. The common emitter gain β is the ratio of collector current to base current change ($\Delta I_C / \Delta I_B$) and can be read from the curves, as shown, for any value of collector–emitter voltage V_1, base current or collector current.

Fig. 13.66 shows the transistor characteristics where the magnitude of the sweep voltage has been increased to the point which results in a rapid collector current increase. This is known as breakdown, and the voltage at breakdown can be read from the curves. It should be noted that base current has only a minor effect on the magnitude of the breakdown voltage. During

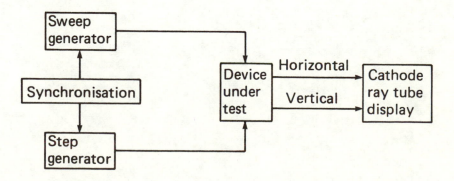

Fig. 13.64. Simplified block diagram of a curve tracer.

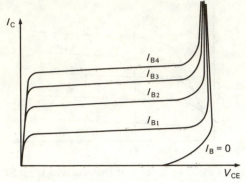

Fig. 13.66. Breakdown characteristics of an NPN transistor as displayed on a curve tracer.

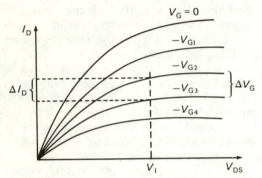

Fig. 13.67. Drain current characteristics of an FET.

this test the curve tracer limits the breakdown current to a safe value, but the test time should be kept as short as possible to prevent over-dissipation in the device.

To make measurements on an FET, and obtain the characteristics shown in Fig. 13.67, both the sweep and step generators must be operated in a voltage mode. The transconductance, or gain, of the FET can be found by the ratio $\Delta I_D / \Delta V_G$ and is in units of mho. The pinch off voltage of the FET can also be found from the characteristics of Fig. 13.67, but is limited in accuracy by the steps possible in the curve tracer step selector. For example suppose the step selector is set to 0.1 V, and gives the entire family of curves, as in Fig. 13.67, from $V_G = 0$, -0.1 V, -0.2 V, -0.3 V and -0.4 V. Now the steps are increased to the next selector setting, say 0.2 V, and suppose that only $V_G = 0$, -0.2 V and -0.4 V are displayed. The curves for $V_G = -0.6$ V and -0.8 V are superimposed at a very low value of I_D, corresponding to pinch off. Therefore the pinch off voltage lies between 0.4 V and 0.6 V. The exact value of pinch off voltage can be found by connecting an external bias voltage to the gate of the FET, and adjusting this to show the pinch off state on the trace.

For two terminal devices, such as diodes and zener diodes, only the sweep generator is used, in voltage mode. A voltage sweep is applied across the device, (forward or reverse connection), and feeds the horizontal input of the CRT, whilst the current through it is monitored on the vertical input. Only a single trace is displayed, rather than a family, and from this the characteristic can be noted, such as voltage drop, reverse breakdown voltage, and zener voltage.

14. Audio and video circuit tests

14.1 Introduction

In this chapter the characteristics and measurement techniques, for audio and video systems used in radio, hi-fi and television, are introduced. This includes amplifiers, with distortion considerations, radio and TV receivers, and hi-fi equipment. Measurements in audio and video transmission are described in Chapter 15.

14.2 Audio amplifiers

Amplifiers can be broadly classified as tuned (r.f.) and untuned, which are also known as audiofrequency or video amplifiers. Audio amplifiers are used in applications such as loudspeaker systems. Wide-band or video amplifiers cover a frequency band well beyond audio, into the megahertz range. A TV video amplifier, for example, needs to have uniform performance over a frequency range from d.c. to greater than 4 MHz.

Perhaps the most obvious characteristic of an amplifier, which needs to be measured, is its *voltage gain*. This can be measured using the circuit of Fig. 14.1. The signal generator output

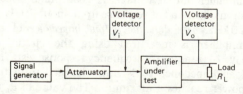

Fig. 14.1. Measurement of amplifier gain, power, sensitivity and frequency response.

amplitude is controlled by the attenuator, and fed into the amplifier under test. The voltage input V_i and output V_o are measured, and the voltage gain is given by the ratio V_o/V_i. Gain varies with frequency, and has a shape similar to

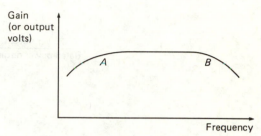

Fig. 14.2. Typical gain (or frequency response) curve of a power amplifier.

that of Fig. 14.2, where A and B are the roll off points.

The *power gain* of an amplifier is equal to the ratio of output power P_o to input power P_i. Output power is found by measuring the output voltage V_o across load resistor R_L and calculating it as V_o^2/R_L. Input power is found by first measuring the input impedance of the amplifier R_i and the input voltage V_i and calculating it as V_i^2/R_i. The power output P_o should be the maximum undistorted output, and it can be obtained by noting V_o on an oscilloscope, and increasing it until the waveform just starts to appear distorted. For a music amplifier a pulse generator, rather than a sine wave generator, should be used, since music and voice have narrow peaks of amplitude.

The *full power bandwidth* of the amplifier is measured by operating it at its full rated power and noting the distortion (say $x\%$) at 1 kHz, as in Fig. 14.3. The power output is then reduced by 3 dB, and the frequency of the input signal is reduced to f_L until the distortion is again $x\%$ (roll off point in Fig. 14.2). The frequency is then increased to f_H until the second $x\%$ distortion point is found. The power bandwidth is the difference in the two frequencies, that is $f_H - f_L$.

The *input sensitivity* of an amplifier is the ratio

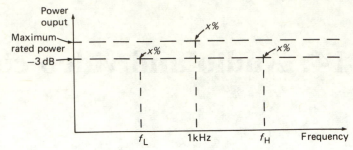

Fig. 14.3. Illustration of adjustments needed to measure full power bandwidth.

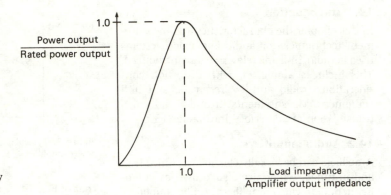

Fig. 14.4. Typical load sensitivity curve of a power amplifier.

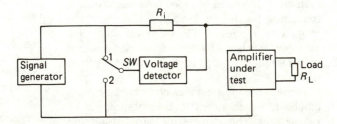

Fig. 14.5. Dynamic input impedance test of an amplifier.

of power output to input voltage (P_o/V_i) and is sometimes specified in place of power gain. Amplifiers are also sensitive to their load, and the *load sensitivity* curve has the shape shown in Fig. 14.4. The output power varies with load impedance, and reaches a peak when the load impedance matches the output impedance of the amplifier. Load sensitivity is measured using the circuit of Fig. 14.1 in which R_L is a non inductive variable resistor. The test should also be repeated over the frequency range of the amplifier.

The *frequency response* curve of an amplifier has the typical shape shown in Fig. 14.2. It is measured by applying constant voltage to the amplifier under test (by adjusting the attenuator in Fig. 14.1) whilst changing the frequency of

the signal generator, and measuring the amplifier output voltage. A good audio circuit should have a flat response from about 20 Hz to 20 kHz. The *dynamic output impedance* of an amplifier is measured using the circuit of Fig. 14.1, as for the frequency response test. The load impedance R_L is now changed until the power reaches a maximum. The value of R_L at this point is the dynamic impedance of the amplifier at this frequency. The test should be repeated at different frequencies.

The *dynamic input impedance* of an amplifier is measured using the arrangement of Fig. 14.5. With the signal generator at a convenient value switch *SW* is set to position 1 and then to position 2, and the value of R_i is adjusted such that the voltage detector has the same reading in both positions. R_i is now equal to the dynamic input impedance of the amplifier under test.

The *common mode rejection ratio* (CMRR) of an amplifier is illustrated by the arrangement of Fig. 14.6. If V_d is the differential input voltage at

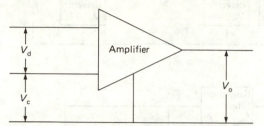

Fig. 14.6. Illustration of common mode rejection ratio.

the input of the amplifier, V_c is the common mode input, G_d is the differential gain and G_c the common mode gain, then the following equations may be derived.

$$V_o = G_d V_d + G_c V_c \qquad (14.1)$$

$$G_d = V_o/V_d \qquad (14.2)$$

$$G_c = V_o/V_c \qquad (14.3)$$

$$\begin{aligned} \text{CMRR} &= G_d/G_c \\ &= V_c/V_d \text{ for constant } V_o \end{aligned} \qquad (14.4)$$

The CMRR is measured by applying a differential input voltage V_d and adjusting it to give a convenient value of V_o. The two differential inputs are then shorted together and V_c is applied to give the same value of V_o. CMRR is then found from (14.4).

Amplifier *slew rate* is measured using a unity gain non inverting follower circuit configuration, as this gives worst case results. A signal is applied in both directions so as to saturate the amplifier, as in Fig. 14.7 and the slew rate is found as the ratio $\Delta V_o/\Delta t$.

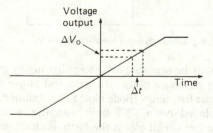

Fig. 14.7. Illustration of slew rate.

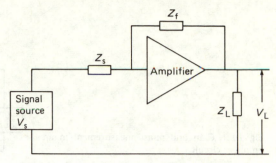

Fig. 14.8. Feedback in amplifier circuits.

Most amplifiers are operated with feedback, as shown in Fig. 14.8. Z_S is the source impedance, Z_L is the load impedance and Z_f the feedback impedance. Depending on the gain being measured, the external impedances should be chosen to be either much greater, or much smaller, than the input Z_i and output Z_o impedances of the amplifier. For example if voltage gain V_L/V_S is being measured then Z_S should be much smaller than Z_i but Z_L should be much larger than Z_o. If current gain I_L/I_S is being measured then the reverse conditions apply. For transconductance (I_L/V_S) measurement, Z_S and Z_L must both be much smaller than Z_i and Z_o respectively.

14.2.1 Network analyser

Voltages are complex values, having amplitude and phase, so that for an input voltage of $V_i \cos \phi_i$ and an output of $V_o \cos \phi_o$ the gain is given by (14.5)

$$G = \frac{|V_o|}{|V_i|} \angle \phi_o - \phi_i \qquad (14.5)$$

This gain can be measured by recording voltage and phase difference using the circuit arrangement shown in Fig. 14.9, which behaves as a crude form of network analyser. The accuracy of this arrangement depends on the accuracy and frequency response of the voltmeters and phase meter, and it is not good. A better method is to use a network analyser, as shown in Fig. 14.10. The amplifier to be tested is connected into one channel and a shorting link is used in the second channel. The frequency of the mixer signal source $(f_i + f_m)$ is adjusted in line with the main signal source (f_i) so that the frequency output from the mixers and band pass filters is fixed (f_m), enabling the meters to operate at constant frequency. For good accuracy both

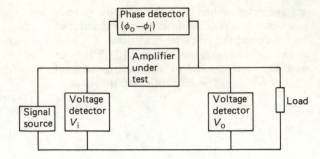

Fig. 14.9. Gain and phase measurement in an amplifier circuit.

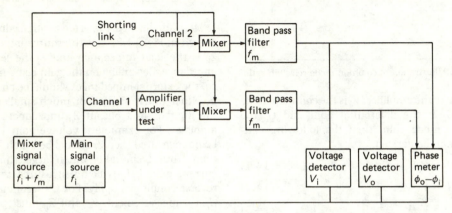

Fig. 14.10. Amplifier gain and phase measurement using a dual channel network analyser.

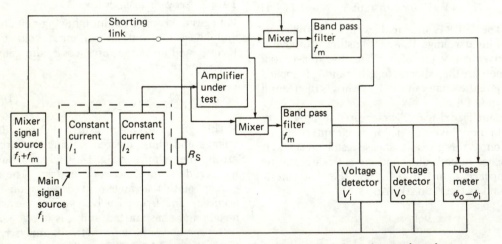

Fig. 14.11. Amplifier input impedance measurement using a dual channel network analyser.

channels should be identical. Small constant offsets can be measured, and compensated for, by replacing the amplifier under test by another shorting link, and then calibrating the instrument.

The signal source of a network analyser can usually be operated in a swept frequency mode. This enables a plot of the gain and phase angle versus frequency (Bode plot) to be obtained, and displayed on a CRT built into the analyser. Another useful plot is the Nichols chart, which shows the variation of log gain with phase shift, at different frequencies: this is useful for characterising closed loop amplifiers.

The network analyser can measure many other parameters, such as the input or output impedance of an amplifier. Fig. 14.11 shows the connection for measuring input impedance. The main signal generator is now operated as two constant current sources, and the analyser measures the complex voltages across a standard resistor R_s and across the input impedance of the amplifier under test. If R_s and I_i are constant then V_i need not be read, and V_o can be calibrated, using known standard resistors in place of the amplifier, to indicate the value of amplifier input impedance.

14.2.2 *Vector voltmeter*

Vector voltmeters can be used in place of network analysers to measure the signal at two points in a circuit, and the phase difference between them. They are useful in VHF applications for measurements on amplifiers, filters and networks, whereas network analysers are more commonly used for microwave measurements.

The two inputs to the vector voltmeter of Fig. 14.12 can be r.f. signals (1 MHz–1 GHz). These are sampled at a low frequency (e.g. 20 kHz) and than fed into tuned amplifiers which are selective at the sampling frequency. Therefore the two r.f. signals are converted to i.f. signals, whose fundamentals have the same amplitude and phase relationship as the fundamentals of the two original r.f. signals. These can then be measured by the voltage detector and phase meter.

14.3 Distortion

14.3.1 *Types of distortion*

The ear is a very good judge of distortion in audio

systems. Although distortion is caused by many devices, an amplifier is considered in all the examples of this section. There are different types of distortion:

(i) Frequency distortion occurs when the amplifier does not have equal amplification for all the frequencies at its input.

(ii) Phase distortion, which results when the signal at the output of the amplifier is displaced in phase relative to the input signal. If all frequencies were displaced by the same amount the distortion would not be noticed, but in practice some frequencies are phase shifted more than others.

(iii) Harmonic distortion occurs due to the fact that the amplifier generates harmonics of the fundamental of the input signal. Harmonics always result in amplitude distortion, for example when an amplifier is overdriven and clips the input signal.

(iv) Intermodulation distortion, which is a consequence of the interaction or heterodyning between two frequencies, giving sum and difference frequencies of the original two frequencies.

(v) Crossover distortion, which occurs in push–pull amplifiers due to incorrect bias levels, as shown in Fig. 14.13.

In the sections which follow, the measurement of total harmonic distortion, intermodulation

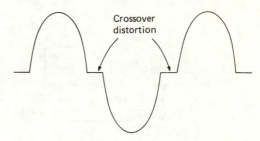

Fig. 14.13. Illustration of crossover distortion.

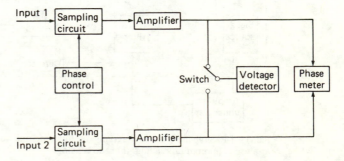

Fig. 14.12. Simplified block diagram of a vector voltmeter.

distortion, and transient intermodulation distortion, are described.

14.3.2 *Total harmonic distortion*

A non-linear system produces harmonics of an input sine wave, the harmonics consisting of sine waves with frequencies which are multiples of the fundamental of the input signal. Total harmonic distortion (THD) is measured in terms of the harmonic content of the wave, as given by (14.6)

$$\text{THD} = \frac{[\sum(\text{Harmonics})^2]^{1/2}}{\text{Fundamental}} \qquad (14.6)$$

In a measurement system noise is read in addition to harmonics, and the total waveform, consisting of harmonics, noise and fundamental, is measured instead of the fundamental alone. Therefore the measured value of total harmonic distortion (THD_M) is given by (14.7)

$$\text{THD}_\text{M} = \frac{\{\sum[(\text{Harmonics})^2 + (\text{Noise})^2]\}^{1/2}}{\{\sum[(\text{Fundamental})^2 + (\text{Harmonics})^2 + (\text{Noise})^2]\}^{1/2}} \qquad (14.7)$$

Equation 14.7 leads to below $\frac{1}{2}$% error, due to the approximation, for values of THD below 10%.

Fig. 14.14 shows a harmonic distortion analyser which is used to measure THD. The signal source has very low distortion, and this can be checked by reading its output distortion by connecting directly into the analyser. The signal from the source is fed into the amplifier under test. This generates harmonics and the original fundamental frequency. The fundamental frequency is removed by the notch filter.

In the manual system of Fig. 14.14(a) the switch *SW* is first placed in position 1 and the total content of fundamental and harmonics (E_T) is measured. Then the switch is moved to position 2 to measure just the harmonics E_H. The value of THD is then found from (14.7), as in (14.8)

$$\text{THD} = \frac{E_\text{H}}{E_\text{T}} \times 100 \qquad (14.8)$$

The meter can be calibrated by putting the switch in position 1 and adjusting the reading for full scale deflection. With the switch in position 2 the meter reading is now proportional to THD. Fig. 14.14(b) shows an alternative arrangement, where the values of E_T and E_H are read simultaneously, and their ratio calculated and displayed as THD on the indicator. For good accuracy the notch filter must have excellent rejection and high pass characteristics. It should attenuate the fundamental by 100 dB or more, and the harmonics by less than 1 dB. The filter

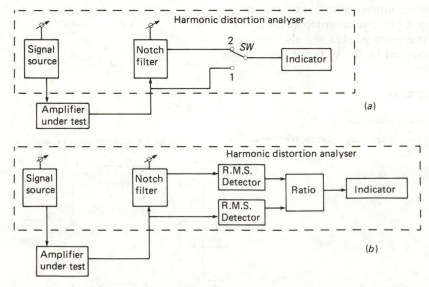

Fig. 14.14. Simplified block diagrams of fundamental suppression harmonic distortion analysers; (a) manual reading, (b) ratio reading.

also needs to be tuned accurately to the fundamental of the signal source. This is difficult to achieve manually, and most distortion analysers do this automatically. A common form of notch filter is a Wien bridge. This balances at one frequency only (see Section 7.7), and at this frequency the output voltage at the bridge null detector is minimum.

14.3.3 *Intermodulation distortion*

When a high frequency (f_1) signal and a low frequency (f) signal are mixed in a linear circuit the output contains the two frequencies only. When mixed in a non-linear circuit, for example an amplifier having distortion, modulation occurs. The output now contains the original frequencies $(f_1$ and $f_2)$, and also the sum and difference frequencies $(f_1 + f_2$ and $f_1 - f_2)$, in addition to several harmonics and their sum and difference frequencies.

Fig. 14.15 shows an instrument for measuring intermodulation distortion. The SMPTE (Society of Motion Picture and Television Engineers) defines the two signal source $(f_1$ and $f_2)$ frequencies as 7 kHz and 60 Hz, having amplitudes in the ratio 1:4 respectively. The lower frequency has a much higher amplitude since a low degree of non-linearity is expected. In Europe the values of f_1 and f_2 are defined by DIN 45403 as 8 kHz and 250 Hz, again having an amplitude ratio of 1:4.

The output from the amplifier under test is fed into a high pass filter, giving a waveform at point A which consists of the high frequency carrier

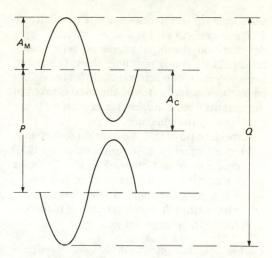

Fig. 14.16. Waveform output from an intermodulation distortion meter.

and its low frequency modulation, as in Fig. 14.16. This waveform can be observed on an oscilloscope, and intermodulation distortion (IMD) calculated from (14.9) or (14.10)

$$IMD = \frac{\text{Amplitude of modulation } (A_M)}{\text{Amplitude of carrier } (A_c)} \times 100 \tag{14.9}$$

$$= \frac{Q - P}{P} \times 100 \tag{14.10}$$

The analyser shown in Fig. 14.15 is capable of calculating (14.9) directly and displaying the value of IMD on an indicator. The r.m.s. value of

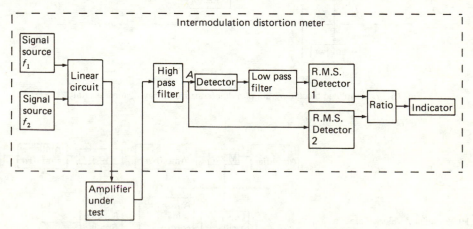

Fig. 14.15. Simplified block diagram of an intermodulation distortion meter.

the waveform at *A* is measured by r.m.s. detector 2. The waveform at *A* is also demodulated by a detector and the high frequency components removed by the low pass filter, leaving the low frequency distortion components. This is measured by r.m.s. detector 1, and the ratio is taken of the outputs from detectors 1 and 2 before being displayed on the indicator.

A special case of intermodulation distortion is difference frequency distortion, also called CCIF intermodulation distortion. Fig. 14.17 shows the construction of a difference frequency analyser. A low pass filter is used after the amplifier under test so that the meter is sensitive to the low (difference) frequencies only.

14.3.4 *Transient intermodulation distortion*

Transient intermodulation distortion occurs because an amplifier is not able to respond rapidly to changing inputs. It can be tested by applying a square wave signal, with a superimposed high frequency sine wave, to the input of the amplifier, as in Fig. 14.18(*a*), and observing the output on an oscilloscope. If transient intermodulation distortion is present then the output will have an initial portion of the high frequency signal missing, as at

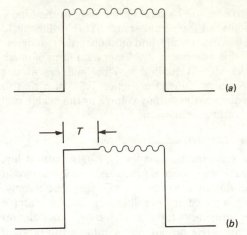

Fig. 14.18. Waveforms to measure transient intermodulation distortion; (*a*) input signal, (*b*) output signal.

Fig. 14.18(*b*), and the value of *T* can be noted.

14.4 **Radio receivers**

14.4.1 *Receiver construction*

Although there have been many different designs for radio receivers the super heterodyne design, illustrated in Fig. 14.19, has been a

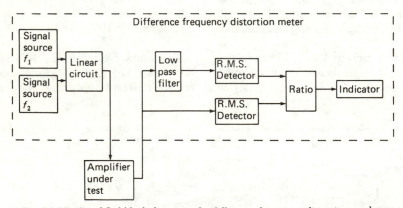

Fig. 14.17. Simplified block diagram of a difference frequency distortion analyser.

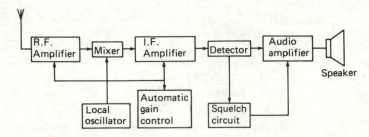

Fig. 14.19. Simplified block diagram of a superheterodyne receiver.

standard over many years. The r.f. signal of frequency f_r is picked up by the antenna, and feeds the r.f. amplifier. This amplifier can be tuned, so as to accept only a very narrow band of frequencies. The output from the amplifier is heterodyned in the mixer, with a local oscillator, whose frequency (f_o) is also variable.

The output from the mixer consists of the two original frequencies (f_r and f_o) plus their sum ($f_r + f_o$) and difference ($f_r - f_o$). The i.f. amplifier, which contains several stages, is usually tuned to a fixed frequency equal to $f_r - f_o$. Because of this the i.f. amplifier can be designed to give high gain without oscillating.

The detector demodulates the audio signal from the r.f. and this is amplified by the audio amplifier before driving the speaker. The squelch circuit monitors the audio, and cuts off the audio amplifier when no audio is present, to prevent noise only being transmitted to the speakers.

Many different parameters are used to define the performance of radio receivers. Only a few of the most common ones are described in this section, together with their measurement method.

14.4.2 Sensitivity

Sensitivity can be specified and measured in different ways, depending on the type of receiver. In all methods, however, a signal from a generator is fed into the receiver and the output signal is detected. Although the power of the output signal can be measured, it is more common to note signal plus noise and distortion (SINAD ratio), which is the ratio of the total output to the output caused by noise and distortion. SINAD meters are available which combine the properties of power and distortion meters.

In making sensitivity measurements care is needed when coupling the signal generator to the receiver. The impedance of the generator should be matched to that of the receiver, and for receivers which are designed to be used with external antennas the signal generator should be coupled via an artificial antenna.

Many different systems of units are used to define sensitivity. A few of these are, voltage across a matched load; open circuit voltage stated in dB relative to $1\,\mu V$; and voltage across a matched load in dB relative to $1\,\mu V$. Although open circuit voltage is the recommended inter-national standard, it is important to know what units the meter is calibrated in, when making measurements. Some meters have a choice of calibrations.

Sensitivity can be stated in several ways, four of these are noise sensitivity, audio power sensitivity, quieting sensitivity, and usable sensitivity.

Noise sensitivity. This is the r.f. signal input which is needed to produce an output with a specified signal to noise ratio. For broadcast a.m. or f.m. receivers the measurements are made using a signal having 30% amplitude modulation depth, or an f.m. of 22.5 kHz deviation. The output signal to noise ratio is 20 dB. For narrow band a.m. or f.m. receivers an f.m. deviation of 60% of rated system deviation is used, and a signal to noise ratio of 10 dB. To make measurements the signal modulation is set to the required value, and the amplitude of the r.f. adjusted until the audio output provides the required signal to noise ratio, as measured when the modulation is switched on and off.

Audio power sensitivity. This is also known as *gain sensitivity* and applies only to a.m. receivers. It is stated as the r.f. input needed to give a specified audio output, usually 50 mW. To test the receiver the r.f. signal generator is set to produce a 30% a.m. depth at a modulating frequency of 1 kHz, and the r.f. level is adjusted until the output is 50 mW at maximum gain.

Quieting sensitivity. When there is no input signal to a receiver the output is mainly all noise. When an unmodulated signal is applied to the input there is gain reduction in an a.m. receiver, since the automatic gain control circuit operates to keep the detector signal level constant, as the received signal level changes. This reduces the output noise. For an f.m. receiver the output noise also decreases as the input signal captures the limiting stages in the receiver. Eventually, in both a.m. and f.m., saturation is reached, where the increase in the unmodulated input signal gives very little decrease in output noise.

Quieting sensitivity is the input at which the output decreases by 20 dB over the zero input value. It is measured by first noting the receiver output with no input signal, and then increasing the unmodulated r.f. input until the output falls by 20 dB.

Usable sensitivity. This is often used to specify the sensitivity of narrowband receivers, and is the r.f. input which produces an output SINAD ratio of 12 dB. To measure sensitivity the signal generator is modulated with 30% a.m., or with an f.m. deviation of 60% of rated system deviation, as appropriate. The input level is then adjusted to give a SINAD ratio of 12 dB, (equivalent to distortion of 25%). Weighted SINAD ratios can also be used with weighting networks such as CCITT P53, for which the SINAD ratio is 20 dB (weighted distortion of 10%).

14.4.3 *Signal to noise ratio*

Sensitivity tests indicate whether a receiver is capable of detecting weak signals. Signal to noise ratio, on the other hand, measures the quality of the output at high signal levels. The test is performed by inputting a signal of 1 mV which has a modulation of 30% a.m., or of 60% rated system deviation. The modulation is then switched on and off to obtain the audio outputs, and their ratio, in the two conditions.

Weighted signal to noise ratio tests can also be performed, using weighting networks such as CCITT P53 for narrowband receivers, and CCIR for broadcast receivers.

14.4.4 *Selectivity*

Selectivity tests indicate the ability of a receiver to reject interfering signals on a channel adjacent to the one to which it is tuned. Two signal generators are used to measure adjacent channel selectivity, and these are coupled together and feed the input of the receiver. One of the signal generators is tuned to the required receiver channel frequency, and is modulated by a 1 kHz tone, to give a 30% a.m. or an f.m. deviation of 60% of the rated system deviation. The second signal generator is modulated by a 400 Hz tone, and with the same modulation conditions as the first channel. It is tuned to a channel frequency which is adjacent (up or down) to that of the first signal generator.

To perform the test, initially the output from both generators is set to zero. The signal from the first generator is then increased to give a SINAD ratio at the output of the receiver, of 12 dB (distortion of 25%), with the receiver set to 50% of rated power. The signal level from the second generator is then increased until the SINAD ratio falls to 6 dB (50% distortion). The ratio of the output levels from the two generators is quoted in dB. Measurements are made of the ratio for both the up and down channels, and the lowest value is used to specify the adjacent channel selectivity of the receiver.

14.4.5 *Blocking*

Blocking measures the effect which a strong signal has on the sensitivity of a receiver, when it is detecting a weak signal. If f_r and f_o are the frequencies of an r.f. signal and receiver local oscillator, and f_{if} is the frequency to which the i.f. amplifier is tuned, then the i.f. stage will accept signals having a frequency $(f_o + f_{if})$ in addition to $(f_r - f_o)$, which can result in blocking or imaging.

To measure blocking, two signal generators are used coupled to the receiver input. One generator is tuned to the wanted frequency, and carries the reference modulation of 60% of rated system deviation. The output level from this generator is adjusted, to give a receiver output SINAD ratio of 12 dB, whilst the level from the second generator is zero. The second generator is unmodulated, and its level is now increased until the SINAD ratio falls to 6 dB. The level at the input to the receiver at this point is called its blocking level. The frequency of the second generator is varied, usually in the range 2–10 MHz from the wanted frequency, to establish the lowest blocking level.

14.4.6 *Bandwidth*

Bandwidth can be found by noting the output from the receiver, usually on an oscilloscope, whilst the input is swept through a range of frequencies. Bandwidth can also be found approximately as follows. The input signal, from a signal generator, is set to 30% a.m. or 60% rated system deviation, and its level adjusted until the signal to noise ratio in the output is 10 dB. The input level is then increased by 6 dB, and is tuned to find two frequencies at which the signal to noise ratio is again 10 dB. The difference between these two frequencies is the approximate bandwidth of the receiver.

Another measure of bandwidth is the *modulation acceptance* bandwidth, which is mainly applicable to narrowband f.m. receivers. To measure this parameter the signal generator is modulated to 60% rated system deviation, and the input i.f. level is adjusted to give a 12 dB SINAD ratio at the receiver output. The signal generator output level is then increased by 6 dB, and its f.m. deviation is also increased until the SINAD ratio is again 12 dB. The deviation of the signal generator at this point is the modulation acceptance bandwidth of the receiver.

14.4.7 *Harmonic distortion*

To measure harmonic distortion of a receiver its input signal is set to 1 mV, with a modulation of 30% a.m. or an f.m. deviation of 60% of rated system deviation. The volume control is then adjusted to give a power output level which is between 10% and 50% of full rated power, and the distortion of the receiver is measured by a distortion analyser (Section 14.3).

14.5 Stereo decoder separation

A fundamental measurement for stereo systems in hi-fi and radio is the separation between the left and right systems. Fig. 14.20 shows the block diagram of a very basic stereo system. The output from the f.m. discriminator goes through a passband filter (23–53 kHz), so that the difference signal $(L - R)$ consists of frequencies within this range. The signal is demodulated in the detector, and the $(L - R)$ signal is fed to one of the mixers. This combines with the $(L + R)$ signal to give a 2L signal to drive the top speaker.

The $(L - R)$ signal from the detector is also phase inverted to get $(R - L)$, and this is mixed with the $(L + R)$ signal in the second mixer. The resulting 2R signal is used to drive the bottom speaker, so that separation between left and right speakers is achieved.

For good separation it is important to have good cancellation in the mixers. This implies a passband circuit with a sharp cut-off frequency, and the absence of phase shift in the $(L - R)$ sideband signal. It is also necessary to control signal levels closely, in order to give an accurate subtractive process. In an ideal arrangement, if only an R signal is applied, then there should be an output from the right channel; and if only the L signal is applied then an output is expected from the left channel only. In practice each channel gives some output on the other channel, although this is usually 20–30 dB lower.

Fig. 14.21 shows the test set up for measuring stereo separation. The system under test is fed from a stereo generator, and the L and R test signals are applied in turn. The output from the left and right channels is measured by a calibrated oscilloscope, or VTVM with a dB scale. This is connected to each channel in turn, and the difference of the two readings is taken to find the separation in dB.

There are two types of separation tests. The first measures the performance of the decoder on its own, and for this an audio composite signal is applied from the generator. In the second the performance of the complete stereo system is measured. An r.f. carrier is now frequency

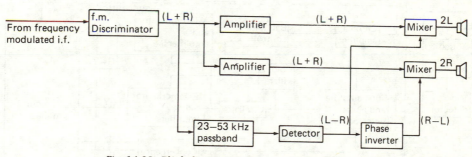

Fig. 14.20. Block diagram of a basic multiplex receiver.

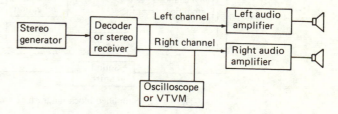

Fig. 14.21. Test arrangement for measuring stereo separation.

modulated by an audio composite test signal, and applied to the antenna input terminals of the stereo receiver. The stereo separation on the complete system is generally less than that of the decoder on its own.

14.6 Television receivers

The television receiver, by virtue of the fact that it has a built in CRT, is able to provide an indication of its own performance. Although repair of a television is often done without use of external instrumentation, this section will look at the pattern generator, which is extensively used for television performance measurements.

Fig. 14.22 shows a simplified block diagram of a colour television. The signal is captured by the antenna, and the required channel selected by the tuner. After amplification the signal is demodulated and then distributed to the other parts of the receiver. The brightness and contrast section handles all aspects of a black and white signal, and it also controls the brightness of a colour picture. The colour control section processes the colour information for display on the colour tube. The raster control provides the horizontal and vertical deflection of the picture tube, and the accelerating voltage to excite the phosphor. It also synchronises the deflection of

the beam on the tube with the video signal. Finally the sound control section amplifies and demodulates the f.m. audio signal, and then provides the power drive to operate the speakers.

Pattern generators can produce signals for a variety of patterns, which are coupled into the television set and displayed on its screen. Fig. 14.23 shows the cross-hatch and white dot

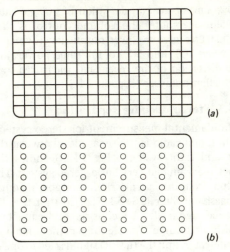

Fig. 14.23. Cross-hatch and white dot generator patterns: (*a*) cross-hatch, (*b*) white dot.

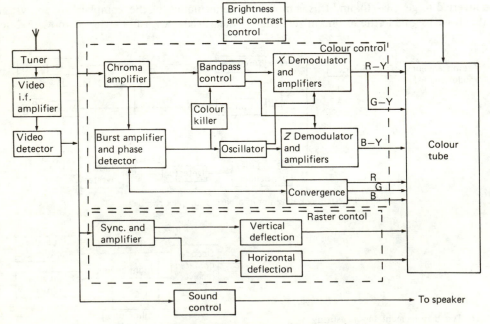

Fig. 14.22. Simplified block diagram of a colour television.

patterns. These are also called convergence patterns since they are used to check picture tube convergence. The white dot pattern is primarily used for dynamic convergence; the beams are in convergence if the dots are small and white, and they are out of convergence if the dots are multicolour.

The colour control section of a television is mainly checked by producing a band of colours from the pattern generator, and specialist instruments which do this are also called colour bar generators. These work on the principle of generating two signals which are phase shifted by the correct angle for each colour, and then

gating to separate them into bars of different colours. Fig. 14.24 shows a colour bar generator connected to test a television receiver, and

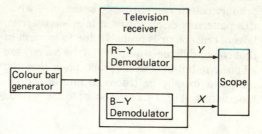

Fig. 14.24. Television test set-up.

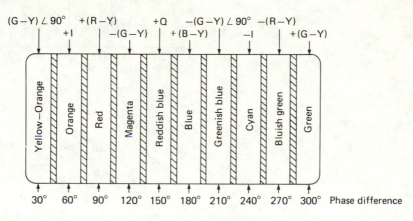

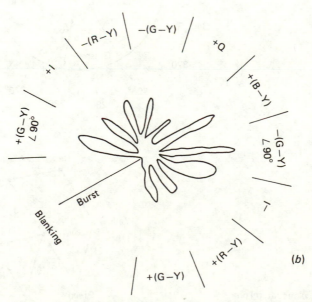

Fig. 14.25. Keyed rainbow pattern; (a) colour bar, (b) vectogram.

Figs. 14.25 and 14.26 show two colour pattern standards.

The keyed rainbow pattern consists of successive bursts which differ in phase by 30°. A missing primary colour can obviously be detected when this pattern is displayed on a television screen. Loss of one colour is probably due to a defect in that colour difference amplifier. If the colours are all present, but in different proportions, then the fault can usually be traced to one of the demodulators (Fig. 14.24). For example if blue predominates in the display then the X demodulator is faulty i.e. a weak red and green colour is being produced. A colour tint in a black and white picture would also indicate that the complementary colour was weak or missing. For example a red and blue tint would indicate a suspect G–Y amplifier, giving a weak green signal.

The NTSC colour pattern consists of seven colours of varying strength and phase, as shown in Fig. 14.26(a) and (b). The waveshapes and

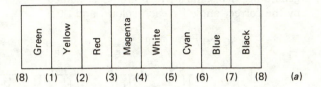

(a)

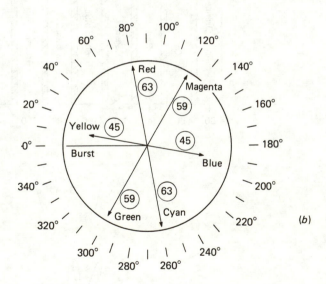

(b)

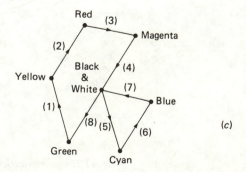

(c)

Fig. 14.26. NTSC pattern; (a) colour bar, (b) phase of colours, (c) vectogram.

amplitudes of the signal, after passing through the receiver, give considerable information on circuit performance. A vectogram can also be used to check on the operation of the chroma demodulator. This will produce the pattern for an NTSC colour signal as in Fig. 14.26(c), where the spot travels around the screen in the sequence numbers indicated. The vectogram for the keyed rainbow pattern should theoretically be symmetrical, but in practice it is distorted as shown in Fig. 14.25(b) due to the characteristics of the receiver.

15. Transmission system testing

15.1 Introduction

In the previous chapter the characteristics, and testing, of audio and video signal receivers, such as radio and television, were described. The present chapter looks at three components used in the transmission of signals; antennas, transmission lines and transmitters.

15.2 Antennas

15.2.1 *Antenna operation*

An antenna emits electromagnetic radiation when an oscillating signal excitation is applied to it. The efficiency of the radiation is greatest when the excitation frequency coincides with the resonant frequency of the antenna. Antennas are used for many different applications, such as H.F., V.H.F., U.H.F. and microwave. In this section emphasis is placed on H.F. and V.H.F. dipole and monopole antennas.

A half wavelength dipole is mounted horizontally to the earth's surface, as in Fig. 15.1(a). It emits an electromagnetic field in which the electrical and magnetic components are at right angles to each other, and the electric field is horizontal to the earth's surface. A quarter wavelength monopole (Fig. 15.1(b)) is mounted vertical to the earth's surface, and the electrical field component, of the electromagnetic radiation, is also perpendicular to the earth's surface. The field strength, picked up by the receiver, is strongest when the transmitting and receiving antennas have the same polarization.

Antennas are usually cut to a length which is a function of the signal wavelength to be transmitted or received. Usually an antenna is cut to half a wavelength or a quarter wavelength, of the centre operating frequency. However, due to capacitance and end effects, the antenna electri-

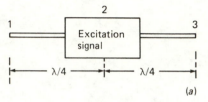

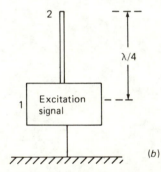

Fig. 15.1. Two types of H.F. and V.H.F. antenna; (a) half wavelength dipole, (b) quarter wavelength monopole.

cal length is greater than its physical length. If f is the resonant frequency of an antenna, then its electrical length E_L in metres is given by (15.1) and (15.2), for half wavelength and quarter wavelength antennas respectively.

$$E_L = \frac{150}{f} \tag{15.1}$$

$$E_L = \frac{75}{f} \tag{15.2}$$

When the antenna is close to the earth's surface, or to that of a conducting object, as would normally happen when an antenna is mounted, then the electrical length is slightly less than that given by (15.1) and (15.2).

The physical length P_L of the antenna is related to its electrical length by (15.3), where constant $K = 0.94$ for resonant frequencies f greater than 30 MHz, and $K = 0.96$ for f less than 3 MHz.

$$P_L = KE_L \qquad (15.3)$$

15.2.2 Antenna impedance

The impedance of an antenna varies along its length. This is illustrated in Fig. 15.2, which

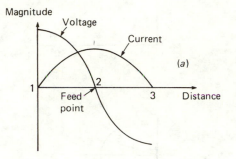

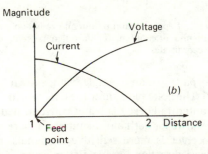

Fig. 15.2. Voltage–current characteristics for a typical antenna; (a) half wavelength dipole, (b) quarter wavelength monopole.

shows the current and voltage distribution for a half wavelength dipole and a quarter wavelength monopole. The point of maximum current and minimum voltage, that is lowest impedance, is called the feed point, and the impedance at this point is the feed point impedance Z_F given by (15.4)

$$Z_F = R_R \pm jX_F \qquad (15.4)$$

R_R is called the radiation resistance of the antenna, and X_F is the reactive part of the feed point impedance. At resonance X_F should equal zero, whilst at frequencies above resonance the impedance is inductive, and at frequencies below

resonance it is capacitive. In practice the reactive component of the impedance varies between zero and ± 100 ohms.

The impedance of an antenna also varies due to other factors, such as the closeness of the earth's surface or of a conducting surface. In the ideal case a half wavelength dipole would have a radiation resistance of 73 ohms and a quarter wavelength monopole a resistance of 53 ohms. In practice these resistances vary between 5 ohms and 120 ohms, and between 5 ohms and 80 ohms, for the two antennas respectively.

The antenna resistance can be measured using an impedance bridge, which is usually based on the Wheatstone bridge (Chapter 7). Radiation resistance measurements are more meaningful. These disregard the d.c. resistance of the antenna, which is small except for long wire construction used in low frequency applications.

Fig. 15.3 shows a method for measuring the

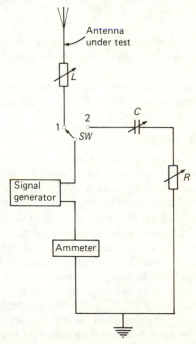

Fig. 15.3. Measuring antenna radiation resistance.

radiation resistance of an antenna. The signal generator is set to the frequency at which the antenna resistance is to be measured, and switch SW is put to position 1. The antenna is tuned to this frequency, by adjusting inductor L, until the ammeter reads a peak value of current, say I_M. The switch is then moved to position 2, and capacitance C is adjusted until the ammeter reading is maximum. The value of resistance R is then changed until the ammeter reading equals I_M. The value of R at this setting is equal to the radiation resistance R_R of the antenna, at this frequency. The radiated power P_R being delivered by the antenna is now given by (15.5)

$$P_R = I_M{}^2 R_R \qquad (15.5)$$

Another way of defining the relationship between radiated power and radiation resistance is to suppose that the antenna is surrounded by a sphere, having a radius which is large relative to the dimensions of the antenna and the radiated wavelength. The waves will therefore be almost normal to the surface of the sphere, and its value, integrated over the whole surface of the sphere, is the total radiated power P_R. The value of the radiation resistance R_R is then given by (15.6), where I is the input current to the antenna.

$$R_R = P_R / I^2 \qquad (15.6)$$

R_R is also equal to the antenna resistance, measured at its input terminals, when there is no loss.

It is usual to measure the antenna impedance over the whole usable frequency range, in order to design the coupling network.

15.2.3 *Antenna directivity*

Antennas are usually designed so that they transmit and pick up signals from a few selected directions only. This property is called directivity, and is important in improving the gain and efficiency of an antenna, as explained in Section 15.2.4. The directivity of an antenna may be plotted using Cartesian or polar co-ordinates, as in Fig. 15.4 which shows only part of the plots, the remainder being symmetrical.

The directivity of a transmitting antenna can be obtained by rotating it, and measuring the strength of the signal, at the transmitted frequency, using a field strength meter of the type shown in Fig. 15.5. This will give a variety of

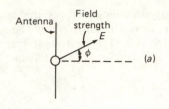
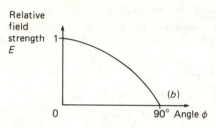
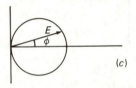

Fig. 15.4. Representation of antenna radiation pattern; (*a*) antenna orientation, (*b*) Cartesian coordinates, (*c*) polar coordinates.

polar patterns, an example being shown in Fig. 15.6. The polar plots show the direction in which the antenna energy is being concentrated, and this is true for both a transmitter and a receiver. A receiver is more sensitive to signals in the forward direction, and is also more effective in rejecting signals from the null direction of a polar pattern.

15.2.4 *Antenna gain*

When the same amount of power is radiated in all directions by an antenna it is called an isotropic source. Good gain can be obtained by concentrating the antenna energy in one direction, for a vertical monopole, or in two directions for a horizontal dipole. The total energy radiated is the same, but it is being directed to the required

Fig. 15.5. Simplified block diagram of a field strength meter.

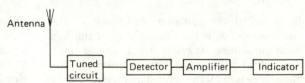

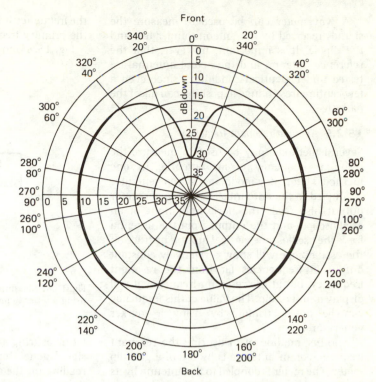

Fig. 15.6. Polar pattern of an antenna.

direction. In this case the directivity gain G_D of the antenna is given by the (15.7)

$$G_D = \frac{\text{Maximum power flux due to the antenna under test}}{\text{Power flux due to an isotropic source radiating the same power}} \quad (15.7)$$

Directivity gain G_D is a measure of the increase in power flux due to the shaping of the antenna radiation pattern. The power gain G_p of the antenna is given by (15.8), and the antenna radiation efficiency η_R by (15.9)

$$G_P = \frac{\text{Maximum power flux due to the antenna under test}}{\text{Power flux due to an isotropic source with the same input power}} \quad (15.8)$$

$$\eta_R = \frac{\text{Power radiated by the antenna under test}}{\text{Power input to the antenna}}$$

$$= \frac{G_P}{G_D} \quad (15.9)$$

It is possible for antennas to have a high directivity gain (G_D) but still have a low power gain (G_P) if the antenna losses are high, giving a low radiation efficiency (η_R).

Antenna gain is usually applied to receiving antennas. The voltage at the terminals of the antenna is compared with that of a thin wire dipole of the same size, and working at the same frequency and the same distance from the transmitter. Usually the gain is expressed in decibels relative to the reference dipole, stated as dB over a dipole or dBd. Fig. 15.7 shows a typical curve of antenna gain variation with frequency.

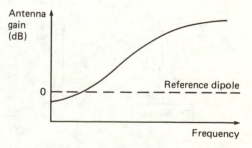

Fig. 15.7. Variation of antenna gain with frequency, referred to a reference dipole.

A wavemeter can be used to measure the signals received by the antenna under test and the dipole. It is important, however, that the reference antenna and the actual antenna are tested under identical conditions, since even a few centimetres change in position can affect the patterns.

15.2.5 *Antenna resonant frequency*

Several methods may be used to measure the resonant frequency of an antenna. In one method a frequency modulated sweep generator is coupled to the antenna. The sweep generator output also feeds the horizontal input of an oscilloscope, and the output from the antenna feeds the oscilloscope vertical input. The trace on the oscilloscope will show a relatively constant output at frequencies far from the resonant frequency, but at the resonant frequency it will display a null or dip. The value of this frequency can be noted, usually by using a marker generator.

Another method of measuring the resonant frequency of an antenna is by use of a grid dip meter. The meter is coupled to the antenna by its tuning coil, or via a pick up coil. The grid dip meter is tuned for a dip, and the resonant frequency noted. Readings should be made with the antenna in the actual operating position. There is also likely to be mismatch between the antenna and the leads from the meter. This will give rise to two reactances which can interact to result in extra resonances, which may be misleading. To make resonant frequency measurements the grid dip meter generator should initially be set at a low frequency, and this frequency gradually increased until a dip occurs. This first dip is the primary resonant frequency of the antenna. Subsequent dips, which occur as

the frequency is increased, are exact multiples of the primary frequency, and are harmonics.

Fig. 15.8 shows the series ammeter method

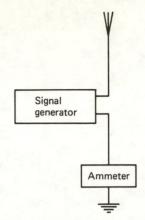

Fig. 15.8. Measuring resonant frequency of an antenna using a series ammeter.

for measuring antenna resonant frequency. The signal generator is tuned to give a maximum reading on the ammeter, which corresponds to the point of maximum transfer of energy from the signal generator to the antenna, since both are at the same frequency. The value of the resonant frequency is read from the setting of the signal generator. Although relatively simple to operate, the series ammeter method suffers from two problems. (i) The series ammeter is limited in operating frequency. (ii) The series ammeter consumes power.

Fig. 15.9 shows the use of a wavemeter to measure resonant frequency. Note that the wavemeter is similar to a field strength meter (Fig. 15.5) but its tuned circuit can be externally adjusted, so that it acts as more than just a peak

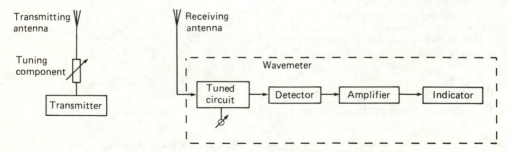

Fig. 15.9. Measuring resonant frequency of an antenna using a wavemeter.

indicator. To make resonant frequency measurements the tuned circuit is adjusted, to give a maximum reading on the wavemeter. The frequency setting is now the antenna resonant frequency. To set up a transmitting antenna the wavemeter is initially tuned to the transmitter operating frequency. The tuning component (inductive or capacitive) on the transmitting antenna, is then adjusted to give a maximum reading on the wavemeter.

15.3 Transmission lines

Transmission lines are used in a variety of applications, one common use being to feed an antenna. Usually one cannot locate the oscillator, or radio transmitter, exactly at the antenna feed point, as indicated in Fig. 15.1. In these instances it is necessary to locate the transmitter some distance away, which could be a few metres to a few miles, and to use a transmission line to carry power to the antenna.

Transmission lines may consist of parallel wires, although coaxial cables are more frequently used. In this section some of the characteristics of transmission lines are described, and the methods by which these can be measured, and this is followed by a description of fault location in cables.

15.3.1 *Transmission line characteristics*

Three of the parameters of transmission lines are introduced here, velocity factor, characteristic impedance and standing wave ratio.

Velocity factor. Radio waves in free space travel at the speed of light, i.e. 3×10^8 m/s. In a transmission line their speed is reduced by a factor F, and this is known as the velocity factor of the transmission line. The value of F varies, depending on the construction of the line. Typical values are: 0.75, for a two wire line with plastic dielectric; 0.67 for a coaxial line with solid

plastic dielectric; 0.85 for a coaxial line with air dielectric; and 0.97 for a two wire open line with air dielectric.

The velocity factor alters the electrical length, given in (15.1) and (15.2), to those given by (15.10) and (15.11) respectively.

$$E_L = \frac{150F}{f} \tag{15.10}$$

$$E_L = \frac{75F}{f} \tag{15.11}$$

F is the velocity factor f the operating frequency in MHz, E_L the electrical length in metres.

Characteristic impedance. The characteristic impedance of a transmission line is the value of load resistance which enables maximum transfer of power from the source to the load. If the load has a different value from the characteristic impedance of the line, then some of the power is reflected back down the line, towards the source.

The characteristic impedance is a measure of the distributed inductance and capacitance of the transmission line. If L_d is the distributed inductance per unit length, and C_d the capacitance per unit length, then the characteristic impedance Z_c of the line is given by (15.12)

$$Z_c = \sqrt{\frac{L_d}{C_d}} \tag{15.12}$$

Practical values of characteristic impedance vary from 100 ohms to 1000 ohms for parallel wire, and 10 ohms to 150 ohms for coaxial cable.

Standing wave ratio. When a transmission line is terminated by an impedance different from its characteristic impedance, some of the forward signal wave is reflected back. This reflected wave mixes with the forward wave, and the resultant amplitude at any point is the algebraic sum of the amplitudes of the two waves, as in Fig. 15.10.

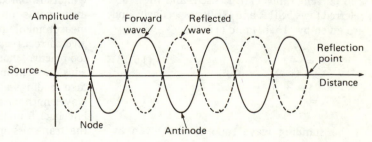

Fig. 15.10. Illustration of standing wave.

The nodes and antinodes do not move relative to the transmission line i.e. they are stationary, and the waves are called standing waves.

An important consideration in transmission line and antenna design is the standing wave ratio, and this is illustrated using Fig. 15.11,

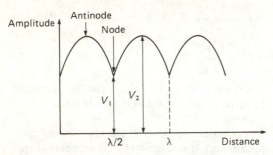

Fig. 15.11. Illustration of standing wave ratio (SWR = V_2/V_1).

which represents the standing wave on a line. If the line is open circuit or short circuit (load impedance being infinite or zero respectively), then all the forward wave is totally reflected back to the source and V_1 is zero. For an open circuit line the antinodes are located at half wavelengths along the line, starting back from the open circuit point. For a short circuited line the nodes are located at half wavelengths, starting back from the short circuit point. Therefore although open and short circuited lines give identical standing wave patterns, they are shifted relative to the positions of the nodes and antinodes.

In most systems some of the forward wave will be absorbed by the load, for example in the case of an antenna some of the forward energy is radiated into space. Therefore the reflected wave has a lower amplitude than the forward wave, resulting in a standing wave similar to that in Fig. 15.11. The standing wave ratio (SWR) is then given by the ratio of the maximum and minimum amplitudes of this wave. Both voltage and current ratios can be used, and these are referred to as VSWR and ISWR respectively, and are given by (15.13) and (15.14)

$$VSWR = \frac{V_2}{V_1} \qquad (15.13)$$

$$ISWR = \frac{I_2}{I_1} \qquad (15.14)$$

The standing wave ratio is also given by

(15.15) to (15.18) using impedance, power, or forward and reflected voltage as the parameters.

$$SWR = \frac{Z_c}{Z_L} \quad \text{(for } Z_c > Z_L) \qquad (15.15)$$

or

$$SWR = \frac{Z_L}{Z_c} \quad \text{(for } Z_L > Z_c) \qquad (15.16)$$

where Z_c is the characteristic impedance of the transmission line and Z_L is the characteristic impedance of the load.

$$SWR = \frac{1 + (P_R/P_F)^{1/2}}{1 - (P_R/P_F)^{1/2}} \qquad (15.17)$$

where P_F is the forward power from the source to the load and P_R is the power reflected back from the load.

$$SWR = \frac{1 + V_R/V_F}{1 - V_R/V_F} \qquad (15.18)$$

where V_F is the voltage component of the forward wave, and V_R is the voltage component of the reflected wave.

When the load impedance equals the characteristic impedance of the transmission line, the forward wave is totally absorbed in the load, and there is no reflected or standing wave. Now V_1 equals V_2 in Fig. 15.11 and the system is ideal, having a standing wave ratio defined as 1:1.

15.3.2 *Transmission line measurements*

The standing wave ratio (SWR) is one of the most fundamental parameters, as it checks for mismatch between the transmitter, transmission line and antenna. The SWR varies with frequency, since the impedance of the source, line and load are unlikely to change by the same amount with frequency; so the SWR should always be measured at several frequencies.

Any of the equations (15.13) to (15.18) can be used to find SWR, and meters are available which are based on all of these. Most low cost instruments use voltage or current as their measurement parameter. These instruments give correct readings only when the load is not too reactive, that is the SWR ratio being measured is low. The meters must also be connected to take readings at the load feed point, or at half wavelength distances along the transmission line, back from the load. This usually means that the transmission line needs to have a length which is an integral value of its electrical length, as given by (15.10) or (15.11).

SWR meters based on power measurements give results which are independent of length. A true r.f. wattmeter is used, fitted with a directional coupler so as to measure the forward and reflected power separately, and the SWR is calculated from (15.17).

The radiation resistance of an antenna can also be measured using a resistance bridge, and the SWR found from (15.15) or (15.16). Fig. 15.12 shows a Wheatstone bridge arrangement in which resistors R_B, R_B and R_x form three of the bridge arms, and the radiation resistance of the antenna forms the fourth arm. The supply to the bridge is a low power r.f. signal generator, which provides the excitation. The value of resistor R_x is chosen depending on the antenna

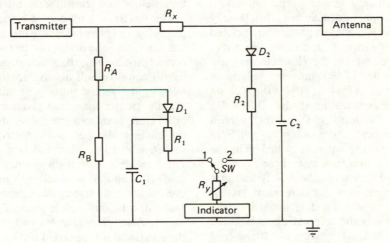

Fig. 15.12. An SWR meter arrangement based on a Wheatstone bridge.

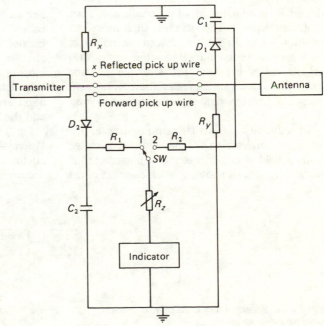

Fig. 15.13. An SWR meter arrangement based on forward and reflected wave measurements.

type, i.e. a 50 ohm or a 75 ohm system. When making measurements switch *SW* is first set to position 1 and the value of R_Y is adjusted to give full scale deflection on the indicator. Having calibrated the instrument, switch *SW* is moved to position 2, and the value of the SWR is read directly from the indicator.

Fig. 15.13 shows an alternative SWR meter construction. The pick up unit consists of three conductors, the centre conductor being the core of the transmission line and the outer two conductors are spaced equally from this. Usually this pick up unit is made as a pattern on a printed circuit board. Resistors R_x and R_y are selected to have values equal to the characteristic impedance of the line i.e. 50 ohms or 75 ohms. In operation any r.f. signal on one of the pick up wires will be rectified by diodes D_1 or D_2, smoothed by capacitors C_1 or C_2 and then applied as d.c. to the indicator. The forward signal, from the transmitter to the antenna, picked up by the reflected wire, is absorbed by resistor R_x, so no forward voltage appears on the pick up line after point *X*. However this signal remains on the forward pick up wire, and is applied to the indicator. Similarly the reflected signal is picked up by the reflected pick up wire, but is absorbed in resistor R_y of the forward pick up circuit.

When making measurements with the meter shown in Fig. 15.13 the switch is set to position 1 and R_Z is adjusted to give a reading at a set calibration point. The switch is then moved to position 2 and the SWR read from the indicator.

A good antenna has a standing wave ratio which varies between 1:1 and 1:1.2. Poor antennas have ratios greater than 1:1.5 and most SWR meters do not give a reading beyond 1:3.

Instruments are usually multipurpose, for example combining a standing wave ratio meter with a field strength meter and power output meter. Fig. 15.14 shows a block diagram of an impedance noise bridge, which is a low cost instrument for measuring several parameters on antennas and transmission lines. The noise source produces a wide spectrum of pseudogaussian noise. This is amplified and then goes through an isolator before being coupled to three outputs. The detector gives an indication of the received signal, and shows a high noise level at all frequencies, which are separate from the resonant frequency of the device under test (usually an antenna or a transmission line).

The electrical length of a transmission line differs from its physical length due to the velocity factor *F*. The noise bridge can be used to find the correct physical length, which gives an electrical length equal to half a wavelength. The transmission line should initially be cut to a length slightly greater than that given by (15.10) or (15.11). R_x is set to a low value, and one end of the cable under test is connected to the test socket, whilst the other end is short circuited. The detector is set to the frequency at which the cable is to be half a wavelength in length, and the frequency is slowly decreased until null is reached. Now the cable is gradually trimmed in stages. After each trim its end is again short circuited, and it is re-tested until eventually the null point occurs at the required frequency.

15.3.3 *Cable fault location*

Several methods are used for locating faults in cables, such as telephone wires or power transmission lines. Usually these faults consist of short circuits between the conductor and ground. Most of the instruments used are based on the Wheatstone bridge. Two methods of fault location are described here, the Murray-loop test and the Varley-loop test.

Fig. 15.15 illustrates the connection for the Murray-loop test. Suppose that the faulty conductor has a length l_a and that a fault has occurred at a distance I_c from one end. This

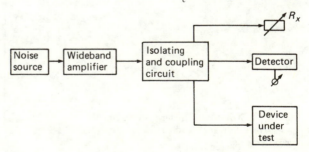

Fig. 15.14. Simplified block diagram of an impedance noise bridge.

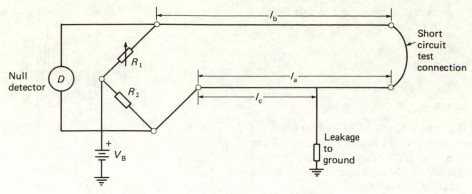

Fig. 15.15. Murray-loop test method for measuring faults in cables.

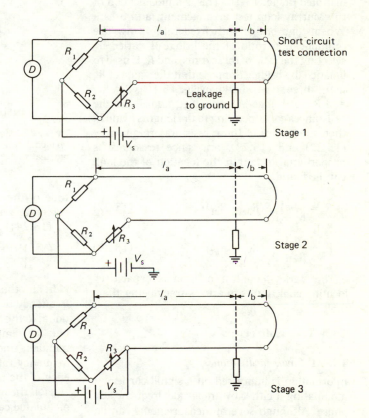

Fig. 15.16. Varley-loop test method for measuring faults in cables.

conductor, and an adjacent good conductor, are connected into the measurement bridge, as shown, and their far ends are short circuited together. The good conductor is assumed to have a length of l_b, although for a multicore cable lengths l_a and l_b will be equal. If R_x is the total resistance of the loop, due to $(l_a + l_b)$, and R_y is the resistance of the faulty conductor, from the bridge terminal, (due to l_c) then at bridge balance the value of R_y is given by (15.19).

$$R_y = \frac{R_2}{R_1 + R_2} \cdot R_x \qquad (15.19)$$

For conductors of uniform cross section resistance is proportional to length, so that (15.19)

can be re-written as in equation (15.20). Further
if $l_a = l_b = l$ the value of l_c is given by (15.21)

$$l_c = \frac{R_2}{R_1 + R_2}(l_a + l_b) \qquad (15.20)$$

$$l_c = \frac{2R_2 l}{R_1 + R_2} \qquad (15.21)$$

The Murray-loop method is capable of locating
low resistance ground faults accurately. If the
ground fault is high resistance then the low
voltage battery V_B is no longer suitable, and a
high voltage source is needed.

Three stages in the Varley-loop test are illustrated in Fig. 15.16. This is a modification of
the Murray-loop test arrangement, and uses a
Wheatstone bridge with fixed ratio arms. The
multiplication ratio of the bridge is varied by
selecting the ratio of these arms, and R_3 is used to
balance the bridge. Suppose that the value of R_3
in the three stages shown in Fig. 15.16 is R_{B1}, R_{B2}
and R_{B3} at bridge balance. If R_x and R_y are the
resistances corresponding to the lengths l_a and l_b,
then the values of these resistors are given by
(15.22) and (15.23), and since resistance is
proportional to length the location of the fault
can be found.

$$R_x = \frac{R_1}{R_1 + R_2}(R_{B2} - R_{B1}) \qquad (15.22)$$

$$R_y = \frac{R_1}{R_1 + R_2}(R_{B3} - R_{B2}) \qquad (15.23)$$

The Varley-loop test method is capable of
locating cable faults to an accuracy better than
0.5%.

15.4 Transmitter tests

15.4.1 *Single modulation*

Information is impressed on a signal carrier by
modulating it either by amplitude, frequency or
phase. Although several measurements can be
made on a transmitted signal the amount of
modulation is the most common. Transmitters
are allocated frequency channels, and the
modulation on a carrier is limited by the need to
prevent interference between adjacent channels.

Fig. 15.17 shows a carrier which is amplitude
modulated (a.m.), frequency modulated (f.m.)
and phase modulated (p.m.). In the a.m. wave
the audio modulating signal varies the ampli-

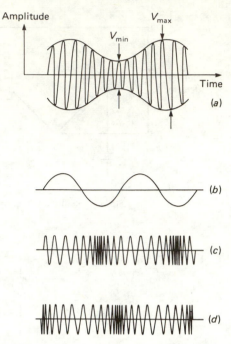

Fig. 15.17. Modulated carriers; (*a*) amplitude modulated,
(*b*) modulating signal, (*c*) frequency modulated, (*d*) phase
modulated.

tude of the carrier between V_{max} and V_{min}. The
percentage amplitude modulation (A_M) is given
by (15.24)

$$A_M = \frac{V_{max} - V_{min}}{V_{max} + V_{min}} \times 100 \qquad (15.24)$$

In f.m. the audio modulating signal varies the
frequency of the carrier, by an amount proportional to the amplitude of the modulating signal,
but the amplitude of the modulated carrier
remains unchanged. The modulated carrier
frequency rate is also proportional to the frequency of the modulating signal.

If f_D is the maximum frequency deviation of the
modulated carrier, and f_M is the frequency of the
modulating signal, then the modulation index M_I
for a f.m. signal is given by (15.25).

$$M_I = \frac{f_D}{f_M} \qquad (15.25)$$

The instantaneous phase deviation of the
modulated carrier, with respect to the phase of
the unmodulated carrier, is seen from Fig. 15.17
to be proportional to the amplitude of the

modulating signal. The peak phase deviation of the carrier, in radians, is the modulation index, and is given numerically by (15.25).

It should be noted that 100% modulation has some physical meaning in a.m. systems, and is obtained when $V_{min} = 0$. This is not the case in f.m. and p.m. systems when 100% modulation has to be defined by standardisation. For example f.m. broadcasting systems have standardised on ± 75 kHz as 100% modulation; for TV sound signals 100% modulation is defined as a deviation of ± 25 kHz; and for two-way radio 100% modulation is ± 5 kHz.

15.4.2 *Measurements with an oscilloscope*

Measurements can be made on a transmitted signal by observing its waveform on an oscilloscope. For example the waveforms shown in Fig. 15.17(a) will be obtained if the modulated carrier is fed into the vertical terminals of the oscilloscope, and the oscilloscope's internal circuit is used to give time along the horizontal axis. From these waveforms measurements can be made of V_{max}, V_{min}, etc. and (15.24) and (15.25) used to calculate the amount of modulation.

An alternative connection, for measurements on amplitude modulated waves, is to connect the modulated carrier signal to the vertical input of the oscilloscope, and the modulating signal to the horizontal input. This will give a display on the oscilloscope screen of the form shown in Fig. 15.18. Because this is linear it is easier to

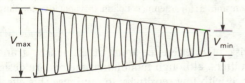

Fig. 15.18. Envelope of an amplitude modulated carrier plotted against the modulating signal.

measure V_{max} and V_{min}. The depth of modulation is again given by (15.24).

Oscilloscope measurements are not very accurate, especially at low modulation depths. Measurement techniques using spectrum analysis are preferred, as described in Section 15.4.3. Because of its high dynamic range, up to about 70 dB, a spectrum analyser on logarithmic display enables modulation depths as low as 0.06% to be measured.

15.4.3 *Measurements using a spectrum analyser*

If a single tone r.f. carrier of frequency f_c is amplitude modulated by a modulating signal of frequency f_m, then the resulting modulated signal consists of three r.f. signals; the original carrier f_c, an upper sideband of frequency $(f_c + f_m)$, and a lower sideband of frequency $(f_c - f_m)$. A spectrum analyser has the capability of breaking the modulated signal into its three frequency components, and displaying these as in Fig. 15.19(a). The amplitudes V_1 and V_2 can

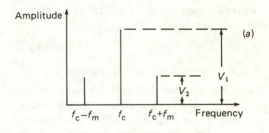

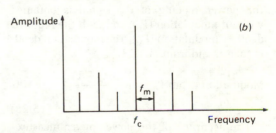

Fig. 15.19. Spectrum analyser plots of modulated signals; (a) amplitude modulated, (b) frequency modulated.

be measured from the spectrum analyser, and the percentage amplitude modulation of the original signal calculated from (15.26)

$$A_M = \frac{2V_2}{V_1} \times 100 \qquad (15.26)$$

This assumes linear modulation, where the sidebands are of equal amplitude and are spread equally from the carrier.

If a frequency modulated signal is displayed on a spectrum analyser, the resulting waveform will be of the form shown in Fig. 15.19(b). The amplitude of the sidebands can vary, although the number of sidebands is directly related to the

Table 15.1. *Relationship between number of sideband pairs and modulation index of a frequency modulated wave*

Number of sideband pairs	2	3	4	6	7	8	11	12	15	18	23
Modulation index I_M	0.5	1	2	3	4	5	7	8	11	14	18

modulation index I_M, and is given by Table 15.1. The spacing between the sidebands is equal to the frequency of the modulating signal. To find the maximum frequency deviation, the value of f_m is read from the spectrum analyser display. The modulation index I_M is determined by counting the number of sideband pairs, and then reading I_M from Table 15.1. The value of f_D is then calculated from (15.27)

$$f_D = I_M \cdot f_m \qquad (15.27)$$

15.4.4 *Measurements with a power meter*

Power meters can be used to measure the modulation depth of a signal. It is essential, however, to use a meter which is based on true power sensing, for example thermal methods, since instruments based on other techniques will result in errors. To measure modulation depth the power reading of the signal is obtained without modulation (P_0), and with the required depth of modulation (P_M). The modulation depth is then found from (15.28).

$$\text{Modulation depth (\%)} = \left[\frac{2(P_M - 1)}{P_0} \right]^{1/2} \times 100 \qquad (15.28)$$

The accuracy of the power meter measurement technique is limited by the resolution of the power meter, and this method is not normally used to measure modulation depths less than 50%. It also suffers from the disadvantage of needing a system for turning the modulation on and off, on the signal being measured.

15.4.5 *Modulation meters*

Probably the commonest method of measuring the modulation of a transmitted signal is to use a modulation meter. There are a variety of these instruments, varying from low cost meters, with limited modulation frequency measurement range, to the expensive microprocessor controlled meters. These can operate over a wide range of modulation frequencies, and can measure other parameters, apart from modulation, such as signal to noise ratio and distortion.

Connection. Modulation meters can be connected directly to low power transmitters only, with power levels less than 1 W. Above this value the input circuit of the meter may be damaged, and attenuators, usually built into the modulation meter, must be used. For signal strengths above about 10 W attenuators can become expensive, and for these power levels it is usual to couple the modulation meter to the transmitter via an antenna, as in Fig. 15.20(a). There is now the

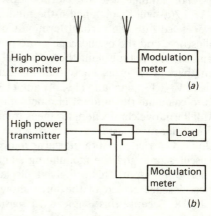

Fig. 15.20. Coupling a modulation meter to a high power transmitter; (a) antenna coupling, (b) capacitive coupling.

risk of interference from other nearby transmitters, although modulation meters are designed with low sensitivity to minimise this effect. Interference is worse with auto tuned meters, since with manual tuning the required transmitter can be carefully selected. Unfortunately modulation meters have a broadband input capability, so interference can still result from adjacent channels.

Capacitive coupling can also be used to connect a modulation meter to a high power load, as in Fig. 15.20(b). This can be used to make measurements on transmitters up to about 1 kW, and results in an isolation between the transmitter and the meter of about 50 dB at 20 MHz.

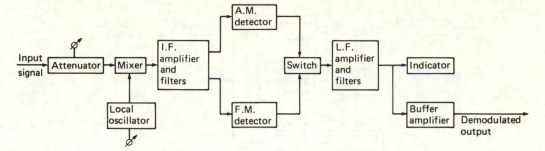

Fig. 15.21. Block diagram of a modulation meter.

Construction. Fig. 15.21 shows a simplified block diagram of a modulation meter. The signal to be measured is passed through a variable attenuator. This protects the input circuitry of the meter against high level signals, and also provides a mixer design with good linearity and low noise, since it does not need to cope with a wide range of signal levels, and so does not require a high dynamic range.

The input signal is mixed with a local oscillator in a mixer stage. An important requirement of a modulation meter is that its local oscillator must have low noise, since this noise will be transferred to the output of the mixer, resulting in errors. The i.f. from the mixer is amplified. When measuring wide bandwidth signals, such as in stereo broadcasting, i.f. filters are not used. For a.m. signals it is usual to insert filters after the amplification stage which limit the frequency to 10 kHz, and even narrower bandbass filters are used in measurements on systems such as mobile radio.

The i.f. signal can be demodulated by operator selectable a.m. or f.m. detectors. Phase modulation is usually derived from the f.m. signal by integration. The output from the detector consists of a signal with an amplitude proportional to the amount of modulation on the signal. This l.f. signal is amplified, and a buffered output is provided for input to other instruments. The signal is also measured (usually peak value), and the modulation shown on the indicator. Low frequency filters may be cut in after the l.f. amplification stage, to reduce the l.f. bandwidth, such as would be needed to minimise the effects of noise.

Automatic Tuning. The modulation meter can be tuned by manually varying the frequency of

the local oscillator, as in Fig. 15.21, until the required value of i.f. frequency is obtained. In order to do this the operator needs to have a knowledge of the approximate value of the input signal frequency, and the tuning method is also very slow.

Modern modulation meters use automatic tuning systems. These can reduce the tuning time down to one to two seconds, which is a fraction of the manual tuning time. The expensive meters use a frequency synthesizer as the local oscillator, and its frequency is controlled by a microprocessor. The instrument can still be tuned manually, to any required frequency, by entering this value via a keyboard. In the automatic tuning mode the microprocessor adjusts the local oscillator frequency until a signal is detected in the i.f. stage. Based on this reading the microprocessor calculates the value of the input signal frequency and sets the local oscillator to the optimum value.

Lower cost instruments use a form of sweep tuning, as shown in Fig. 15.22. The sweep control sweeps the frequency of the local oscillator until a signal of sufficient strength is detected at the output of the mixer. This stops the frequency sweeping so that the local oscillator frequency is fixed. The automatic frequency control circuit is then operational, and fine tunes the local oscillator frequency to give an optimum output from the mixer.

Extended frequency range. Fig. 15.23 shows a method of extending the frequency range of a modulation meter, beyond that obtained from its local oscillator. An external mixer and oscillator are used, and the external oscillator is tuned to give an output from its mixer, which is within the frequency range of the modulation meter.

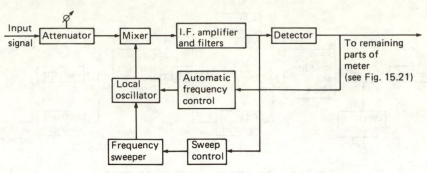

Fig. 15.22. Part of a modulation meter showing automatic tuning.

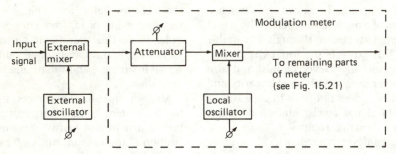

Fig. 15.23. Part of a modulation meter showing the method for extending its frequency range.

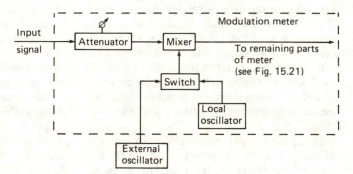

Fig. 15.24. Part of a modulation meter showing the connection of an external oscillator.

Errors. Errors occur due to several reasons when using modulation meters. Noise is the most important since it gives false readings, and limits the dynamic range of the instrument. The main causes of noise errors are noise on the input signal, noise due to detector operation, and noise in the local oscillator. Choosing the narrowest l.f. filter possible minimises the effect of noise. Ultimately, however, it is the noise in the local oscillator which limits the performance of the modulation meter.

Fig. 15.24 shows a method in which an external low noise oscillator can be connected to reduce the errors, caused by noise, in the modulation meter. The internal local oscillator is switched off. If W_i is the noise level of the instrument and W_R is the required measurement noise level, then the permitted noise level of the external oscillator is given by (15.29), where N is an integer.

$$W_o < \frac{W_R - W_i}{N} \qquad (15.29)$$

In order to reduce the stringent noise requirements placed on the external oscillator, N should be chosen to be a low value. If f_s is the signal frequency and f_i is the i.f. frequency, then the frequency of the external oscillator is given by (15.30), where N is the integer used in (15.29).

$$f_o = \frac{f_s + f_i}{N} \tag{15.30}$$

The residual a.m. noise in a modulation meter is due to pick up of stray signals by the audio wiring; by power supply hum; and by thermal noise in amplifier stages. It results in a variation of the amplitude of the r.f. signal envelope, when the a.m. or f.m. input is removed from the transmitter. For an a.m. signal, residual noise can be measured by noting the modulation meter reading, when the transmitter modulation is at 0% and at 100%, and expressing the ratio in decibels. For an f.m. signal the residual a.m., with no f.m. signal, is the ratio between the peak value of the signal envelope and the average detected d.c. voltage.

The residual f.m. noise is the ratio of the output from the meter for full system deviation and for no audio input. For an a.m. transmitter the reference is the f.m. detector output, with an f.m. signal having a known deviation, such as 1 kHz, and the f.m. detector output is measured with and without a.m.

An a.m. signal will contain some f.m. and an f.m. signal will have some a.m., and this is referred to as incidental modulation. The modulation meter will give a reading which is equal to the sum of the two modulation effects, and this will result in an error. An a.m. source produces f.m. due to the variation in capacitive effects within the modulator or the amplifier stages, which results in phase variations in the signal. An f.m. source produces a.m. when filters are used which do not have a flat frequency response, resulting in a.m. effects.

Because there is no fixed phase relationship between the main signal and the incidental modulation components, it is difficult to compensate for this. Most modulation meters specify the level of incidental modulation which can occur, and this should be taken into account when making measurements. Section 15.4.6 shows

how incidental modulation can be measured, if required.

A modulation meter will also give a false reading if there is more than one modulated input signal, since they will interfere with each other. In the worst case the meter can even lock onto the wrong signal, especially if auto tuning is used. It is therefore important to reduce the chance of multiple input signals, by coupling the modulation meter effectively to the required signal source.

The input circuit of a modulation meter is not tuned, so it will respond to harmonics in the input signal. These harmonics will be mixed with the local oscillator frequency, and harmonics of the local oscillator frequency. This will result in several outputs from the mixer, many of which may fall within the bandwidth of the i.f. stage, and so will be detected, resulting in errors. This can be minimised by using filters which stop the harmonics in the signal from reaching the modulation meter, and by using an external oscillator with low harmonic content.

15.4.6 *Other measurements with a modulation meter*

Usually a modulation meter has a built in capability to make several types of measurement, apart from modulation. A few of these are described here, with the basic modulation meter used in the measurement circuit.

Signal to noise ratio. The arrangement for measuring this is shown in Fig. 15.25. The demodulated signal from the modulation meter is passed through a band pass filter, and then, for f.m., through a de-emphasis circuit. This has values of about $75 \mu s$ for broadcast f.m. and $750 \mu s$ for narrowband radio. Two common forms of noise weighting networks are in use. For noise measurements on narrowband radio the telephone psophometric weighting filter (CCITT) is used, and for f.m. broadcast systems the broadcast psophometric weighting network (CCIR) is used. To make a measurement the signal is modulated with the test frequency, and

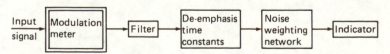

Fig. 15.25. Connection of a modulation meter to measure signal to noise ratio.

the indicator is set to read zero dB. The test modulation is now removed and a second reading taken, which gives the residual modulation due to noise etc., and from this the signal to noise ratio is obtained.

Incidental modulation.   As mentioned in Section 15.4.5 incidental modulation is the unwanted component of modulation, that is a.m. on an f.m. signal or f.m. on an a.m. signal. It can be measured as for normal modulation, but now an a.m. or f.m. signal generator is used to set the reference level.

Fig. 15.26 shows the test set up for measuring

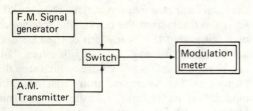

Fig. 15.26. Connection of a modulation meter to measure incidental f.m. on an a.m. transmitter.

incidental f.m. modulation on an a.m. transmitter. The f.m. signal generator is used as the reference, and this sets the zero dB level on the meter. The meter is then switched to the a.m. transmitter, and a reading can be taken of the incidental f.m. modulation on this source.

Readings of incidental modulation are affected by the characteristics of the modulation meter. For example some a.m. is produced on an f.m. signal when it passes through the filters within the meter.

Residual modulation.   A test modulation signal, consisting of a deviation of about 10 kHz for f.m. and a modulation depth of 30% for a.m., is applied to the meter, and it is set for a reading of zero dB. The modulation is then switched off and the residual modulation is read, which is usually expressed as an equivalent f.m. deviation or a.m. depth.

Distortion.   This can be measured by passing the demodulated output from a modulation meter into a distortion meter. Sometimes the distortion meter functions are built into the case of the modulation meter, to form one instrument.

Frequency response.   Measurement of frequency response, of a transmitter, is made by varying the modulation frequency and noting the effect on the modulation, as measured by a modulation meter. It is important to choose the l.f. filter carefully, so that the readings are not affected by its characteristics. When making measurements on an f.m. transmitter it is usual to switch in an appropriate de-emphasis circuit, which corresponds to the pre-emphasis introduced by the transmitter.

16. Digital circuit analysis

16.1 Introduction

This chapter describes the equipment used to analyse digital circuits. These cover the range from the simple logic pulsers and probes, which are used for hardware testing one node at a time, through to the more comprehensive testers such as signature analysers, logic analysers and development systems for software based products.

16.2 Probes, pulsers and clips

A logic probe is a hand held device, very similar in shape to a pencil with a very fine point. It detects the voltage level at any point of a circuit to which it is connected. Fig. 16.1 shows the block diagram of a probe. A switch on the probe can be set to measure different logic levels, such as for TTL or CMOS. Usually the maximum voltage is 20 V and input protection circuitry is used to shield the probe from high voltage or current levels.

The probe detects whether the input voltage is above or below the threshold levels for the logic family chosen. Some probes have a single indicator light, and this is on or off for the two logic levels. 'Bad' logic levels cause the light to be at half brightness. Some probes use multicolour indicators to show the different logic states. A pulse stretcher is usually also available which can stretch pulses of down to about 10 ns, so that they can be indicated. Pulse trains up to 50 MHz are also stretched to give an indicator blink at about 10 Hz.

Logic clips are designed to fit over integrated circuits, and they have indicators for each pin of the device. The logic clip draws power from the pin supply, and it contains its own gating logic for finding the supply pins automatically.

A logic pulser is a pulse generator, which is also shaped as a hand held pen with a fine point. It is capable of inserting voltage and current signals at any node of the circuit, and is usually used in conjunction with logic probes and logic clips. The pulser is capable of producing a high current output, greater than half an ampere, and this is sufficient to override existing logic levels at the nodes to which it is applied. Because the current pulse is applied for less than 300 ns its energy is low, and it does not damage any circuit components. Controls on the body of the pulser enable it to operate in several modes, such as single shot, pulse train or pulse burst.

The logic current tracer is similar to the logic probe in appearance, but it senses pulsing current flow in a circuit, rather than voltage levels. Its narrow tip senses the magnetic field caused by the current, and it can cover a typical range of 1 mA–1 A. An indicator light turns on when current flow is detected. The device can be used to sense current in multilayer boards, and its sensitivity is adjusted by a control on its side.

Another device, which is mainly used to detect faulty components in a logic circuit, is known as a logic comparator, shown in Fig. 16.2. It works on the principle of applying an identical series of stimuli to a reference component and the compo-

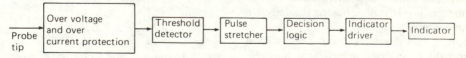

Fig. 16.1. Block diagram of a typical logic probe.

Fig. 16.2. Block diagram of a logic comparator.

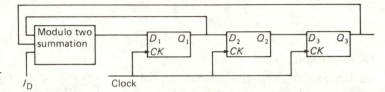

Fig. 16.3. A three bit pseudo random binary sequence generator.

nent under test, and indicating a fault if the output response from both is not identical. The reference device is usually plugged into the body of the logic comparator, and a probe clip is available for fixing over the device under test.

16.3 Signature analysis

Signature analysis techniques, and signature analyser equipment, were mainly developed for use in the repair of faulty equipment, especially in the field. When trouble shooting an analogue system the engineer usually has a circuit diagram, on which the voltage levels and waveforms at various nodes are indicated. By tracing through the equipment, and checking the nodes, the source of the fault can be located and repaired.

In digital circuits the signal at the nodes consists of logic ones and zeros. The data stream at the test points can be very complex and the faults may be caused by nodes being stuck at logic one or logic zero, and by timing errors. Analysis can now be done by applying a suitable stimulus to the circuit, and then recording the data output in a convenient form. Signature analysers usually convert the bit stream at the different nodes into a small number of hexadecimal digits, which forms the 'signature' of the circuit at that node. If the circuit diagram is annotated with the signatures expected from a good circuit, then the equipment nodes can be tested in turn, until the node having the defective signature is found.

Signature analysis should not be confused with transition analysis, which is a long established method of debugging faulty digital equipment. In this latter method the changes in the logic states at the nodes are counted for a fixed time period and then compared to the value expected from a good circuit. It will be seen in Section 16.3.2 that signature analysis is much more effective in detecting faults than transition counting.

16.3.1 *Principle of signature analysis*

The signature analysis technique is based on the use of cyclic redundancy check (CRC) code produced by a pseudo random binary sequence generator. A pseudo random binary sequence is a sequence of logic ones and zeros which look random, but which in practice repeat after a period. A pseudo random binary generator having a bit size of q will produce a sequence of $(2^q - 1)$ bits before repeating. These will cover all the possible states with q bits, except all zeros.

Fig. 16.3 shows a three bit pseudo random sequence generator. The modulo two summation gate produces an output of logic zero only when the modulo two sum of all its inputs is zero. Fig. 16.4 shows the operation of this circuit,

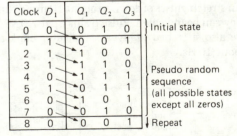

Clock	D_1	Q_1	Q_2	Q_3	
0	0	0	1	0	Initial state
1	1	0	0	1	
2	1	1	0	0	
3	1	1	1	0	
4	0	1	1	1	Pseudo random
5	1	0	1	1	sequence
6	0	1	0	1	(all possible states
7	0	0	1	0	except all zeros)
8	0	0	0	1	Repeat

Fig. 16.4. Sequence table for the pseudo random binary sequence generator of Fig. 16.3, with no data input.

assuming that the input data I_D is not connected. The initial state is assumed to be $Q_1 = 0$, $Q_2 = 1$ and $Q_3 = 0$. There is feedback from Q_1 and Q_3 to the modulo two summation gate, and since both its inputs are zero, its output at D_1 is also zero. Therefore on the first clock pulse Q_1 goes to zero and the states all shift right.

At the end of the first clock pulse one of the inputs to the modulo two summation gate is at logic one, so the output at D_1 is also one. Therefore on the second clock pulse this logic one is fed through to Q_1. The sequence proceeds until after the third clock pulse when Q_1, Q_2 and Q_3 are all at logic one. Now the output from the modulo two summation gate is logic zero and this is clocked through to Q_1 on the fourth clock.

It is seen by examining Fig. 16.4 that a sequence of $(2^3 - 1)$ or 7 patterns is produced, and that these are random in appearance, but repeat after every 7 clock pulses.

If a stream of input data I_D is applied to the pseudo random binary sequence generator of Fig. 16.3, then the data sequence is modified depending on the characteristics of the input data. Suppose that the input data consists of a repetitive bit stream 0101111000. Its effect on the pseudo random binary sequence generator is shown in Fig. 16.5. In the initial state the data

Clock	I_D	D_1	Q_1	Q_2	Q_3	
0	0	0	0	1	0	} Initial state
1	1	0	0	0	1	
2	0	0	0	0	0	
3	1	1	0	0	0	
4	1	0	1	0	0	Pseudo random
5	1	1	0	1	0	sequence modified
6	1	1	1	0	1	by input data
7	0	1	1	1	0	
8	0	0	1	1	1	
9	0	1	0	1	1	
—	—	—	1	0	1	} Remainder (signature of input data)

Fig. 16.5. Sequence table for the pseudo random binary sequence generator of Fig. 16.3 with data input sequence (I_D) as shown.

input I_D and feedbacks Q_1 and Q_3 are all at logic zero, so D_1 is also at zero. This is clocked through to Q_1 on the first clock pulse. Now Q_3 goes to logic one, but since the input data I_D is also at one the output at D_1 is at logic zero. This is now fed through to Q_1 on the second clock pulse, which is a different sequence to that of Fig. 16.4. In fact

checking the sequences of Figs. 16.4 and 16.5 they are seen to be totally different. The value remaining in the registers Q_1, Q_2 and Q_3 after a specified number of clock pulses, in the present instance nine pulses, is called the signature of the input data bit stream. It can be shown that a single bit change in this data will result in a very different signature, which can therefore be used to measure the performance of the circuit.

The data in signature analysis can be of any length, although the signature length will be determined by the number of bits in the shift register. The data can also be in true or complement state. It is important that, to get consistent signatures on each sample, the system under test is doing the same operation on every sample. This is usually done by using the system under test to provide its own stimulus, and by feeding the system clock into the logic analyser for synchronization.

16.3.2 *Error detection*

There are two primary requirements in signature analysis; stimulating the circuit under test to produce a data stream, and compressing this data stream so as to form a signature for the node being measured. As mentioned in the previous section two main techniques have been used to compress the data stream, transition counting and using a linear feedback shift register to generate a pseudo random binary sequence. Both these methods are considered further here. The test of any method is how well it detects an error in the data bit stream.

If a data sequence has a bit length of p, and it is fed into a linear shift register with feedback of length q, then the percentage probability P_E of detecting an error in the data sequence is given by (16.1).

$$P_E = [1 - K(p-q)(2^{p-q} - 1)(2^p - 1)^{-1}] \times 100$$
$$(16.1)$$

In this equation K is a constant which makes P_E equal to 100% when $p < q$, since in this instance all the data is stored in the shift register and an error will be detected.

If only one error occurs in the input data sequence then it will always be detected, i.e. $P_E = 100\%$, since there is never another error bit to cancel the feedback generator. If more than one error occurs then there is a possibility that a later error will cancel out the effect of an earlier one,

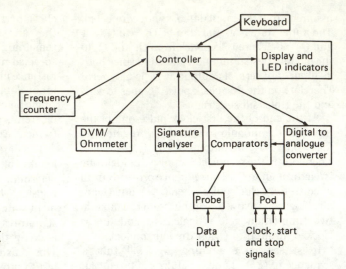

Fig. 16.6 Block diagram of a signature analyser incorporating a DMM and a frequency counter.

so both errors go undetected. Under these circumstances the probability of detecting multiple errors is given by (16.2), and is seen to be dependent on the length of the shift register, but independent of the length of the data stream.

$$P_E = (1 - 2^{-q}) \times 100 \qquad (16.2)$$

Therefore, for example, if a 16 bit shift register is used the probability of detecting multiple errors in a data stream is $(1 - 2^{-16}) \times 100 = 99.998\%$, which is high.

For transition counting, assuming r transitions, the percentage probability of detecting an error in the data sequence is given by (16.3).

$$P_E = \frac{1 - \sum_{r=0}^{p}\left[\dfrac{p!}{(p-r)!r!}\right]\left[\dfrac{p!}{(p-r)!r!} - 1\right]}{2^p(2^p - 1)} \times 100$$

$$(16.3)$$

Now for single bit errors, the probability of detecting the error is given by (16.4)

$$P_E = [1 - (p - 1)(2p)^{-1}] \times 100 \qquad (16.4)$$

Therefore for long data sequences there is only a 50% probability that a single bit error will be detected. This compares with 100% probability of detecting single bit errors, for signature analysis using a pseudo random binary sequence generator.

For multiple bit errors the percentage probability of detecting an error with transition counting varies with the length of the data sequence p, and is found from (16.4) to vary from about 80%

for $p = 5$ to 95% for $p = 100$. Therefore again the signature analysis technique is superior in detecting data errors.

16.3.3 *Signature analysers*

A signature analyser is an instrument which is primarily intended for testing and trouble shooting faulty digital equipment. Most modern instruments incorporate other features to avoid the need for multi instruments. For example the Hewlett Packard H.P. 5005A is a signature analyser with a $4\frac{1}{2}$ digit DMM and a 50 MHz counter timer, all working via the same probe. It is therefore possible to check analogue circuits and asynchronous parts, which a signature analyser cannot test on its own.

Fig. 16.6 shows the block diagram of a typical multi purpose signature analyser and Fig. 16.7 shows its waveforms in data sample mode. The input data stream is fed via an active probe into the comparator. The high and low threshold setting on this can be programmed by the central controller using the digital to analogue converter. This comparator technique digitises the probe readings, and results in a fast set up time. The threshold levels for the comparator can be varied between ± 12.5 V in 50 mV steps, and can be preset to several levels, for example to test TTL, CMOS and ECL circuits.

A single pod also connects into the circuit under test to pick up a clock, and a start and stop pulse which provides the measurement window. Since the same clock is used for the analyser and for the system under test, synchronism is en-

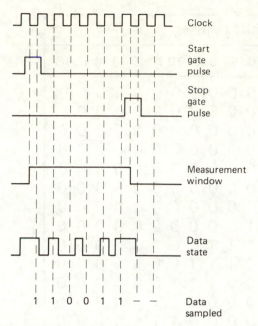

Fig. 16.7. Illustration of data sampling in a logic analyser.

sured. The active edge of the clock pulse can be selected from the instrument front panel; Fig. 16.7 shows an active rising edge. The clock samples the input data stream at each active edge, and data changes in between clock pulses are ignored.

The maximum clock frequency for a typical instrument, such as the H.P. 5005A is 25 MHz with a minimum pulse width of 15 ns in the high or low state. The minimum gate length is one clock cycle, i.e. one data bit between the start and stop pulses. There is no maximum gate length. The data probe has a set up time, which is the time that the data needs to be present and steady before arrival of the selected gate pulse, of 10 ns. The time for which the data needs to remain steady after occurrence of the clock edge, called

the hold time, is zero. The start and stop pulses have a set up time of 20 ns and also no hold time.

A hold feature on the signature analyser allows one shot signatures to be noted, such as for power-on transients. This will display the signature of the first window only, and will hold this until the reset button is pressed. A flashing 'gate' light on the instrument front panel indicates detection of valid start, stop and clock conditions. If there is a difference between two successive signature readings then a flashing 'unstable' light comes on, and this indicates a possible intermittent fault.

The H.P. 5005A uses a 16 bit feedback shift register to generate the pseudo random binary sequence, as in Fig. 16.8. The tap points on the shift register are chosen to scatter the missed errors as much as possible. Evenly tapped points, for example at 4 or 8 bits, are avoided since most bus oriented systems, based on microprocessors, tend to repeat patterns at 4 or 8 bit intervals.

Fig. 16.9 shows the sequence table for the 16 bit generator of Fig. 16.8 when the input data sequence is 1110000111000011011. Its operation is similar to that shown in Fig. 16.5. Note the dissimilarity between the input bit sequence, and the signature left after the 20th clock pulse. This is displayed using a non standard hexadecimal code shown in Fig. 16.10, and forms the signature of the input data sequence.

The DVM/Ohmmeter mode can be selected in Fig. 16.6 by I/O signals from the controller. The ohmmeter has several facilities, such as displaying 'open' when the circuit is open circuit, rather than giving an overload display as on most meters. The controller automatically switches to higher ranges when the maximum of any range is reached, and it goes through a self calibration check every ten readings.

The frequency counter can be used for frequency measurements up to 50 MHz, and for time interval measurements with up to 100 ns reso-

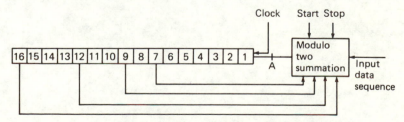

Fig. 16.8. Sixteen bit pseudo random binary sequence generator.

Clock	Input data	Input at A	Shift register															
			1	2	3	4	5	6	7	8	9	10	11	12	13	14	15	16
0	1	1	0	0	0	0	0	0	0	0	0	0	0	0	0	0	0	0
1	1	1	1	0	0	0	0	0	0	0	0	0	0	0	0	0	0	0
2	1	1	1	1	0	0	0	0	0	0	0	0	0	0	0	0	0	0
3	0	0	1	1	1	0	0	0	0	0	0	0	0	0	0	0	0	0
4	0	0	0	1	1	1	0	0	0	0	0	0	0	0	0	0	0	0
5	0	0	0	0	1	1	1	0	0	0	0	0	0	0	0	0	0	0
6	0	0	0	0	0	1	1	1	0	0	0	0	0	0	0	0	0	0
7	1	0	0	0	0	0	1	1	1	0	0	0	0	0	0	0	0	0
8	1	0	0	0	0	0	0	1	1	1	0	0	0	0	0	0	0	0
9	1	1	0	0	0	0	0	0	1	1	1	0	0	0	0	0	0	0
10	0	1	1	0	0	0	0	0	0	1	1	1	0	0	0	0	0	0
11	0	1	1	1	0	0	0	0	0	0	1	1	1	0	0	0	0	0
12	0	1	1	1	1	0	0	0	0	0	0	1	1	1	0	0	0	0
13	0	1	1	1	1	1	0	0	0	0	0	0	1	1	1	0	0	0
14	1	0	1	1	1	1	1	0	0	0	0	0	0	1	1	1	0	0
15	1	1	0	1	1	1	1	1	0	0	0	0	0	0	1	1	1	0
16	1	1	1	0	1	1	1	1	1	0	0	0	0	0	0	1	1	1
17	0	0	1	1	0	1	1	1	1	1	0	0	0	0	0	0	1	1
18	1	0	0	1	1	0	1	1	1	1	1	0	0	0	0	0	0	1
19	1	1	0	0	1	1	0	1	1	1	1	1	0	0	0	0	0	0
20			1	0	0	1	1	0	1	1	1	1	1	0	0	0	0	0

9   *H*   7   0   Display reading

Fig. 16.9. Sequence table for the sixteen bit pseudo random binary sequence generator of Fig. 16.8.

Bit position				Code
1	2	3	4	
0	0	0	0	0
1	0	0	0	1
0	1	0	0	2
1	1	0	0	3
0	0	1	0	4
1	0	1	0	5
0	1	1	0	6
1	1	1	0	7
Q	0	0	1	8
1	0	0	1	9
0	1	0	1	A
1	1	0	1	C
0	0	1	1	F
1	0	1	1	H
0	1	1	1	P
1	1	1	1	U

Fig. 16.10. Hexadecimal display code used in the HP5005A signature analyser.

lution. It can be used to test one short circuits by making time interval measurements. Asynchronous circuits can also be tested by totalizing the pulses between start and stop states. The totalizer can even be used as a form of transition counter.

16.3.4 *Measurement with signature analysers*

When making measurements on a circuit using signature analysers a large number of bit sequences, greater than 20, should be used. It is also often necessary to design the system such that feedback paths within circuits can be broken easily, when testing, to stop faulty sequences being fed back and affecting good nodes. It is then possible to trace the fault back to the defective component.

Microprocessor based systems are especially suitable for testing using signature analysis. The circuits are designed with a small part of the program memory containing a special diagnostic program, which can be activated to simulate nodes. This causes bit sequence patterns to be generated at various nodes, whose signature can be measured and compared to that of a good circuit.

The microprocessor itself can also be tested by isolating it from its memory, usually by removing test jumpers in the circuit. The microprocessor is then made to free run by its data lines to logic zero or one states, so as to give an operation such as CLR (clear) or NOP (no operation). The processor now reads the same data at each address, and so free runs around all possible memory addresses. The signatures for the processor bus in this condition can be verified against the expected signature for a good device.

16.4 Logic analysers

The instruments described in earlier sections, logic probes and signature analysers, are primarily used to test electronic circuit hardware. Logic analysers represent a step increase in complexity from signature analysers. Logic analysers were initially primarily intended to test hardware, but with the shift towards microprocessor systems the emphasis is now on using logic analysers for testing software. The change in emphasis from hardware to software has also resulted in a modification of some of the features of a logic analyser. For example hardware analysers need to run at relatively high speeds, and sampling rates in excess of 500 MHz are often required. However most microprocessor systems operate at speeds below 20 MHz, so for software analysis lower logic analyser speeds are adequate.

Many of todays systems are based on bus structures, and for effective testing it is necessary to monitor several lines simultaneously. For example an 8 bit microprocessor would probably have 16 address, 8 data and 8 control lines, requiring an analyser with 32 inputs. This would be increased to 48 lines for a 16 bit microprocessor. A conventional oscilloscope does not have as many input channels as a logic analyser; several analysers provide up to 72 input channels.

Analogue signals are usually repetitive, and can be shown as a trace on an oscilloscope. Many digital states only occur once during the execution of a microprocessor program. In order to observe these they need to be captured or stored in digital form in memory, and then displayed either as a data stream or as an analogue waveform. Because the signals are all digital the output does not indicate any amplitudes, only logic zero and logic one states.

The size of the memory available with a logic analyser is an important feature, since with multichannel operation, and a high sampling rate, even a large memory will soon be full. Selective triggering is commonly used to conserve memory, by storing data for only that portion of the operation which is of direct

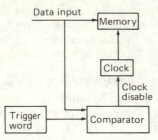

Fig. 16.11. Simplified representation of the trigger function within a logic analyser.

interest. Fig. 16.11 shows in simplified form how this works. A trigger word is programmed via the operator panel into the logic analyser. The data input is shifted through the analyser's memory, and is compared with the trigger word. When the data input matches the trigger word, recording of data occurs. Usually this is done by disabling the memory clock after a programmable delay called the trigger delay.

Logic analysers are complex instruments and it is important that they are designed such that they can be easily used. Usually this is done by setting up the instrument's functions using menus and prompts which appear on the screen. One manufacturer has also provided his instrument with a help button. This can be used to call up onto the screen the entire operating manual for the instrument, which is stored in an internal read only memory.

16.4.1 *Logic analyser construction*

A basic block diagram of a typical logic analyser is shown in Fig. 16.12. The large number of input lines are usually organised into groups onto data pods. The design of the pod is important. Because of the large number of lines involved the pods must be easy to use. Usually they fix onto the body of the integrated circuit (microprocessor) and pick up the signals from its pins. It is also important to avoid ringing and crosstalk, and to have good bandwidth and sensitivity. This is especially important when operating at high clock rates above about 100 MHz. Usually most probes used with logic analysers have a field effect transistor input stage, with an isolating resistor of 1 MΩ and a balancing capacitance of 5 pF. Lower values of resistance and capacitance may be used for higher operating frequencies. Other problems which need to be avoided include skewing, due to different propagation delays of the probe connections.

Usually each pod has some memory associated with it, which communicates with the main back up memory. Most logic analyser manufacturers also provide a family of probes for different microprocessors. These probes plug into the microprocessor socket of the unit under test, and contain special circuits which format the information on the data bus to a convenient form. This allows additional facilities, such as the display of software instructions in mnemonics, so as to make it easier for debugging.

The data inputs to the pods are compared against a threshold level which sets the logic zero and logic one states. These levels are controlled from the instrument's front panel and it is

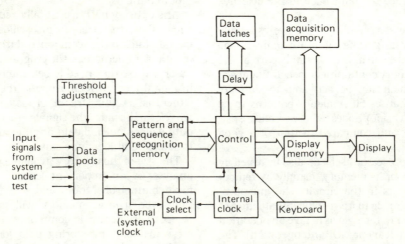

Fig. 16.12. Basic block diagram of a logic analyser.

important that they match the levels which are expected from the type of logic family under test. After passing through the pods the data is compared with the trigger information stored in the pattern and sequence recognition memory. As will be seen in the next section the trigger word can consist of a complex combination of many patterns.

The logic analyser may be operated either on its internal clock or on the clock obtained from the system under test via a data pod. Often the input may consist of many external clocks, all clustered onto one pod. These can be used in any number of logical combinations, and either the rising or falling edge of the clock may be programmed as the active edge. It will be explained in Section 16.4.2 that both the external and internal clocks are used under different operating mode conditions.

The input to set up the logic analyser is usually via keys, and many instruments use a full ASCII keyboard. Touch sensitive keys are also used on the screen to select items such as menus, and soft keys, programmed with short routines, help in making the operator interface easier. The control section, usually based on microprocessors, co-ordinates the overall operation of the instrument, storing the required data in memory and displaying it in the required format on the display screen, which is usually a cathode ray tube.

16.4.2 *Operation of a logic analyser*

In this section some of the basic operating features of a logic analyser are introduced. The next section gives simple examples showing the use of the logic analyser.

There are two operating modes for logic analysers, asynchronous and synchronous. In the asynchronous sampling mode the internal clock of the logic analyser is used to sample and transfer the input data to memory. The frequency of the sampling clock should be at least 4–10 times greater than that of the signal being sampled. The output from the asynchronous sample mode is usually shown as *timing diagrams*.

Fig. 16.13 shows a simple example, the timing diagram of a 4 bit binary counter. The counter operates on the falling edge of its system clock. If the asynchronous clock operates at a high frequency then in the asynchronous mode the time difference between sections of the display can be found by showing them on an expanded scale, as in Fig. 16.14, and noting the time difference based on the asynchronous clock frequency. It should be noted that since the clock and the data are independent there can be a difference of ± 1 clock cycle between successive samples.

In the synchronous sampling mode the clock

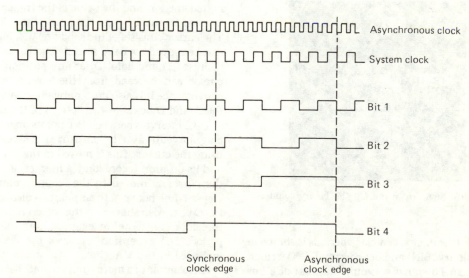

Fig. 16.13. Logic waveforms for a four bit binary counter.

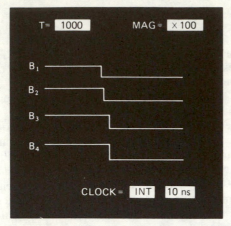

Fig. 16.14. Expanded trace of the four bit binary counter of Fig. 16.13.

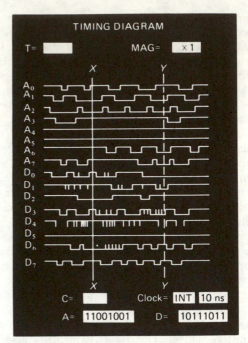

Fig. 16.16. A timing diagram.

of the system under test is used as the logic analyser clock. The objective now is to store as many logic states as possible, so a large memory is important rather than a high speed clock. The recording is made by the clock edge at which the circuit is not triggered, which would be the rising edge of the clock in Fig. 16.13. This allows for system set up and hold times. Since logic states are of prime interest in this mode of operation the timing diagram is not the most appropriate method of displaying the results. A *state list* is preferred, as shown in Fig. 16.15. The state of

CURS	BIN	HX
TRIG	0000	00
501	0001	01
502	0010	02
503	0011	03
504	0100	04
505	0101	05
506	0110	06
507	0111	07

T= 500

Fig. 16.15. State list of the four bit binary counter of Fig. 16.13.

the counter after seven clock periods is shown in both binary (BIN) and hexadecimal (HX) format.

Fig. 16.16 shows a timing diagram of a bus structured system having 8 address and 8 data

lines. Logic analysers have the capability of displaying all input channels and the full depth of memory simultaneously, so the total system can be studied.

Selective areas can then be isolated and magnified for closer scrutiny. In Fig. 16.16 the position of the trigger word is shown by the dotted line YY, and the value of the trigger word is given by the T box at the head of the display. The cursor consists of the solid line XX, and this can be moved over the screen. The value of the address (A) and data (D) at any position of the cursor can be read from the screen, but for convenience it is also automatically displayed in the A and D boxes at the bottom of the screen.

Time intervals between two points on a timing diagram can also be measured conveniently using the cursor. This is moved to the first point and its position recorded by a marker, and then the cursor is moved to the second point. The elapsed time between these points is displayed in the C box. Also shown on the screen is whether the clock is internal or external, and the clock period. The expansion factor of the display is indicated in the MAG box.

An example of a more complex state list to that given in Fig. 16.15, is shown in Fig. 16.17. This

STATE LIST

T=   C=

LABEL	A	D	F
BASE	OCT	BIN	BIN
09	01011	11111101	1
10	01004	11111011	0
11	01002	11001010	0
12	01034	10001110	1
13	01201	11010110	0
14	01000	10001111	1
15	01040	10101011	0
16	02002	01100111	1
17	02006	00010010	1
18	02124	10110110	0
19	02126	11001110	1

Fig. 16.17. A state list.

is sometimes also called a *data list* or a *data domain list*. It is more appropriate for data obtained in synchronous mode, and for microprocessor systems, since it is difficult to show the flow of a microprocessor program as a timing diagram. In a state list the information on the top of the list is usually the data which is recorded first. It is also usual to show only part of the list at a time, and to scroll through the stored memory using the cursor.

The trigger word is given in the T box at the head of the display of Fig. 16.17. The time interval between sequences can be measured using the cursor and marker, and this is displayed in the C box in terms of the clock period. The channels which are displayed, in this instance A, D and F, can be selected, and also the base in which they are to be shown, the most common being octal (OCT), binary (BIN), hexadecimal (HX) and decimal (DEC). Usually the

user would decide after the information had been recorded whether to display it as a timing diagram or a state list. It is therefore possible to flip from one to the other, and moving the cursor on one display would automatically move it on the other.

In both the synchronous and asynchronous operating modes data is stored in the memory of the logic analyser on an active clock edge only. Therefore short duration glitches, which occurred in between these clock pulses, would not normally be recorded. To catch these glitches most logic analysers are capable of operating in a latch mode. In this mode if more than one signal edge occurs between two clock edges, then this will be assumed to be a glitch and will be stored. This will only work if the sampling frequency is greater than the data frequency. The glitches are stored in separate glitch latches, and are displayed in the next clock period. This is illustrated in Fig. 16.18 which also shows that glitches are lost in a usual sample mode. Although the presence of glitches is displayed in the latch mode the actual position, amplitude or length of the glitch cannot be obtained from the display. However if a circuit is malfunctioning it is usually sufficient to know that this is being caused by glitches, which can then be suppressed.

Some instruments have a special system just to capture glitches. These are stored in a separate memory and then displayed. The glitches can now be used as a trigger parameter.

The logic analyser is usually set up using a series of menu driven *format specifications*, an example of which is shown in Fig. 16.19. In this the start point, clock slope and clock period are input at the top of the screen. The threshold voltages are then assigned to each input data

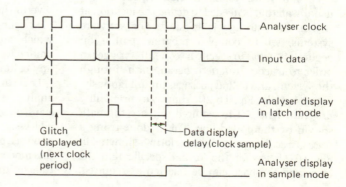

Fig. 16.18. Displaying glitches on a logic analyser.

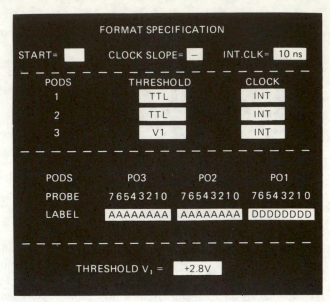

Fig. 16.19. Format specification for allocating pods and setting threshold voltages.

pod, and these may be preset values such as for TTL and ECL, or they can be adjustable, and the set value is given at the bottom of the page. In this example each pod has eight lines, and a label can be assigned to each line; Pods 2 and 3 are labelled address (A) and pod 1 is labelled data (D).

Triggering in a logic analyser is important since it allows the suspect part of the operation to be stored in memory, which is usually limited in size, and it also avoids the need for the engineer to analyse irrelevant data. As mentioned earlier there are three stages to triggering: (i) Storing pre-trigger information, for example the pre-trigger word and the amount of pre- and post-trigger data, which are determined by the trigger delay. (ii) Searching for the trigger word as the data goes through memory. (iii) Storing the post trigger data and stopping after the trigger delay.

For hardware systems a single trigger word is usually all that is needed to define the location at which data storage is to start. For software systems, where complex program paths can occur, it is usual to specify a sequential trigger in order to trace a unique program path through the system, and to find passages unambiguously. For example Fig. 16.20 shows a branching program where it is required to trace the path shown by the dotted line. This is done using a *trace specification* of the form shown in Fig. 16.21. As for the format specification the label and the base can be selected as required.

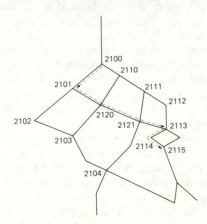

Fig. 16.20. Example of a branching program.

The count facility allows each item to occur a number of times (usually up to a maximum of several thousand times) before going on to the next item in the sequence. The number of words which must be found in sequence is usually limited to between three and ten. The crosses in any column indicate a 'don't care' state.

Logic analysers are capable of carrying out an analysis of stored data such as histograms of the elapsed time between two events or of the address space used. Fig. 16.22 shows the logic analyser operating in a graph analysis mode. In this mode it can plot any parameter with magnitude on the Y axis and sequence of occurrences

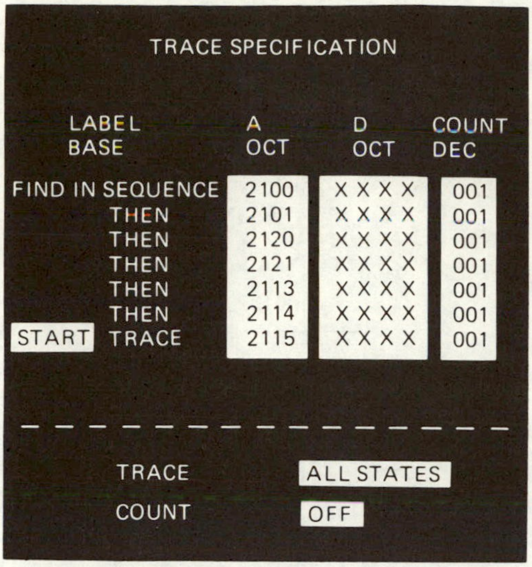

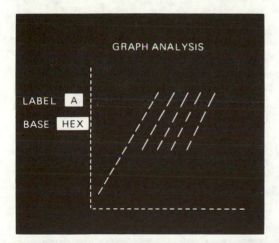

Fig. 16.21. Trace specification for the program of Fig. 16.20.

on the *X* axis. In Fig. 16.22 the *Y* axis is labelled addresses (A). This means that if the program were to operate in sequence through its address locations the graph would be a straight line. In Fig. 16.22 the program is seen to loop through a group of addresses, and if this is not expected then this part of the program can be studied in more detail using, for example, a state list output.

16.4.3 *Using a logic analyser*

Before making measurements using a logic analyser its variables are set up with a format specification menu, and then the section to be recorded is defined within the trace specification menu. As an example suppose that it is suspected that a fault is occurring within a section of program starting at memory location 00110. Fig. 16.23 shows the trace specification set up for this. The starting address is specified in A and the D content is unimportant. Triggering starts at the first occurrence of the trigger word (location 00110) hence the count is one.

The state list corresponding to the specification of Fig. 16.23 is shown in Fig. 16.24. It is assumed that a special pod for the target microprocessor is in use, so that a mnemonic listing

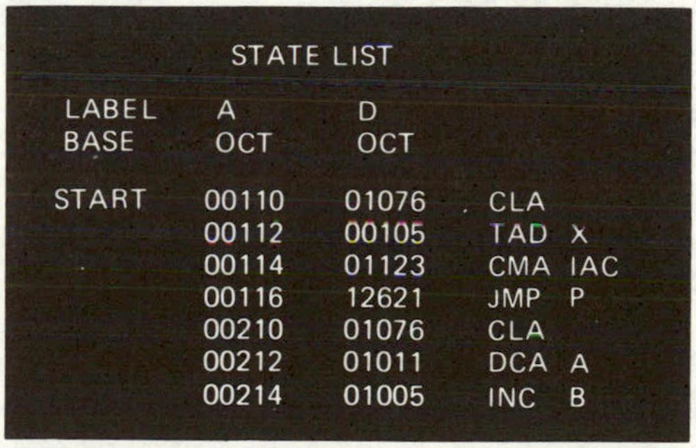

Fig. 16.22. A graph of address sequences produced by a logic analyser.

STATE LIST

LABEL BASE	A OCT	D OCT		
START	00110	01076	CLA	
	00112	00105	TAD	X
	00114	01123	CMA	IAC
	00116	12621	JMP	P
	00210	01076	CLA	
	00212	01011	DCA	A
	00214	01005	INC	B

Fig. 16.24. State list corresponding to the trace specification of Fig. 16.23.

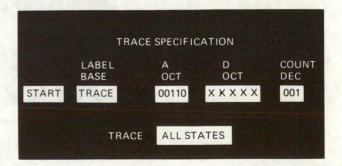

TRACE SPECIFICATION

LABEL BASE	A OCT	D OCT	COUNT DEC
START TRACE	00110	X X X X X	001

TRACE ALL STATES

Fig. 16.23. Trace specification to print out a section of program starting at location 00110.

can also be obtained. Examination of this list shows that there is a jump instruction in location 00116, which was not intended and can now be corrected.

An alternative way of locating this discontinuity in the program, which would be easier to spot, especially in a large program, would be to use the graph analysis facility. This is shown in Fig. 16.25. Each state is represented by a dot

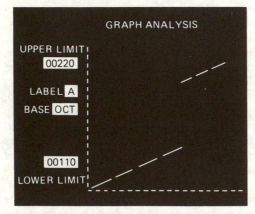

Fig. 16.25. Graph analysis for the example of Fig. 16.23.

(dash shown in the figure), although for a large program several million states can be condensed by first displaying every 1000th state (say) only, and then expanding the scale around the suspect area. Fig. 16.25 shows visually the discontinuity in the program around address 00116, and this can now be examined using a state list output.

As another example consider Fig. 16.26 which illustrates the flow diagram for a system

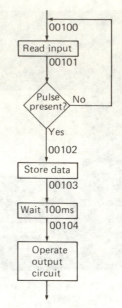

Fig. 16.26. A system flow diagram.

used to capture input data. As long as a pulse is not present it waits in a continuous loop of monitoring the input. When a pulse is sensed it is captured and stored. The system then waits 100 ms before operating an output circuit. Fig. 16.27 shows the trace specification for Fig. 16.26, assuming that the sequence shown is to be traced, where the numbers in Fig. 16.26 indicate A locations. Suppose also that the elapsed time is to be indicated so that the count field is programmed with TIME. The time can be absolute (ABS) or relative (REL).

If the time is programmed as REL in Fig. 16.27

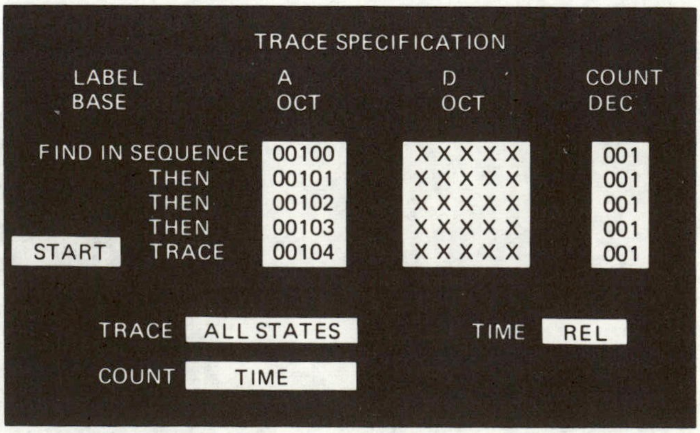

TRACE SPECIFICATION

		A	D	COUNT
LABEL BASE		OCT	OCT	DEC
FIND IN SEQUENCE		00100	X X X X X	001
	THEN	00101	X X X X X	001
	THEN	00102	X X X X X	001
	THEN	00103	X X X X X	001
START	TRACE	00104	X X X X X	001

TRACE ALL STATES   TIME REL

COUNT   TIME

Fig. 16.27. Trace specification for the system shown in Fig. 16.26.

STATE LIST			
LABEL	A	D	TIME
BASE	OCT	OCT	DEC
	00100	01035	–
	00101	00214	35µs
	00102	10123	10µs
	00103	10056	15µs
	00104	01365	100ms

Fig. 16.28. State list for the example of Fig. 16.26 using relative time indication.

the state list would be as in Fig. 16.28. The value of A is traced through as specified in Fig. 16.27 and the stored D in each location is also indicated. The time given in this list is the elapsed time between individual states, and it is not accumulative. If the time in Fig. 16.27 was ABS, then the state list would show the accumulated time from the trigger word, in Fig. 16.29, where the start point is at time zero.

STATE LIST			
LABEL	A	D	TIME
BASE	OCT	OCT	DEC
	00100	01035	–1.8ms
	00101	00214	–1.765ms
	00102	10123	–1.750ms
	00103	10056	–1.735ms
	00104	01365	0µs

Fig. 16.29. State list for the example of Fig. 16.26 using absolute time indication.

Fig. 16.30 shows a program with 8 nested loops. Suppose that it is suspected that a fault occurs in this program on the 40th occurrence of

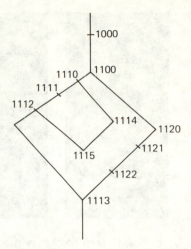

Fig. 16.30. Example of a nested loop program.

address 1115 during the 6th execution of the major loop. To find this state the trace specification can be written as in Fig. 16.31. The count column is now used to specify the number of times any of the required addresses is to occur before the analyser is triggered.

The logic analyser is also a useful tool for locating intermittent faults in a system. To do this a good trace of the suspect area is first stored in the internal memory of the analyser, as in Fig. 16.32(a). The analyser is then put into the compare mode. In this mode there are several options, such as 'stop if equal' and 'stop if not equal', the former mode being shown in Fig. 16.32(c). The logic analyser will now continuously cycle through the required section of the program, and will display all zeros as long as the measured trace equals the stored trace. Suppose now that the measured trace momen-

TRACE SPECIFICATION			
LABEL	A	D	COUNT
BASE	OCT	OCT	DEC
FIND IN SEQUENCE	1000	X X X X	001
THEN	1100	X X X X	006
START TRACE	1115	X X X X	040
TRACE	ALLSTATES		
COUNT STATE	1115	X X X X	

Fig. 16.31. Trace specification for the nested loop Fig. 16.30.

STATE LIST		STATE LIST		TRACE COMPARE	
				COMPARE MODE STOP NOT EQ	
LABEL	A	LABEL	A	LABEL	A
BASE	OCT	BASE	OCT	BASE	OCT
START	001100	START	001100	START	000000
01	002617	01	002617	01	000000
02	713581	02	713581	02	000000
03	210456	03	210456	03	000000
04	103167	04	103167	04	000000
05	001076	05	001076	05	000000
06	323567	06	323567	06	000000
07	753217	07	753217	07	000000
08	325612	08	325612	08	000000
09	124700	09	124700	09	000000
10	007431	10	007602	10	000233
(a)		(b)		(c)	

```
007431 = 0000 0000 0111 0100 0011 0001
007602 = 0000 0000 0111 0110 0000 0010
000233 = 0000 0000 0000 0010 0011 0011
```
(d)

Fig. 16.32. The use of an analyser to locate an intermittent fault; (a) required trace, (b) faulty trace, (c) analyser display in compare mode, (d) derivation of the error code.

tarily changes to that in Fig. 16.32(b), with an intermittent fault occurring which affects item 10. The logic analyser will now display the trace shown in Fig. 16.32(c), with a non zero in item 10, and will stop. The figure shown in item 10 indicates which bits of the two displays in Fig. 16.32(a) and (b) are not equal, as illustrated in Fig. 16.32(d).

16.5 Microprocessor development systems

The digital instruments discussed in previous sections have been used primarily to analyse hardware systems, although logic analysers are becoming increasingly accepted for software development. In the present section the techniques used for software development and analysis are discussed, primarily those associated with microprocessor based systems.

Fig. 16.33 shows the flow diagram for product development of a microprocessor based system. The customer requirements are translated into a system specification, and this is followed by system design, which enables the hardware and software contents of the product to be defined.

The hardware is split into modules, usually consisting of one or more printed circuit board assemblies, and these are then specified. Hardware design is initially done on a module basis, and is debugged using computer based engineer-

ing work stations, as in Fig. 16.34. These enable the schematic diagram to be captured, compiled to check for correctness, and then simulated to check for functional performance and for correct timing. Data required for this analysis is obtained from a data base of component characteristics, which is stored within the engineering work station.

The engineering work station is also used by the test engineer to ensure that the test points provided are adequate for manufacturing test. Test patterns are input to the work station and the system run to determine the amount of test coverage. Faults can be simulated at circuit nodes (stuck at zero and stuck at one) to see if they are detectable.

Simulation on an engineering work station eliminates the costly and time consuming requirement for system breadboarding, although this can still be done, if required, to verify the operation of critical parts of the circuit. When the engineer is satisfied with his design it is passed through a packager, which assigns gates to integrated circuit packages. The design then moves from the circuit design to the physical design phase, where the system is laid out onto a printed circuit board. This may result in a redefinition of package numbers and pins, and this back annotation information is fed back to

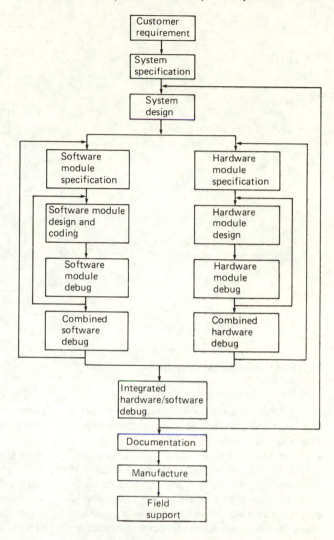

Fig. 16.33. Product development flow.

modify the initial schematic drawing. Conventionally the circuit design and physical design aspects of hardware design have been done on different work stations, usually supplied by different vendors, which has given rise to interface problems. The trend today is to perform all hardware design on a single work station.

Software development, shown in Fig. 16.33 follows a similar route to hardware development, but uses a very different set of tools. The individual software modules are initially specified. Very few tools are currently available for this phase, although several are being developed. One such system, known as structured analysis, allows the system operation to be represented

graphically, as data flow diagrams. These are easy to change, using computer based tools, which also check for omissions and consistency.

Following module definition is module design, coding and debug. The tools available for these operations are described in Section 16.5.1. The software is then combined with the hardware for system test, and it is here that emulation, described in Section 16.5.2, is especially useful. Finally documentation for both the hardware and software is prepared for manufacture.

The design phase of a software development project is particularly important since it determines the software quality, reliability and maintainability. This is especially important since

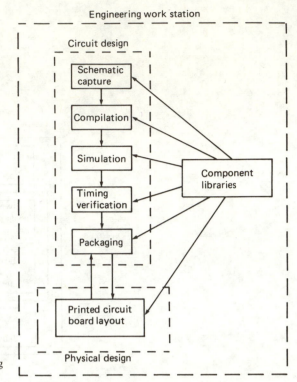

Fig. 16.34. Hardware design using an engineering work station.

about 70% of the cost of software occurs after it has been released, in the form of changes and bug fixes. A well designed and documented software package is much easier to maintain.

Software development costs predominate in the costs of developing a microprocessor based system, accounting for 60% to 90% of the overall cost of an average project. The software development is also most likely to cause programme slippage. Time to market is usually the most important consideration for any project. A 50% increase in development costs can usually be expected to reduce the profits obtained from the product by about 5%. On the other hand a six month delay to market can reduce profits by as much as 30%. It is because of the importance of software development within a product that much emphasis has been put, in recent years, on improving the productivity of software engineers.

16.5.1 *Software development and debug*

Several generic software facilities are required for any development system, and these are described in this section. An *editor* is essential in order to

enter code, and to modify it as required. Editors also perform many tasks which are transparent to the user, such as text compression prior to disc storage to minimise space.

Programs are rarely written in machine code, microprocessor mnemonics being the lowest level normally used. For example it is easier and more meaningful to write 'DCA A' rather than '1325' in order to decrement the accumulator. The mnemonic code is translated on a one to one basis into machine code by a utility called the assembler, hence the code used is also referred to as *assembler language*. The assembler also provides other facilities, such as checking for common coding errors, and jumping to labels in text rather than locations in memory.

Assembler languages are tied to the instruction set of particular microprocessors, and cannot therefore be ported from one microprocessor to another. Because of the large investment made in software development this means that companies are usually tied to using a family of microprocessors from one vendor, to ensure software compatibility. *Higher level languages*, such as Pascal and C, are largely microprocessor independent, and therefore more portable. They are also much easier to code and to understand. This is especially important during the debug and maintenance phases, since changes are often made long after the original code was written, and a program which is easy to understand can be changed with less new errors occurring.

Another advantage of higher level languages is that they require fewer lines to code. For example to multiply A and B together could be coded 'MULT A AND B' whereas the same operation in assembler language could take up to 20 lines of code. Most programmers usually write the same number of lines of code (non commentary source statements, NCSS) in a given time period. Therefore programmer productivity is increased with the use of higher level languages.

Table 16.1 gives a comparison of some of the most commonly used higher level languages. Although these languages significantly reduce programming time, they suffer from the disadvantage that the *compiler* produces about twice as much equivalent machine language code compared to that produced by an assembler. This means that more memory is required, and execution time is also longer due to the extra code. It is therefore usual to write the bulk of a program in higher level language, but to use assembler code for time critical parts.

Most higher level languages (the exception is C) also take the programmer away from the internal workings of the microprocessor. They are therefore not suitable for use in applications which need access to the internal parts of a microprocessor, for example bit manipulation, such as reading from or writing to input and output ports respectively. Again assembler language code can be introduced within a higher level language program, to handle these tasks.

Once the different modules of higher level language and assembler code have been written they are passed through a utility called a *linker*. The compilers and assemblers often do not assign fixed memory addresses to codes, that is they can be reallocated anywhere in memory. The linker's function is to link together these modules into a sequence of memory, so as to produce a single homogeneous program code.

Programs are usually written on development systems, which have the facility for compiling, assembling, linking and then storing the code within the system. After the code has been checked it is transferred from the development system into the memory of the actual target system, so that it can be tested further. This process is known as downloading, and the utility which does it is called a downloader.

The *operating system* provides the interface between the user and the development system, and is the heart of the software development system. Operating systems are usually interactive, and use default conditions to reduce the amount of information which has to be put in by the user. They also tend to be file oriented, in which the user operates on the programs by using their file names.

Operating systems carry out many control tasks, such as calling up compilers and editors, as shown in Fig. 16.35. The control part of the operating system works off a user source file which specifies items such as memory locations, interrupts etc. The control reacts to the user set up requirements by generating code and command files, so as to automatically handle details

Table 16.1 *Comparison of higher level languages*

Language	Easy to learn	Many data types	Many operators	Good flow control	Generates reentrant code	Type checking	Structured
Basic	√	×	×	—	×	—	×
C	×	√	√	√	√	×	√
Fortran	×	—	—	×	×	×	×
Pascal	√	√	√	√	√	√	√
PL/X	√	×	×	×	—	—	√

√ = Good   × = Weak

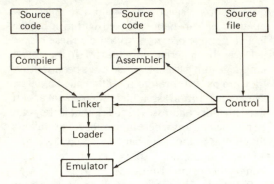

Fig. 16.35. The control exercised by an operating system.

which will execute the system program as specified. Examples of this are code generation for initialization and reset; creating linker command files; setting up emulator command files, which enable the linked object code to be downloaded and executed in the emulator.

Many other important facilities are available from operating systems. The *multi-tasking* feature enables several tasks to be performed simultaneously. The system could therefore be compiling or assembling a source code file, whilst editing another file and printing a third file. Multi-tasking is especially important for centrally based development systems having many users connected to them. The overall system software is written and debugged as smaller modules, and the operating system allocates memory and processing power to each module, as needed to execute the programs.

The *multiuser* facility of an operating system allows several users simultaneous access to the development system. It utilises the multi-tasking facility for allocating resource to each user, and then protecting this resource from other users. Operating systems also enable a *hierarchical file system* to be used in which files are arranged at different levels, and each file has files or directories which point to other files at lower levels.

In large software systems it is useful to have a facility whereby a simple command will cause all interdependent source code modules, which have been changed during a debug session, to be automatically recompiled or reassembled, and then linked with the total system. This facility of an operating system is known as *automatic software update*. Another useful facility is *electronic mail* which can be used to keep mem-

bers of a project team in touch, and is specially useful for broadcasting messages on problems, etc.

Two other features available from some operating systems are *virtual input–output* and *command indirection*. Virtual input–output allows extra peripherals to be added easily to a system by treating each device independently of its characteristic. Command indirection enables commands to be read from a file, without operator input, and is useful for process control applications.

The use of software analysis utilities was introduced in Section 16.4. Performance analysis gives an overview of program activity and allows areas of the software to be optimised. Examples of performance analysis utilities available are: (a) Breakdown of the memory used throughout, which enables more efficient memory allocation and utilization. (b) An analysis of the time distribution between successive accesses to different memory modules, which enables queuing problems to be identified. (c) Plot of the time needed to run separate parts of a program including best and worst case conditions. Areas which take more time than expected can be re-written. (d) A count of the number of occurrences of different instructions.

As software systems start to grow, involving large teams of programmers, and multi microprocessor based systems with over 0.5 Mbyte of code, software project management and control tools become vital. The problems associated with project management arise not only because of the large number of programmers involved, but also because of the need to make many changes and to keep several versions, and to develop different variations of the same software, for example for disc based systems, main memory systems, etc.

Fig. 16.36 shows some of the project management tools in use on a typical project. Each individual project member is allocated his own data base, which interacts with a central data base, and is under control of the management tools. These tools perform the following functions: (a) Control the access to individual program modules. Once a module is booked out from the central data base to a programmer no other engineer has access to it. (b) Automatic log of all changes made to a program module, when it is returned to the central data base. This

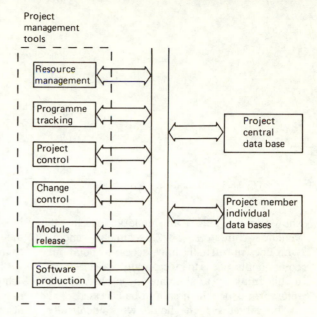

Fig. 16.36. Project management tools.

specifies when the change was made, who made the change, what the change was and why the change was made. (c) Automatic update of all individual project data bases with the latest information, to ensure that everyone is working with up to date program modules. (d) Segregation of information by revision, version, type etc., so that a project member can quickly get to the exact module he requires. (e) An up to date record of all data held and their status. This is useful in providing incremental compilation facilities, so that if a module is changed then only those modules which are affected need to be re-compiled and re-linked prior to re-test.

16.5.2 *Emulation*

When testing conventional hardware systems it is possible to understand the operation of the system by studying the circuit diagram. Fault diagnosis of microprocessor based systems is much more difficult since the circuit performance is determined by its software. Emulation is used in these circumstances. This is done by replacing the circuit under test (also called the target system) by a device which emulates its behaviour. This device is called an emulator, and it can be subject to controlled stimulus in order to study the operation of the target system. This concept of emulation is not new since, for example, when testing hardware circuits it is

usual to replace part of the system with external equipment, such as power supplies and function generators, which in effect act as emulators.

When used with microprocessor development systems the emulator gives control over the target microprocessor, and a view of the activities taking place in its registers. The emulator may be part of the development system, so that it is tightly coupled to the overall system, or it can be a stand alone dedicated emulator linked onto the main bus, when software development is done using a large host mainframe computer.

Emulators can be obtained from microprocessor manufacturers or from instrument makers. When a new microprocessor is developed the initial support for this is available from the semiconductor vendor, and it is only later, sometimes one to two years later, that the instrument vendor has a support tool available. Usually the emulator available then has more features, since operational experience has been gained on the microprocessor, and the same equipment can support devices from several different vendors. One of the problems of microprocessor emulation is that often it is not possible to access all the required lines on the device, since they are not brought out to pins. The microprocessor vendor gets around this problem by bringing out these lines onto special pins. These devices, called bond-out versions, as

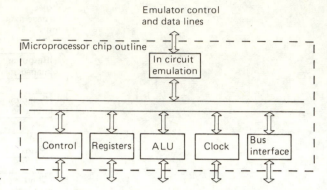

Fig. 16.37. A bond-out microprocessor.

shown in Fig. 16.37, are made for use in the vendor's emulators. The instrument vendor, who does not usually have access to bond-out components, needs to incorporate extra circuitry in his emulator to simulate those parts of the microprocessor which cannot be accessed.

The target system should work identically with and without the emulator, and this property is known as having transparency. There are two types of transparency. Electrical transparency is when the two systems have the same electrical characteristics, such as operating time and machine cycles, and functional transparency when the same resources, such as memory and interrupts, are used by the two. The emulating microprocessor also needs to be mounted close to the target system to minimise pick up effects, and it is usually built into an external pod which is plugged in, in place of the target microprocessor. The cables connecting the emulator to the rest of the system should be of good quality and have impedance compensation. Usually the emulating microprocessor needs to be buffered to drive cables and the target system, so performance variances are introduced from the target system, such as extra delays.

Several features are required from an emulator based development system, to test the target equipment. It should be possible to single step through the program, examining register contents or modifying memory content at each step, and to set special stop points, called breakpoints. These can be software breakpoints, where the program instruction is replaced by an emulator call to stop the program at specific points, or hardware breakpoints, which can also be set by comparing parameters such as memory address or input–output datalines. Hardware

breakpoints can also generate triggers for instruments, such as logic analysers and oscilloscopes, to make further readings.

Changes can be made to a program during the debug stage using an emulator. These changes may be patched into the system for assessment. A disassembler facility should be available to observe the microprocessor operation in mnemonics, and an assembler facility for patching. It is also useful to be able to carry out debugging at a higher level language (source code), and to be able to set breakpoints at this level. The complexity of the system is now increased since the emulator needs to track the source code program.

Logic analysers are often teamed with emulators within the development system. The logic analyser is primarily a monitoring instrument, and is very useful in collecting timing data using the asynchronous clock mode. Emulators operate in synchronous mode from the target system clock, and cannot do this effectively. Fig. 16.38 shows an arrangement in which an emulator is used within a development system and is connected to the target system via its microprocessor socket and the emulator interface.

The emulator shown in Fig. 16.38 has access to its internal logic state circuit which provides analysis of functions such as software performance. It also has access to timing and state analysis within the development system. Software performance analysis does not affect the program execution, so information can be collected and displayed in real time. The emulator memory must be wide enough for the microprocessor address, data, status and control lines, which can be up to 96 bits wide. It must also be

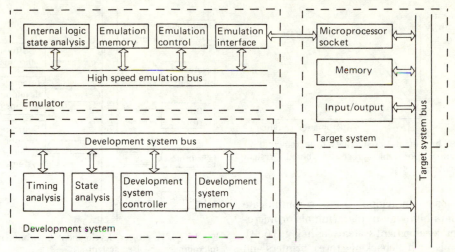

Fig. 16.38. Use of an emulator with a development system.

deep enough to capture between one and two thousand steps of data. In spite of this memory size careful setting of breakpoints is essential to optimise use of trace memory.

Fig. 16.38 shows separate bus structures for the emulator and the development systems. This allows both systems to operate independently making it useful for emulation of multiprocessor based products, for examining logic performance whilst the microprocessor is running, and for updating the emulator trace memory without stopping the processor.

Fig. 16.38 shows the emulator connecting into the microprocessor socket of the target system. This in effect replaces the target's microprocessor with the emulator's microprocessor, which is under control of the development system. This form of emulation is known as in circuit emulation, and is the one most commonly used. It cannot be employed for bit slice processors since these operate at high speeds, and their microcoding feature allows each device to have its own instruction set. In these circumstances memory or ROM emulation is used.

In memory emulation the target's microprocessor is functional, but interface with the emulator is made via the target's program memory. A fast RAM within the emulator replaces the program memory, giving the emulator control over the system's microcode. The emulator can stop the clock at the breakpoints needed. However although this may be satisfactory for static systems it is not suitable for dynamic systems, since stopping the clock causes loss of data. If the emulator operates at a much higher speed than the microprocessor cycle time then it can force the processor into a 'do nothing' loop whenever a breakpoint is detected. Therefore the clock is not stopped but the operator can view the internal registers of the system.

The third type of emulation is known as input–output emulation, and is used when the system needs to be tested before the design of the input–output circuit has been completed. The emulator now interfaces with the input–output part of the target system. Pattern and word generator signals are sent to the microprocessor and timing and state analysis is used to check its response. The system can be highly programmed, for example particular signals can be sent to the microprocessor corresponding to particular patterns on the address bus.

16.5.3 *The development process*

The ideal software development process consists of a single software engineer supported by comprehensive development tools. As the size of the project team increases so also does the need for detailed specifications and rigid control systems, which add to the development overheads. Therefore the incremental productivity to resource curve has the shape shown in Fig. 16.39.

Most large software development projects are based around a central mainframe computer, which provides the power needed for many of the tasks, and the management tools to support the

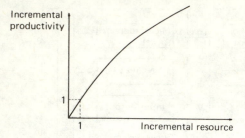

Fig. 16.39. Incremental productivity to incremental resource curve for microprocessor based system development.

project team. Individual users communicate over a data highway using terminals or microprocessor development systems, as in Fig. 16.40. Peripherals such as backing stores, printers and

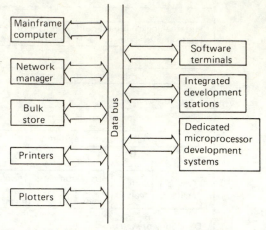

Fig. 16.40. A project development environment for a microprocessor based system.

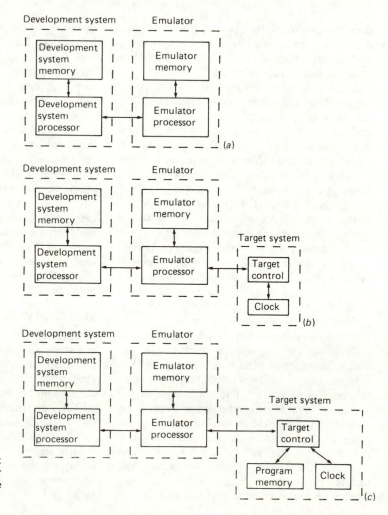

Fig. 16.41. Three development modes used for microprocessor based systems; (*a*) mode 0, (*b*) mode 1, (*c*) mode 2.

plotters are usually shared between users. Cross compilers are currently available for most micro-processors.

The software development process starts with specification, design and coding, as in Fig. 16.33. Most of the program is written in a higher level language, which is subsequently compiled, with critical sections written in assembler language. After linking the software is simulated on the development system using software debug tools. However this simulator can reveal first order faults only, since the behaviour of the target system is difficult to predict. Therefore after the initial software debug emulation is used.

Emulation can occur in three modes, as in Fig. 16.41. In Mode 0 the emulator and development system simulate the target system, without any target hardware. The program for the target system is run in real time whilst resident within the emulator's memory. The development system clock is used as the clock for the emulator microprocessor, and the input–output is simulated. Mode 1 emulation is used when some of the target hardware is available. The clock used is that of the target system. Input–outputs can be simulated, or those from the target system used if available. The memory employed may be that within the emulator or the target system, or even a combination of the two. In Mode 2 emulation, which is the final hardware–software integration stage, all the hardware is available so the target system's memory, clock and input–output

circuits are used. As in the previous two modes the emulator still replaces the target microprocessor, and gives the engineer control over the target system.

Two types of problems can arise when using emulators. In the first the target system operates satisfactorily when the emulator is used, but not when running from its own microprocessor. The most common cause is that the target microprocessor is overloaded, since the emulator is buffered and can therefore support a greater load. Timing problems during power up can also result in this discrepancy in performance. For example if the power-on reset pulse is too short then it will affect the system when it is running from the target microprocessor, but not when it is running from the emulator, since the emulator execution commences after power-on has occurred.

If the target system works satisfactorily when operating from its own microprocessor, but not when connected to the emulator, the fault could be due to timing problems, caused by buffers, power supply noise, etc. Often these problems also occur when the target system is running off its own microprocessor, but they are masked by other effects and show up as intermittent faults.

After the target has been satisfactorily checked out, using the emulator, the software is programmed into EPROM by transferring the files from the development system to an EPROM programmer. This then forms the program memory for the target system.

17. Optoelectronic measurements

17.1 Introduction

Optoelectronics covers that area of science which combines optical and electronic technologies. It spans a very wide range of topics. Most electronic engineers are familiar with the subjects of electronic displays and optical sensors, but because optoelectronics is not a pure electronic science, many find the terminology used to be confusing.

This chapter first introduces the basic concepts used in optoelectronics and the terminology. The units of measurement are then described and the measurement techniques. The chapter concludes with a description of fibre optic parameters and measurements, since these are being used to an increased extent within the electronics industry.

17.2 Some basic concepts

17.2.1 *The optical spectrum*

The electromagnetic spectrum is shown in Fig. 17.1(a). It is seen from this figure that the optical spectrum occupies a small part of the electromagnetic spectrum, and the visible spectrum covers an even smaller band. It is usual to specify the electromagnetic spectrum in terms of wavelength rather than frequency, although both are given in Fig. 17.1. The conversion between frequency and wavelength is given by (17.1).

$$f = c/\lambda \tag{17.1}$$

f is the frequency is Hz, λ the wavelength in metres, and c is the speed of light, which is

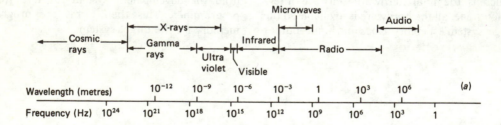

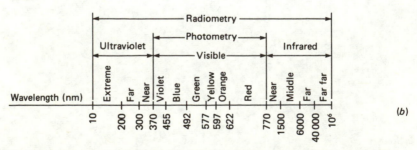

Fig. 17.1. The electromagnetic spectrum; (a) electromagnetic spectrum, (b) optical spectrum.

approximately equal to 2.99×10^8 metres per second.

The optical spectrum extends from 10 nm to 10^6 nm, and is divided into three main regions, ultraviolet, visible and infrared, as in Fig. 17.1(b). The ultraviolet region extends between the visible band and gamma rays. It is further subdivided into extreme ultraviolet (10–200 nm), far ultraviolet (200–300 nm) and near ultraviolet (300–370 nm).

The visible region extends from about 370 nm to 770 nm. This is the band which can be seen by an average human eye. It is divided into colours which can be perceived by a standard observer. The principal colours are violet (370–455 nm), blue (455–492 nm), green (492–577 nm), yellow (577–597 nm), orange (597–622 nm), and red (622–770 nm).

The infrared band extends from the visible region to microwaves. It is divided into four main bands, near infrared (770–1500 nm), middle infrared (1500–6000 nm), far infrared (6000–40 000 nm), and far far infrared (40 000–10^6 nm).

17.2.2 *Interaction of light with matter*

The velocity of light through space is constant for all types of electromagnetic radiation. Different types have different frequencies, and therefore wavelengths, given by (17.1). Usually a beam of light is made up of a number of frequencies, but if the beam has a single frequency, or in practice a very narrow band of frequencies, then it is said to be *monochromatic*. If further the individual waves making up this beam all have the same phase relationship they are said to be coherent.

Optical waves may be considered to move back and forth, at right angles to the direction of travel of the beam of light. Usually these waves are not pointed in any one direction in this plane, and the light is said to be *unpolarised*. If the beam is changed such that the wave motion is in one direction only, in a plane perpendicular to the direction of motion, then it is called linearly polarised light.

When light passes across the junction of two media it is bent, provided the angle of incidence does not equal 90°, and this is known as *refraction*. The *refractive index* of the medium is given by the ratio of the sines of the angles of incidence and refraction, and varies from about 1 to 3, depending on the material composition

and the wavelength of the light.

When a light beam strikes a surface some of the light is reflected, some is absorbed and some is transmitted. The amount of *reflection* varies, depending on the properties of the surface and the wavelength of the incident light, from greater than 98% for visible light on a manganese oxide surface, to less than 1% for visible light on a lampblack surface.

Reflection can also be specular or diffused. *Specular reflection* occurs when the surface is smooth, compared to the wavelength of the light falling on it. The beam is now reflected with no change in divergence, and obeys the well known laws of reflection. *Diffused reflection* occurs from a surface which is rough compared to the wavelength of the light striking it. The light is scattered, and in practice the reflection is proportional to the cosine of the viewing angle. In a perfect diffuser, given by Lambert's law, as in (17.2), the radiance is constant and independent of viewing angle.

$$L = M/\pi \qquad (17.2)$$

L is the radiance in watts per steradian square metre, and M is the radiant exitance in watts per square metre, the units used being explained in Section 17.3.

Some of the light falling on a surface is *absorbed*, the amount of absorption varying from almost zero, for pure silica glass, to greater than 99% for lampblack. The speed of light is less in a material than in air, and as light travels through the media its intensity gradually decreases. The speed of light through the media also varies depending on the wavelength of the light, and this is called *dispersion*. Some of the light is scattered and some is absorbed, the absorbed light being converted to heat.

Transmittance refers to the effect where some of the light striking a surface is transmitted into the media, the penetration depth in a practical substance varying from zero to greater than 75%. The sum of the proportion of reflected, absorbed and transmitted light for any substance is equal to unity.

17.2.3 *Radiometry and photometry*

Radiometry is the name given to the science of measuring the optical radiation covering all wavelengths, as shown in Fig. 17.1(b). It provides an indication of optical energy, and is measured by causing the energy to be absorbed within the instrument and then measuring the temperature change. The instrument should have the same response to all the frequencies within the optical band.

Photometry is the science concerned with measuring visible light, with respect to the response of the human eye. It is used in applications where the eye is the main sensor, for example in room lighting and video screens. The eye has two types of optical sensors, called cones and rods. At normal illumination levels the cones are the most active, and they are primarily responsible for differentiating between colours. Rods are more sensitive than cones in weak illumination, but they do not have colour response, so the eye has difficulty in separating colours at low light levels.

The peak spectral sensitivity between rods and cones is displaced by between 40 nm and 50 nm. In order to obtain consistent photometric measurements the response of an average eye was established by the CIE in 1924, as shown in Fig. 17.2. The response under normal light levels is called *photopic vision*, and it peaks at 555 nm. The effect of the different colours on the eye is indicated clearly by this figure, for example the eye has almost ten times as much response to green as it has to blue. The response at low light

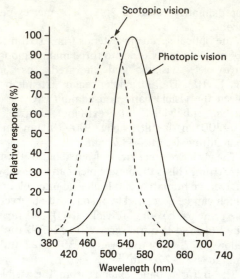

Fig. 17.2. Curves showing the response of an average eye at different wavelengths.

levels is known as *scotopic vision* and it has a peak at 507 nm.

17.3 Units of measurement

Different units of measurement are used for radiometric and photometric systems, and these are summarised in Fig. 17.3. Other terms and

Parameter	Radiometric			Photometric		
	Symbol	Name	Unit (and abbreviation)	Symbol	Name	Unit (and abbreviation)
Wave energy	—	Radiant energy	Joule (J)	—	Luminous energy	Lumen second (1m·s)
Wave energy per unit time	ϕ_c	Radiant power or Radiant flux	Joule/second (J/s or Watt)	ϕ_v	Luminous power or flux	Lumen (lm)
Energy from a point source	I_c	Radiant intensity	Watt/steradian (W/sr)	I_v	Luminous intensity	Lumen/steradian (lm/sr) or Candela (cd)
Energy on a surface (directional)	L_c	Radiance	Watt/steradian metre² (W/sr·m²)	L_v	Luminance	Lumen/steradian meter (lm/sr.m²) or Candela/metre² (cd/m²)
Energy falling on a surface	E_c	Irradiance	Watt/metre² (W/m²)	E_v	Illumination or Illuminance	Lumen/metre² (lm/m²)
Energy leaving a surface	M_c	Radiant exitance	Watt/metre² (W/m²)	M_v	Luminous exitance	Lumen/metre² (lm/m²)

Fig. 17.3. Summary of radiometric and photometric units.

units are also used to describe optical systems. For example the *Kevin Scale* is often used to describe the colour content of a light source. It is based on the concept that a black body radiates when heated. As the temperature of the body is increased it reaches a threshold after which the body begins to glow and to emit a dull red light. As the temperature of the body increases further the colour of the emitted light changes through bright red, orange, yellow, white and blueish-white. Typical values on the Kevin scale are 6500 K for daylight, 3500 K for light output from an incandescent lamp, and 1900 K for candlelight.

Another term used in optical measurements is *visual acuity*. This is the ability of the eye to resolve fine details. Visual acuity is dependent on the level of illumination; it improves with illumination but saturates at high levels.

17.3.1 *Photometric units*

The basic unit of power in the photometric system is the *lumen*. It is called the *luminous flux* or *luminous power*, and is a measure of the rate of flow of luminous energy from a source. Like all photometric measurements it is based on the photopic response of the eye, which peaks at 555 nm.

The energy of the light wave in lumen seconds is called the *light energy* or *luminous energy*. The light energy emitted from a point source is measured as the luminous power passing through a unit solid angle in a given direction from the source. The *steradian* is a measure of the solid angle formed at the centre of the sphere of radius r by the area of r^2 on the surface of the sphere, as illustrated in Fig. 17.4. The energy emitted from the point source is its *luminous intensity* and is measured in *lumens per steradian* or in *candelas*.

If a point source radiates equally in all directions it is known as an *isotropic source*. Since there are 4π steradians in space around a point the total flux from a source having a luminous intensity of one candela is 4π lumens. If the source is not isotropic then a plot of luminous intensity versus angle from a reference can be used to indicate its directional property, this being referred to as a *polar diagram*, as in Fig. 17.5.

The density of light energy incident on a surface is known as *illumination* or *illuminance*,

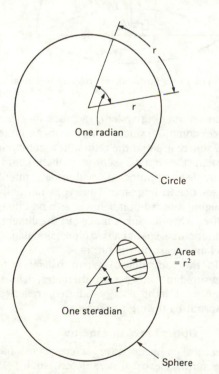

Fig. 17.4. Illustration of radian and steradian.

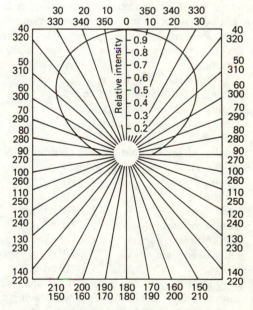

Fig. 17.5. A polar diagram of light output from a non isotropic point source.

and is measured in lumens per square metre. Other units used to measure illumination are lux, phot and foot candle, and these are related to lumens as follows:

1 lux   = 1 lm/m²
1 phot   = 1 lm/cm²
1 foot candle = 1 lm/ft²

As a guide, the approximate values of illuminance under natural lighting conditions are as follows:

Bright sunlight 10^5 lm/m²
Dull sunlight 10^3 lm/m²
Sunset 10 lm/m²
Moonlight 10^{-1} lm/m²
Starlight 10^{-3} lm/m²

In practice no source of light is a point source. However if the source has a diameter less than 10% of the distance between it and the detector, then it can be considered to be a point source. For an area source of light the luminous intensity in any direction is due to many points on the surface of the source. *Luminance* is now used as a measure of the light intensity leaving, passing through or falling on a surface in a given direction. The surface area is the projected area as seen from the specified direction. For example the surface of area A_e has a projected area of $A_e \cos \phi$ in the direction at an angle ϕ to the original area. Luminance is also sometimes referred to as photometric brightness, and is measured in units of lumens per steradian square metre, or in candelas per square meter. Other units of photometric brightness, in common use, are as follows:

1 footlambert = $1/\pi$ cd/ft²
1 lambert   = $1/\pi$ cd/cm²
1 stilb   = 1 cd/cm²
1 nit   = 1 cd/m²
1 apostilb   = $1/\pi$ cd/m²

Examples of approximate levels of luminance from various sources are as follows:

Atomic bomb 2×10^{12} cd/m²
Lightning 7×10^{10} cd/m²
Carbon arc 2×10^7 cd/m²
Tungsten lamp 9×10^5 cd/m²
Sun (from the earth) 2×10^4 cd/m²
Fluorescent lamp 7×10^3 cd/m²
Moon 3×10^3 cd/m²

Energy leaving a surface is defined by *luminous exitance*, which is the luminous power leaving a unit surface area, and measured in lumens per square metre.

7.3.2 *Radiometric units*

The basic measure of power in the radiometric system is the watt. It is an indication of the energy of the wave per unit time. The wave energy is defined as *radiant energy* and is measured in joules. The energy emitted from a point source is called radiant intensity, and is the radiant power per unit solid angle is watts per steradian.

The light flux density falling on a surface is the *irradiance*, and is measured in watts per steradian square metre. The radiant intensity per unit area leaving, passing through or arriving at a surface in a given direction is the *radiance*. The surface area is the projected area as seen from the specified direction, as in Fig. 17.6.

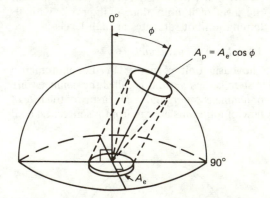

Fig. 17.6. Illustration of projected surface area.

In an imaging system, radiance (and luminance) dominate since the eye casts an image of the source area on the retina. In a non imaging system the optical sensor, which could for example be a photodiode, will respond mainly to irradiance in its plane. The eye is normally an imaging sensor but it behaves as a non imaging sensor when it cannot make out the dimensions of the source, so that a two dimensional image is no longer cast on the retina.

In the radiometric system *radiant exitance*, measured in watts per square metre, defines the density of light energy emitted, reflected or transmitted through a surface.

17.4 Optical measurements

Electrical engineers have become accustomed to measuring electrical parameters, such as voltages and currents, with an accuracy of a few

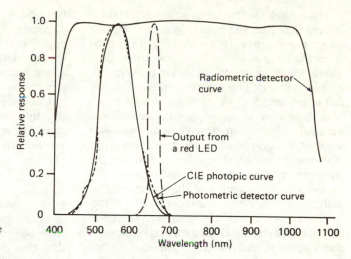

Fig. 17.7. Comparison of detector response curves.

parts per million. However optical measurements, performed with skilled people and good equipment, rarely give errors below 10–20%. Several factors contribute to the measurement errors, and these are discussed in this section.

The prime components for measurements are optical sources and detectors. The spectral emission of the source and the spectral response of the detector are important parameters. Using a detector with the wrong spectral response for the source being measured, results in the greatest errors. For spectrometric measurements the detector performance is specified by its closeness to the CIE photopic response curve, as in Fig. 17.7. However these specifications are on the basis of area under the curve, and can lead to large errors unless matching is planned at specific frequencies. Therefore for the example shown in Fig. 17.7, the photometric detector will give about 25% error when making measurements on a red LED, even though it is relatively closely matched with the CIE response curve over most of the range. The radiometric detector, on the other hand, would give an error below 5%.

Apart from errors due to spectral mismatch, optical measurements also suffer from ageing of the detector and filters, stray light, and non-linear detector characteristics.

17.4.1 Optical sources

Sunlight produces a continuous spectrum of every colour, covering the band from infrared to ultraviolet, with a peak emission in the blue region at about $0.46\,\mu m$.

The most commonly used *incandescent source* is the *tungsten lamp*. It consists of a tungsten filament enclosed in a glass envelope, and gives off light when a current is passed through it to raise it to incandescence. The filament can be designed to run from a wide range of a.c. or d.c. supplies.

The tungsten lamp is often used as a secondary standard. It has a well defined continuous spectrum output, as in Fig. 17.8. The light can be

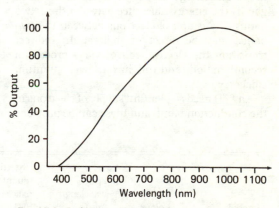

Fig. 17.8. Radiant output curve for a tungsten lamp.

filtered to produce any colour, and is of predictable intensity. As the temperature of the lamp increases the wavelength at which peak emission occurs falls, and the intensity of the output increases. Control of the lamp voltage and current allows calibration at any given colour temperature.

The spatial distribution of a tungsten lamp is almost spherical, and from a distance it appears like a point source. Less than 15% of the light from the lamp is in the visible spectrum. Only about 2% of the electrical input is converted to visible radiation, and if filters are used this can fall to below 0.5%.

A *fluorescent source* produces a continuous spectrum light output, which has superimposed on it peaks at certain wavelengths. These peaks are due to electrical discharge through the mercury vapour in the lamp, and occur mainly in the ultraviolet region of the spectrum. The output from a fluorescent source has a low infrared content, and since more of its spectrum is at wavelengths to which the eye is sensitive, it has a higher overall efficiency than incandescent lamps.

Light emitting diode (LED) operation is based on the phenomenon of *electroluminescence*, which is the emission of light from a semiconductor under the influence of an electrical field. Electrons exist in a solid in one of two states, either bound to the nucleus or free in the lattice. The energy in the free state is greater than that in the bound state. The conduction band in a semiconductor is the energy band for free electrons. The valence band is the energy band for bound electrons or positive holes. The energy gap, or band gap or forbidden gap, is the energy difference between these two bands. Transition of electrons between the two bands may be direct or indirect. In indirect transition the electron moves via a process of recombination, and this has a lower quantum efficiency.

An LED needs a plentiful supply of electrons in the conduction band, and these can be provided by raising the energy of the valence band electrons. These electrons then give up their energy in the form of photons of light and return to the valence band. The excitation can be conveniently obtained by forward biasing the pn junction of the diode.

The wavelength of light emitted by an LED is directly related to its energy gap E. If h is Planck's constant, and c the velocity of light, then the wavelength of the emitted radiation is given by (17.3).

$$\lambda = hc/E \qquad (17.3)$$

Since electrons often move in stages within the forbidden gap they also give up their energy in steps, so the emitted wavelength may be longer than that given on the basis of the band gap.

Several different materials have been used for LEDs, a few of these being illustrated in Fig. 17.9. Many considerations govern the choice of material; the energy gap must be suitable for the wavelength required, and the efficiency of emission must be high. For infrared emitters gallium arsenide is the most popular material, having a direct energy gap of 1.4 eV and a wavelength of emission of around 900 nm. Gallium phosphide is mainly used for emissions in the visible spectrum. It has a gap energy of 2.26 eV and can be doped with zinc and oxygen to give red light, or with nitrogen to give green. As the concentration of nitrogen increases the emission moves from a peak of 560 nm (green) to 585 nm (yellow).

Gallium arsenide and gallium phosphide combine to form a solid solution of gallium arsenide phosphide. The light emission from this material peaks at about 660 nm (red). Nitrogen doping

Material	Colour	Peak wavelength (nm)	External quantum efficiency (%)	Band structure	Lumens per watt
GaP:Zn, O	red	695	15.0	indirect	20
GaP:N	green	570	0.5	indirect	600
GaP:NN	yellow	590	0.1	indirect	450
$GaAs_{0.6}P_{0.4}$	red	649	0.5	direct	75
$GaAs_{0.35}P_{0.65}$:N	orange	632	0.5	indirect	190
$GaAs_{0.15}P_{0.85}$:N	yellow	589	0.2	indirect	450

Fig. 17.9. Properties of some visible LED materials.

increases the conversion efficiency of the phosphorus, and also the wavelength of the emission.

In Fig. 17.9 the external quantum efficiency is a measure of the power output. Luminous efficiency in lumens per watt considers the sensitivity of the eye. It is more sensitive to green than to red hence the increase in luminous efficiency in this region.

Electroluminescent diodes emit radiation in relatively narrow spectral bands in the range from infrared to green (900–550 nm). They can therefore be designed to give a range of outputs without the need for filters. The intensity of the light decays exponentially with time, so that it is convenient to quote the life of the diode as its half life, which is the time for the luminance level to fall to half its original value.

The effect of temperature on an electroluminescent diode relates primarily to variations in its band gap. As the temperature increases the wavelength at which the light peaks also increases. This causes a shift in the colour, and since this affects the eye response the luminous efficiency also changes. For red and green diodes the efficiency decreases with temperature, but for yellow diodes it increases since the emitted wavelength moves more into line with the eye response.

Since the spectral output of the most commonly used LEDs peaks in red, they present problems in measurement. In this region the photoptically corrected detector deviates from the CIE photopic curve, as in Fig. 17.7. Therefore sensors are used which are matched to the curve in the red region, but which deviate considerably in the other areas.

A light source which is finding increasing commercial application is the *laser* (Light Amplification by Stimulated Emission of Radiation). The properties of a laser beam are that it has high radiant intensity, it can be polarised, and it is monochromatic and coherent. The laser can generally be looked on as an optical oscillator, that is an amplifier with positive feedback. Its three main parts are an active material which gives gain, a method of exciting the active material, and a resonant structure to produce feedback.

Many materials are used in lasers, and these may be solid, liquid or gas. A semiconductor laser consists of a forward biased pn junction which is excited by an electric current. Near the junction recombination of holes and electrons takes place releasing photons. If the bias current is large enough then many holes and electrons are concentrated in a small area of the junction, and the photons which are released will stimulate more emission. Relatively high current densities are required to produce lasing, and many laser structures have been developed to reduce this current requirement, and to increase life. Lasers may have d.c. or pulsed output. High power pulsed lasers need special measurement techniques, as described later in this section.

17.4.2 *Optical detectors*

Photoresistive material, photoemissive material, and junctioned semiconductor devices such as photodiodes and phototransistors are all used as optical detectors. These have been described in Section 4.8. In addition thermal sensors may also be used.

Thermal sensors measure the change in temperature, resulting from the absorption of radiant energy, by giving an electrical signal. They usually have a flat response to a band of frequencies, but they are more suited to measuring sources which have a relatively high intensity output, and usually for wavelengths above about 1000 nm. There are several different thermal sensors, a few of these being described here.

The *thermopile* consists of a series connection of many thermocouples. The hot junction is usually blackened to give better radiant energy absorption, and the spectral sensitivity of the sensor depends on the characteristics of this black absorber. The thermocouples are sealed in a transparent envelope. Thermopiles have a slow response and are sensitive to ambient temperature. Their characteristics can be made flat over the desired portion of the spectrum, and they are used in radiation pyrometers and as infrared detectors.

The *bolometer* consists essentially of a strip of blackened metal sealed in a glass film. Radiant energy is absorbed by the strip, raising its temperature and changing its resistance. This change in resistance is measured in a bridge circuit to indicate the value of the radiant energy. In another method a bias is applied to the bolometer. The value of this bias is then decreased as the radiant energy increases, in order to keep the resistance of the bolometer constant.

The change in bias is a measure of the radiant energy.

A *pyroelectric* detector consists of a crystal whose electric polarization is dependent on temperature. It is sensitive to changes in temperature only and it can therefore be used to measure pulsed and chopped light, and to ignore any steady state background. The change in polarization results in a current variation in an external circuit. Pyroelectric detectors have a fast response and are mainly used to measure the pulsed power of lasers.

17.4.3 *Measurement techniques*

Photometry presents difficulties for precise measurements. The original standard was a candle of special construction, which was observed by the human eye and compared to the source being measured. Modern measurement methods use special lamps, and sensors with filters to simulate the standard eye response. Errors still occur due to stray light, non uniform response, sensor drifts and colour temperature changes. The lamp used must be at a known colour temperature; standard lamps usually run at a colour temperature of $2854\,K$.

A photometer normally uses a silicon photodiode, photomultiplier tube or a photoconductive device as a sensor. Filters are added which correct the response to that of a standard photopic eye. The source is coupled to the optical sensor, usually with a microscope having a fixed diameter aperture. Light from the source is focused onto the aperture, and measured by the sensor. When the area of the emitter is smaller than that of the aperture only a part of the emitter is in focus, and therefore consistent measurements are difficult to obtain.

Radiometric instruments use photodiodes, photomultiplier tubes, photoconductive cells and thermal detectors. The sensitivity of the detector varies with wavelength so the instrument must be calibrated over the wavelength to be used. For monochromatic sources the reading of the instrument is multiplied by the sensitivity of the detector at that wavelength, to get the reading of power or energy. For broadband sources measurements need to be made in narrow frequency bands using filters. A spectroradiometer is designed for broadband working. It contains a monochromator in front of the detector. This scans continuously over the detector, isolating all the wavelengths required, in an action very similar to that of a filter.

Spectroradiometers usually have many accessories. The input optics can consist of cosine receptors, LED receptors, imaging optics, integrating spheres, microscopes, telescopes and fibre optic probes. Several detector modules are also available, for different optical sensitivity levels and spectral ranges. Spectroradiometers can make measurements on sources or detectors. A standard source is used to calibrate the instrument when it is set up for measurements on sources, and a standard detector is used when the instrument is to measure detectors. Calibration sources should have similar intensity and optical characteristics to those of the test source.

Optical power meters give an indication of power level using specially designed chopper sensors, which convert light to an electrical signal. Several interchangeable sensors are used with any instrument, each classified by power level and wavelength. There are also special shapes for special applications, such as taking measurements in confined spaces.

Several problems arise when using power meters to make measurements on lasers. The high power density causes non-linearity in the photosensors. Diffusers may be used in front of the sensor to reduce this problem, but these result in reflection and absorption which modifies the spectral response. Non uniform illumination of the sensing elements, such as thermopiles, also results in errors due to non uniform temperature distribution. In laser power measurements it is common to use mirrors to reflect some of the power. Since lasers have narrow beams the reflected beam is as good as the incident beam, and a mirror with about 90% reflection would absorb measurable amounts of heat.

Measurements of the spectrum of light sources is usually done using an optical spectrum analyser. In this instrument the light spectrum is usually resolved by interference or diffraction methods. The interference method usually uses a Fabry–Pérot or Michelson interferometer. The diffraction method resolves the spectrum using a monochromator with a diffraction grating or a prism. A servomotor usually drives the grating, giving good resolution, and the angle of rotation of the grating is displayed on a cathode ray tube. It is clearly important that the monochromator covers a wide frequency band. The signal to noise ratio must also be good to make low level measurements, and this is done by using a synchronous optical chopper in the optical detector.

When using instruments to make measurements of the energy falling on a surface, i.e. illumination or irradiance, the optical sensor must be located on the surface. The detector should also be placed at right angles to the incident light, otherwise compensation needs to be made for angular sensitivity. Since flux density is being measured it is important that the whole area of the detector is illuminated. Therefore in Fig. 17.10 the source should extend at

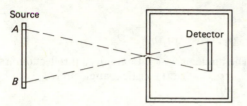

Fig. 17.10. Definition of source area when making measurements of energy on a surface.

least from A to B. If this does not occur then it is necessary to make corrections by taking the actual viewing angle of the probe. To measure irradiance of a point source no special input optics is required. However for large area sources a diffuser input is required, for example an integrating sphere, with a response which varies as the cosine of the angle of incidence.

In making spectral radiance measurements an image of the source is formed at the input slit of a monochromator, as in Fig. 17.11. Low distor-

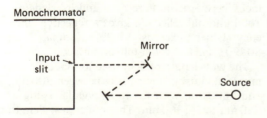

Fig. 17.11. Radiance measurement.

tion mirrors are used and the angle of rotation of the mirrors noted. The system is calibrated using an optical standard. For small area sources a microscope input adaptor may be required, and

for distant sources a telescopic input adaptor. An achromatic lens is necessary for measurements in the visible range, and a fused silica lens for ultraviolet and infrared use. The telescope–spectroradiometer system is calibrated using a point source at the same distance as the source to be measured.

The illumination (E) due to a point source is proportional to the intensity of the source (I) and inversely proportional to the distance (d) between the source and the detector. Therefore the intensity can be calculated from (17.4) once the illumination has been measured, as described earlier.

$$I = Ed^2 \qquad (17.4)$$

It is important that the diameter of the source is at least 50 times smaller than the distance between the source and detector, so that error due to the finite size of the source is low.

17.5 Fibre optics

Fibre optics have gained in popularity over the last few years, especially in the field of communications, and this has been done against competition from conventional and coaxial cables, waveguides and radio. The advantages which fibre optic cables have over conventional copper cables is their much smaller size and weight, and their immunity from noise pick up due to electrical or magnetic fields, caused by effects such as lightning and machines. They also do

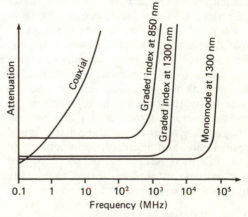

Fig. 17.12. Typical attenuation–frequency curves for coaxial and fibre optic cables.

not suffer from cross talk, since even if light leaves a damaged cable it cannot enter an adjacent cable. Optical fibre cables are difficult to tap, and the intrusion can be easily detected, unlike coaxial cables which can be easily tapped and are difficult to detect. Therefore fibre optic cables are increasingly used in secure applications.

Optical fibres have a much higher bandwidth than conventional cables, and lower attenuation, as illustrated in Fig. 17.12. This means that fewer repeaters are needed in a long haul communication system. It is important at this stage to note that there is a difference between optical and electrical decibels (dB). If optical power P_o is converted to electrical current I in a sensor, then the current is usually proportional to optical power. Therefore optical attenuation is given by (17.5)

$$\text{dB(optical)} = 10 \log \frac{P_{o1}}{P_{o2}}$$

$$= 10 \log \frac{I_1}{I_2} \qquad (17.5)$$

If electrical power P_e is being considered then the power is proportional to the square of the current. Therefore electrical attenuation is given by (17.6).

$$\text{dB(electrical)} = 10 \log \frac{P_{e1}}{P_{e2}}$$

$$= 20 \log \frac{I_1}{I_2} \qquad (17.6)$$

Comparing (17.5) and (17.6) shows that dB (electrical) is numerically twice dB (optical).

Optical fibres are made from low cost, and readily available, material. However they need light sources and detectors in order to interface the signal to the electronic equipment. Light emitting diodes and semiconductor lasers are most frequently used for the light sources, and semiconductor photodiodes for the detectors. The coupling of light sources and detectors to the optical fibre, and interconnection between sections of a fibre, have presented problems in the past, but, as described in the next section, these have now been overcome.

17.5.1 *Principle of optical fibres*

If a light beam crosses the boundary between two

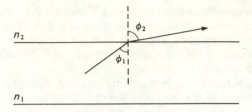

Fig. 17.13. Illustration of refraction.

materials, of refractive indices n_1 and n_2 then it will be refracted, as in Fig. 17.13. As the angle of incidence ϕ_1 is increased, a point is reached when the beam is totally internally reflected. For this to occur inequalities (17.7) and (17.8) must both be true.

$$n_2 < n_1 \qquad (17.7)$$

$$\phi_1 > \cos^{-1}(n_2/n_1) \qquad (17.8)$$

The angle at which total internal reflection first starts is called the critical angle and is given by (17.9)

$$\phi_c = \cos^{-1}(n_2/n_1) \qquad (17.9)$$

The principle of total internal reflection is used to transmit light along optical fibres, as in Fig. 17.14. This is known as a step index fibre

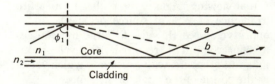

Fig. 17.14. Light transmission in a step index fibre.

since there is an abrupt change in refractive index between the fibre core and its cladding. Total internal reflection gives a reflective efficiency which exceeds 99.9%, compared to 80–90% for a silvered mirror surface.

The wavelength of light used for optical communication using fibres varies between 800 nm, and 1600 nm, with the most popular being at 850 nm and 1300 nm. The longer wavelength gives lower loss and higher bandwidth, but the sources and detectors for these wavelengths are also more expensive.

Typical fibre diameters vary in steps such as 50 μm, 100 μm, 200μm and 300 μm, with single or multi strands. Light waves can travel through the fibre in several modes, the number of modes N being given approximately by (17.10); λ is the wavelength of the light in vacuum, r is the radius of the fibre core, and n_1 and n_2 are the refractive indices of the core and cladding respectively.

$$N = \frac{2\pi^2 r^2}{\lambda^2}(n_1{}^2 - n_2{}^2) \qquad (17.10)$$

The effect of multimode transmission paths is to give signal distortion. If 'a' and 'b' are two modes of travel for light waves in Fig. 17.14 then clearly beam 'a' takes a longer path then 'b' resulting in pulse stretching as in Fig. 17.15.

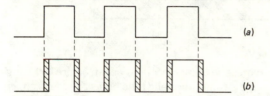

(a)

(b)

Fig. 17.15. Distortion of a light pulse passing through a fibre; (a) input light from source, (b) output light from fibre.

This distortion can be minimised by using graded index fibres. In this cable the refractive index of the core is gradually decreased from the central axis to the edge using techniques such as chemical deposition. The difference in refractive index over the core is about 1%. Fig. 17.16

shows the refractive index profile for step index and graded index fibres.

The graded index fibre may be considered as a series of finite layers, resulting in refractive index changes at each boundary. As the light crosses each boundary it is refracted, but the grazing angle is successively reduced until eventually the beam is totally internally reflected. This results in a parabolic light wave path, as in Fig. 17.17. Since the light wave travelling in the outer regions of the fibre sees lower refractive index material, it also travels faster than the light wave at the centre of the core. This partly compensates for the longer path which needs to be traversed by the outer waves, and reduces the pulse distortion illustrated in Fig. 17.15. The optimum index distribution used in a graded index fibre, to give an almost parabolic refractive index profile, is given by (17.11). n_d is the refractive index at a distance d from the axis of the core; n_a is the refractive index at the fibre axis and n_e the index at the edge; r is the radius of the fibre core and k is a constant (approximately equal to 2) which depends on the composition of the core and the wavelength of the light.

$$n_d = n_a[1 - 2(n_a - n_e)r^{-k}d^k]^{1/2} \qquad (17.11)$$

Step index fibres use a large core so that more light may be coupled into it. They are not good at low temperatures, and it is difficult to fit connectors to them. They also suffer from pulse distortion, which reduces the bandwidth over which

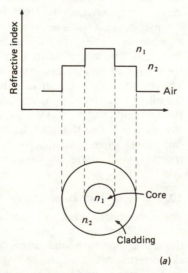

(a)

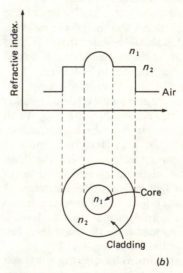

(b)

Fig. 17.16. Refractive index profile of optical fibres; (a) step index, (b) graded index.

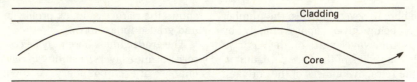

Fig. 17.17. Light propagation in a graded index fibre.

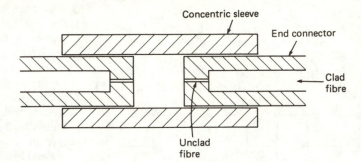

Fig. 17.18. A simple connector for optical fibres.

the fibre can be used. Graded index fibres have about 200 times better pulse distortion characteristics than step index fibres, and therefore higher bandwidth. However for very high bandwidth operation it is necessary to eliminate all propagation modes except one, by use of monomode fibres.

Referring to (17.10) it is seen that the number of propagation modes of light waves within the fibre decreases, as the radius of the core decreases, and the wavelength of the light increases, until eventually N reaches unity. The value of the refractive index differential $(n_1 - n_2)$ is usually chosen such that the radius of the monomode fibre is not less than $5\,\mu m$ to allow easy alignment. It should be noted that (17.10) is true for step index fibres, and that for graded index the value of N needs to be halved.

Glass fibres can consist of plastic core and cladding, silica core and plastic cladding or glass core and cladding. Plastic fibres are rugged and easy to terminate, but have high attenuation. Glass core and glass clad fibres have the best attenuation characteristic, but are the least rugged. Optical cladding on the fibre cores prevents loss of light due to surface scratches, or contaminants such as grease. For total internal reflection the external media needs to have a lower refractive index than the glass core, but grease has a similar index to glass.

Glass fibres are formed by applying a first soft coat, called first buffer, to the newly drawn glass

core, and then a secondary coat, before jacketing in a strong tube called the buffer. The first buffer must adhere to the core but be strippable. It must also protect the fibre from microbends, which can induce losses. The buffer may be loose or tight. In loose buffering the fibre is free to move to allow for flexing, and low friction material such as polyvinylidene fluoride is used. In tight buffering low thermal expansion materials, with low viscosity, are necessary to avoid stresses during extrusion.

When connecting fibres together the two ends are cleaned and then held concentrically together. For joins which need to be removable, say for maintenance purposes, connectors are used, for example as in Fig. 17.18. The two ends of the uncoated fibre are fixed in terminations, and then screwed into a concentric aligning tube. The loss for such an arrangement is about 0.5 dB. If permanent joins are acceptable then the two ends can be welded together, and the overall loss is less than 0.1 dB. Several defects can arise when two cables are connected together, and some of these are shown in Fig. 17.19.

17.5.2 *Fibre optic characteristics*

The main characteristics of optical fibres which are considered in this section are dispersion, numerical aperture and attenuation. Dispersion determines the maximum length of the fibre, or the maximum distance between two re-

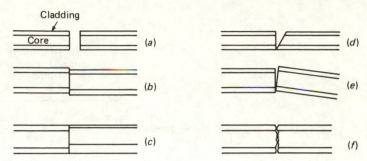

Fig. 17.19. Some types of defects in fibre optic joins; (a) surfaces not mating, (b) fibres not coaxial, (c) cores not concentric, (d) fibre ends not cut perpendicular, (e) fibre axes run out, (f) dirt at fibre ends.

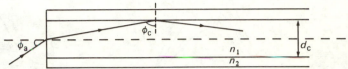

Fig. 17.20. Illustration of numerical aperture.

peaters, as a function of frequency. Numerical aperture and attenuation set the maximum fibre length in terms of available light power, that is they determine the length at low signal frequencies.

Numerical aperture

When a ray of light enters a fibre optic cable, as in Fig. 17.20, only those rays which result in a low grazing angle at the core–cladding interface will be transmitted through the cable, by total internal reflection. If ϕ_c is the critical angle then ϕ_a is known as the acceptance angle, and its value is determined by the refractive index of the core and cladding. Rays of light at angles greater than ϕ_a are not transmitted by total internal reflection within the fibre.

Sin ϕ_a measures the ability of the optical fibre to collect light, and is called its numerical aperture (NA). The value of the numerical aperture for step index fibres is given by (17.12), and for graded index fibres by (17.13)

$$NA = \sin \phi_a = [n_1^2 - n_2^2]^{1/2} \tag{17.12}$$

$$NA = \sin \phi_a = n_1 \left[\frac{2(n_1 - n_2)}{n_1} \right]^{1/2} \tag{17.13}$$

The numerical aperture determines the optical power which can be coupled from the source into the fibre. If d_s and d_c are the diameters of the source and the core of the fibre cable, then when the two are in contact the power transferred is given by (17.14) or (17.15) for $d_s \leqslant d_c$ and $d_s > d_c$

respectively.

$$P = \tfrac{1}{4}\pi^2 d_s^2 R_a (NA)^2 \tag{17.14}$$

$$P = \tfrac{1}{4}\pi^2 d_c^2 R_a (NA)^2 \tag{17.15}$$

R_a is the on axis radiance of the source. In most cases there is about 2 dB loss due to poor coupling between the source and the optical fibre. Typical fibres have diameters which vary from 30 μm to 600 μm, and numerical apertures of 0.3 to 0.4 for step index, 0.2 for graded index, and 0.1 for monomode fibres.

The number of light wave propagation modes within a fibre, given by (17.10), can also be expressed in terms of the numerical aperture using (17.12), and is given in (17.16).

$$N \simeq \frac{2\pi r^2}{\lambda^2} (NA)^2 \tag{17.16}$$

Provided that the diameter of the core is greater than about ten wavelengths it is possible to have many modes of light within the fibre. However if the radius is less than a critical value, given by (17.17), then light can propagate in only a single mode.

$$r_c = \frac{1.2\lambda}{\pi NA} \tag{17.17}$$

This is called the HE11 mode and is not subject to dispersion; fibres made to this criterion are called monomode, as explained in Section 17.5.1.

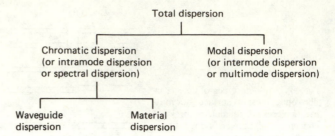

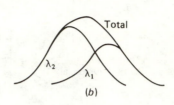

Fig. 17.21. The dispersion 'Family Tree'.

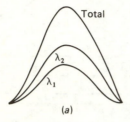

Fig. 17.22. Illustration of pulse stretching due to material dispersion; (*a*) input pulse, (*b*) output pulse.

Dispersion

There are several types of dispersion which occur in optical fibres, and these are shown in Fig. 17.21. The total dispersion in the fibre is given by (17.18)

$$\text{Total dispersion} = \left[\left(\frac{\text{Chromatic}}{\text{dispersion}}\right)^2 + \left(\frac{\text{Modal}}{\text{dispersion}}\right)^2\right]^{1/2} \tag{17.18}$$

The speed of light waves travelling in a fibre changes with its wavelength. Therefore a pulse having a finite frequency linewidth will have its different wavelengths propagated at different speeds, resulting in pulse broadening. This effect is called *chromatic dispersion* and is as a result of *material dispersion* and *waveguide dispersion*.

Material dispersion is caused by the variation of the group velocity of light waves with wavelength, as in Fig. 17.22, since the refractive index of the glass fibre changes with wavelength. In silica fibres there is a unique wavelength, determined by the composition of the core, at which material dispersion is zero. This usually occurs at the longer wavelengths, between 1200 nm and 1800 nm, with 1300 nm being that for most fibre materials, as shown in Fig. 17.23.

Material dispersion is a problem when the

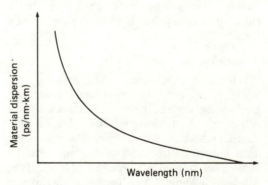

Fig. 17.23. Variation of material dispersion with wavelength in a typical optical fibre.

source used has a high spectral linewidth. The optical bandwidth due to material dispersion is given by (17.19) where K is the electrical bandwidth (3 dB point) due to material dispersion in a fibre of 1 km length using a source with 1 nm spectral linewidth. For silica K is typically 3.3 GHz/km.nm, L is the length of the fibre in km and $\Delta\lambda$ the spectral spread of the source, between its half power points, in nm.

$$f_D = \frac{K}{L \cdot \Delta\lambda} \tag{17.19}$$

Lasers, which have typical spectral bandwidths of 2–4 nm are used where material dispersion needs to be minimised. Light emitting diodes have spectral bandwidths of 30–60 nm.

Waveguide dispersion has a similar effect to material dispersion, but arises due to the variation of phase and group velocities of each mode of the light wave with wavelength. Generally waveguide dispersion is small compared to other distortion effects which occur in multimode large core fibres. For monomode fibres its effect starts to become more important. This can be used to compensate for the effects of material dispersion, so giving a lower overall dispersion. For monomode fibres waveguide dispersion is typically 2 ps/km.nm at 1300 nm.

Modal dispersion has already been introduced and explained with reference to Fig. 17.14 and 17.15. In a graded index fibre, modal dispersion is minimum when the core refractive index profile is a parabola. As seen from (17.16) the number of modes in which light can travel in a fibre increases with its numerical aperture, and this leads to an increase in modal dispersion. For long haul telecommunication networks, operating at high speeds, the numerical aperture is limited to about 0.2 to minimise modal distortion. For short haul local networks the numerical aperture can be increased to about 0.5. Modal dispersion decreases as the number of modes is reduced, and material dispersion then becomes increasingly important in determining the bandwidth of the cable. Generally commercial data given for fibre cables assume a zero linewidth source, so that only modal dispersion is considered.

Attenuation

The output from an optical fibre is less than the amount of light input due to losses within the cable. This loss is proportional to the cable length, and it has an exponential decay with length so that it is usual to specify it in units of dB per unit length. For optical fibres practical values of attenuation vary from 0.2 dB/km to 10 dB/km.

Losses within fibres are due to three prime causes. These are as follows:

(a) *Absorption* of the light due to the impurities within the glass core. These impurities are mainly metallic ions e.g. Fe, Cu, Ni, Cr, Co and Mn, and hydroxide ions (OH) resulting from the presence of water in glass.

Fig. 17.24 shows a typical attenuation curve

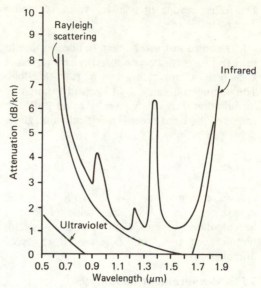

Fig. 17.24. Typical attenuation curve for an optical fibre.

for an optical fibre. The theoretical limit is caused by Rayleigh scattering, described below. The peaks in the curve are due to absorption by the metallic or hydroxide ions. Because of this they are sometimes called water peaks. From Fig. 17.24 it is seen that the attenuation reaches a dip between 800 nm and 900 nm, and because of this the first fibre optic systems operated at this frequency. The sources and light sensors were readily available for this frequency. However a lower attenuation is reached at the longer wavelengths of 1300–1500 nm, and this is now being used. Above 1700 nm glass begins to absorb light due to molecular resonance of silica.

(b) The second cause for light loss within a fibre is scattering. There are two main scattering mechanisms within a fibre. The first is Rayleigh scattering caused by inhomogeneities of the dielectric material due to random molecular distribution within the amorphous glass material. Rayleigh scattering varies as λ^{-4} and in high grade glass it has typical values of 0.9 dB/km at 900 nm and 0.4 dB/km at 1300 nm.

The second scattering mechanism is due to irregularities in the core–cladding interface. This causes rays which are incident on the interface at the same angle to be reflected at different angles, a phenomenon known as mode mixing.

This often results in a lower value of modal dispersion.

(c) The third loss mechanism in fibres is due to the bending of the fibres housed in a cable, and is known as microbending loss or radiation loss. The microbends cause light loss in the fibre due to radiation, but it is generally small until the radius of the bend exceeds a critical value given approximately by (17.20)

$$R_c \simeq \frac{3n_1^2 \lambda}{4\pi(n_1^2 - n_2^2)^{3/2}} \qquad (17.20)$$

As an example for $\lambda = 1300$ nm, $n_1 = 1.5$, $n_2 = 0.99n_1$ i.e. 1.485, the value of R_c can be calculated to be 74μm. Therefore careful handling of the fibres is essential to avoid microbending.

17.6 Measurement techniques

Fibre optics are complex, a fibre with n modes of operation, for example, having $2n$ ports. Measurement standards are therefore very important to ensure uniformity of measurements, for repeatable results. Several standards have been introduced by bodies such as DIN, IEC, VDE and CCITT. In this section the measurement of fibre attenuation, bandwidth, numerical aperture, and fibre dimensions are considered. Also introduced is the optical time domain reflectometer, which is extensively used for a variety of measurements on optical fibres.

17.6.1 *Attenuation*

Fibre attenuation is fundamental in determining the loss along a cable, including that in connectors and splices, and it is therefore an important consideration when planning communication routes.

Three methods are used to measure attenuation, the cutback method, the insertion loss method, and the backscatter method. In the cutback method a constant optical power is fed into the fibre under test and the output power (P_a) is measured. A short length of about 2 metres of the fibre is then cut from the input end, and the optical power at that point (P_b) measured. The fibre attenuation (A) is then given by (17.21)

$$A = 10 \log (P_b/P_a) \qquad (17.21)$$

The cutback method suffers from the disadvant-

age of requiring about 2 metres to be cut from the fibre, so that it cannot be used on an installed cable system. It is also difficult with this method to get reproducible results, even with careful handling.

In the insertion loss method the optical source and detector are first connected by a short length of a reference fibre, and the power output (P_b) is measured. The reference fibre is then replaced by the fibre under test and the power (P_a) again measured. The attenuation (A) of the fibre is given by (17.21), as before. It is important in both these measurement methods to have a stable light distribution within the fibre, so that the attenuation is not influenced by light distribution at the launch point into the fibre. This is obtained by either shaping the light input with a lens system, or by using a mode scrambler, as described in the next section.

The backscatter method of measuring attenuation is used extensively in fibre measurement instruments, such as the optical time domain reflectometer described in Section 17.6.5. In this method light is launched into end of the fibre, and that light power which is returned to the launch end, due to Rayleigh scattering, is measured, together with the time delay. This gives a curve of the form shown in Fig. 17.25 and from this the attenuation over the whole fibre, or over sections of the fibre, can be found. For example the attenuation A_1 over length l_1 in Fig. 17.25 is found by measuring P_{a1} and P_{b1} and then using (17.22).

$$A_1 = 5 \log (P_{b1}/P_{a1}) \qquad (17.22)$$

The factor 5 is used in (17.22) rather than 10, as in (17.21), since now the light travels twice over the length of the fibre.

The backscatter method provides a footprint of all the flaws, and attenuations due to non uniformities, within a fibre. It also has the advantage of not needing the fibre to be cut, and of only using one end of the fibre. However, much higher source power levels are required to cover long fibres, or attenuations due to splices and connectors. Errors can also arise due to inconsistencies, in the numerical aperture, core diameter or scattering coefficient, along the length of the fibre.

17.6.2 *Bandwidth*

Bandwidth is a complex function of the modulat-

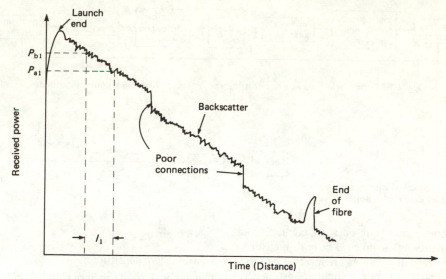

Fig. 17.25. Power received back from a launched pulse.

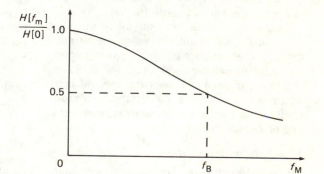

Fig. 17.26. Transfer function of an optical fibre.

ing frequency $H[f_m]$ and is given by the frequency f_B at which (17.23) is satisfied.

$$H[f_m]/H[0] = 0.5 \qquad (17.23)$$

Two basic methods are used for bandwidth measurements, time domain and frequency domain. In the time domain method pulses with fast rise time and short duration are passed through the fibre under test, and the distortion at the far end is measured. Mathematical processing is then used to convert from the time domain to the frequency domain. The disadvantage of time domain measurement methods is that the bandwidth of the signal detector needs to be as wide as that of the fibre, which reduces the length over which bandwidth measurements can be made. It is also necessary to trigger a

sampling oscilloscope during measurements, so accurate synchronization is necessary between the pulse generator at one end of the fibre and a sampling oscilloscope at the other end.

Frequency domain measurements use a swept frequency to establish the fibre's transfer function, as in Fig. 17.26. Since the transfer function, is found directly in the frequency domain no further transformations are needed, as was necessary for time domain measurements.

Fig. 17.27 shows a set up to measure bandwidth using the two point frequency domain method. A mode scrambler is used at the launch end, which overfills the fibre's core. This ensures that the output is independent of the spatial properties of the light source. First output measurements are made at point Y. The launch end

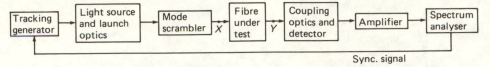

Fig. 17.27. Bandwidth measurement using the two point frequency domain method.

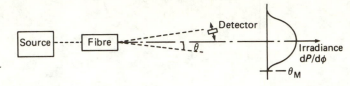

Fig. 17.28. Numerical aperture measurement using far field evaluation.

of the fibre is then cut back to point X and measurements again taken. For both measurements neutral density filters are inserted into the coupling optics, so that in neither instance is there saturation or non-linear operation of the detector. The difference between the two readings gives the bandwidth of the fibre.

Frequency domain measurement methods have good signal to noise ratios since the transmitted power covers a narrow frequency band, requiring a small bandwidth receiver. However chromatic dispersion causes the spectral linewidth of the source to affect the bandwidth reading; narrow bandwidth sources must therefore be used.

17.6.3 *Numerical aperture*

Numerical aperture of a fibre is found by evaluating its far field. If a light source is placed at the centre of a sphere, then the radiant intensity on the inner surface of the sphere is its far field intensity. The near field gives the intensity on the surface of the source of light, which could for example be a laser or the end of an optical fibre.

Fig. 17.28 shows the set up for measuring the numerical aperture of a fibre. A short fibre length, of about 2 metres is used, and the light source overfills the fibre to produce a large spot size. Mode stripping is also used so as to remove transmission modes from the cladding of the fibre. The source shown in Fig. 17.28 has its spot size settable, and the detector can receive light over a small angle, which it scans to measure optical power P as a function of solid angle ϕ. The figure shows the far field distribution for a graded index fibre. From this measurement the numerical aperture (NA) can be found using (17.24).

$$NA = \sin \theta_m \qquad (17.24)$$

17.6.4 *Fibre dimensions*

The dimensions of the fibre cladding can usually be measured under a microscope, but since the core is much smaller it is usual to derive its measurement from the refractive index profile. Different techniques are used for this, such as the transmitted near field method and the refracted near field method. The transmitted near field method is based on the fact that the transmitted near field intensity of a uniformly filled short fibre, of about 2 metres, is proportional to the refractive index profile of the fibre.

In making measurements with the transmitted near field a set up similar to that in Figure 17.28 is used. Now, however, the light profile on the surface of the fibre is measured. Usually imaging techniques are necessary to measure the near field since the fibre has a small emitting area. Mode scramblers or mode strippers are used to remove cladding modes, and an LED to overfill the fibre. Fig. 17.29 shows a typical refractive index profile for a graded index

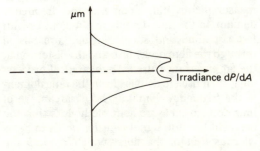

Fig. 17.29. Refractive index profile of a graded index fibre obtained by measurement of the near field.

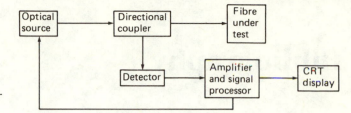

Fig. 17.30. Operating principle of an optical time domain reflectometer.

fibre, obtained by measurement of the near field. The small dip at the centre of the fibre relates to the manufacturing process, and is typical for a graded index fibre. The core diameter is obtained as the intersection points between this curve and the line P given by (17.25), where n_1 is the maximum core refractive index, and n_2 the cladding index.

$$P = n_2 + (n_1 - n_2)/40 \qquad (17.25)$$

17.6.5 *The optical time domain reflectometer (OTDR)*

The OTDR is an instrument which is extensively used for a range of measurements on fibres. Fig. 17.30 shows the operating principle of an OTDR. The optical source consists of a high power laser which is repeatedly pulsed to give an output of 5–100 ns wide, power level from 100 mW to several watts and a repetitive frequency of several kHz. This light is launched into the fibre, and some of the light is reflected or scattered back from the fibre. The directional coupler, or beam splitter as it is also called, prevents the light pulse from reaching the sensor until it has gone through the cable. The detector usually consists of an avalanche photo diode.

Fig. 17.25 shows the profile of the light received back from the fibre. It consists of two components: (i) Discrete reflections, called Fresnel reflections, due to abrupt changes in the refractive index of the fibre, such as occur at connectors, splices, breaks in the cable, or the end of the cable. (ii) Rayleigh backscatter, due to inhomogeneities in the fibre refractive index. The amount of backscattered light is dependent on several factors, such as the source power, the fibre's numerical aperture and scattering cross-section, and the wavelength of the light. Breaks in the fibre result in a reduction or a total loss of backscatter, and knowing the velocity of a propagation it is possible to calculate the distance to the break.

The backscattered signal is about 50 dB below the transmitted power, and is much weaker than the discrete reflections. Signal processing techniques which perform signal averaging, are therefore used to improve the instrument range. The results are displayed on a CRT or a chart recorder. Sometimes pseudo-random sequence pulses are launched into the fibre instead of a single high power pulse. This increases the total amount of power in the fibre and gives a larger backscatter signal, but correlation techniques are now necessary to analyse the signal.

The slope of the backscatter curve is proportional to the attenuation coefficient of the fibre. However the OTDR is not a precise instrument; for example splice loss is found from the relative magnitude of the signal from either side of the splice. The OTDR also gives the mean of the outward and reflected losses, which could be very different from a unidirectional loss. Many techniques have been used to improve OTDR performance, such as using trade offs which reduce the source noise bandwidth, and an increase in the pulse width which improves the backscatter signal. Different OTDR methods are also sometimes used, like photon counting direct detection and coherent detection (S. Barber, optical fibre testing, *Telecommunications*, Jul. 1985).

Commercial OTDR's often have interchangeable laser light sources having wavelengths typically of 850 nm and 1300 nm. Several problems occur at the longer wavelengths. For example two orders of magnitude less power can be launched at 1300 nm into a single mode fibre, compared to 850 nm and a multimode fibre. Since Rayleigh backscatter is inversely proportional to the wavelength raised to the fourth power, the backscatter is about 7 dB less at 1300 nm compared to 850 nm. Rayleigh backscatter is about 6 dB lower for single mode fibres compared to multimode, since it is proportional to the square of the numerical aperture.

Bibliography

Chapter 1

H.S. Hvistendahl, *Engineering Units and Physical Quantities*, Macmillan and Co. Ltd., 1964.

National Bureau of Standards (US), Monograph 56, Aug. 1963

M.L. McGlashan, *Physicochemical Quantities and Units*. The Royal Institute of Chemistry, London, 1971.

G.W.C. Kaye & T.H. Laby, *Tables of Physical and Chemical Constants*, 13th ed. Longmans, Green & Co. Ltd., London, 1966.

D.G. Fink & W. Beaty, *Standard Handbook for Electrical Engineers*, 11th ed. McGraw–Hill, New York, 1979.

Institute of Electrical and Electronics Engineers, *Standard Metric Practice*, IEEE Standard 268, New York, 1976.

Institute of Electrical and Electronics Engineers, *Recommended Practice for Units in Published Scientific and Technical Work*. IEEE Standard 268, New York, 1973.

Institute of Electrical and Electronics Engineers, *IEEE Standard Dictionary of Electrical and Electronics Terms*. IEEE Standard 100-1972. Wiley, New York, 1975.

Institute of Electrical and Electronics Engineers, *Graphic Symbols for Electrical and Electronic Diagrams*. IEEE Standard 315, New York, 1975.

Institute of Electrical and Electronics Engineers, *Letter Symbols for Units in Science and Technology*. ANSI Standard Y-10.19. IEEE Standard 260-1969. New York.

Institute of Electrical and Electronics Engineers, *Letter Symbols for Quantities Used in Electrical Science and Electrical Engineering*. ANSI Standard Y10.5-1968. IEEE Standard 280-1968. New York.

E.A. Mechtly, *The International System of Units*, NASA SP7012, NASA, 1969.

ICSU-CODATA Central Office, *Recommended Consistent Values of the Fundamental Physical Constants*. CODATA Bull. 11, Dec. 1973.

American National Standards Institute. SI Units and Recommendations for the Use of their Multiples, and Certain Other Units, International Standards. LSO-1000, New York, 1973.

F.F. Mazda, *Electronic Engineers Reference Book*, 5th Ed. Butterworths, London, 1983.

Chapter 2

Philco Technological Centre, *Electronic Precision Measurement Techniques and Experiments*, Prentice–Hall Inc., N.J., 1964.

M.B. Stout, *Basic Electrical Measurements*, Prentice–Hall Inc., N.J., 1960.

D. Hartke, A VLF Comparator for Relating Local Frequency to U.S. Standards, *Hewlett Packard J.*, Oct. 1964.

W.G. Cady, *Piezoelectricity; an Introduction to the theory and Application of Electrochemical Phenomena in Crystals*. Dover Publications Inc., New York, 1964.

V.E. Bottom, *The Theory and Design of Quartz Crystal Units*, MacMurray Press, Ill, 1968.

D.C. Hammond, C. Adams, & L.S. Cutler, Precision crystal units, *Frequency*, Jul.–Aug. 1963.

L.S. Cutler, Some aspects of the theory and measurement of frequency fluctuations in frequency standards, *Proc. IEEE* **54**, no. 2, Feb. 1966.

F.B. Silsbee, *Establishment and Maintenance of Electrical Units*, NBS Circ., 475, Jun. 1949.

F.B. Silsbee, *Extension and Dissemination of the Electrical and Magnetics Units by the NBS*, NBS Circ., 531, Jul. 1953.

R.D. Cutkosky, Evaluation of the NBS unit of resistance based on a calculable capacitor, *J. Res. NBS, A,* **65**, 1961.

F. Wenner, Methods, apparatus and procedures for the comparison of precision standard resistors, *J. Res. NBS*, **25**, Aug. 1940.

C.W. Oatley & J.G. Yates, Bridges with coupled inductive ratio arms as precision instruments for the comparison of laboratory standards of resistance or capacitance, *Proc. IEE*, **101**, Mar. 1954.

J.J. Hill & A.P. Miller, An a.c. double bridge with inductively coupled ratio arms for precision platinum resistance thermometry, *Proc. IEE*, **110**, no. 2, 1963.

A.M. Thompson, An absolute determination of resistance based on a calculable standard of capacitance, *Metrologia*, **4**, no. 1, Jan. 1968.

J.L. Thomas, *Precision Resistors and Their Measurements*, NBS Circ. 470.

B.V. Hamon, A 1 to 100 ohm build-up resistor for the calibration of standard resistors, *J. Sci. Instr.* **31**, no. 12, 1964.

L. Julie, *Ratio Metrics: A New, Simplified Method of Measurement Calibration and Certification*, IEEE 1964 Conv. Rec., Pt8, 1964.

J.J. Hill, Calibration of dc resistance standards and voltage ratio boxes by an ac method, *Proc. IEE, Pt1*, **112**, no. 1, Jun. 1963.

M.C. McGregor *et al.*, New apparatus at NBS for absolute capacitance measurement, *IRE Trans. Instr.* **1-7**, Dec. 1950.

D.L. Hillhouse & J.W. Kline, A ratio transformer bridge for standardization of inductors and capacitors. *IRE Trans. Instr.* **1-9**, No. 2, Sep. 1960.

R.D. Cutkosky, Four-terminal pair networks as precision standards, *IEEE Trans. Commun. and Electron.* No. 20, Jan. 1964.

S.L. Howe (Ed.), *NBS Time and Frequency Dissemination Services*, NBS Spec. Publ. 432, 1976.

R.D. Cutkosky & J.Q. Shields, Precise measurement of transformer ratios, *Trans. IRE*, Dec. 1960.

B.F. Field *et al.*, Volt maintenance at NBS via 2 *e/h*: a new definition of the NBS volt, *Metrologia*, **9**, 1973.

B.F. Field & V.W. Hesterman, Laboratory voltage standard based on 2 *e/h*, *IEEE Trans. Instrum. Meas.* Dec. 1976.

R.D. Cutkosky, New NBS measurements of the absolute farad and ohm, *IEEE Trans. Instrum. Meas.* Dec. 1974.

D.G. Lampard, A new theorem in electrostatics with application to calculable standards of capacitance, *Proc. IEEE*, **104**, Pt. C, 1967.

R.E. Beehler, A historical review of atomic frequency standards, *Proc. IEEE*, **55**, No. 6, Jun. 1967.

C.F. Coombs (Ed.), *Basic Electronic Instrument Handbook*, McGraw-Hill Book Co. 1972.

H. Hellwig, *Frequency Standards and Clocks: A Tutorial Introduction*, NBS Technical Note 616, USGPO, Apr. 1972.

A.O. McCoubrey, A survey of atomic frequency standards, *Proc. IEEE*, **54**, no. 2, Feb. 1966.

D. Keppner *et al.*, Theory of the hydrogen maser, *Phys. Rev.* **126**, Apr. 1962.

R.E. Beehler *et al.*, Cesium beam atomic time and frequency standards, *Metrologia*, **1**, no. 3, Jul. 1965.

D.H. Throne, A rubidium vapour frequency standard for systems requiring superior frequency stability, *Hewlett Packard Journal*, **19**, no. 11, Jul. 1968.

D.W. Allen *et al.*, The National Bureau of Standards atomic time scales: generation, dissemination, stability and accuracy, *IEEE Trans. Instrum. and Meas.* IM-21 no. 4, Nov. 1972.

J.A. Barnes, The development of an international atomic time scale, *Proc IEEE*, **55**, no. 6, Jun. 1967.

A.R. Chi & H.S. Fosque, A step in time, *IEEE Spectrum*, **9**, no. 1, Jan. 1972.

J.T. Henderson, The foundation of time and frequency in various countries, *Proc IEEE*, **60**, no. 5, May 1972.

Fundamentals of time and frequency standards, Hewlett Packard, Application Note 52-1.

Timekeeping and frequency calibration, Hewlett Packard, Application Note 52-1.

B.E. Blair (Ed.), *Time and Frequency: Theory and Fundamentals*, NBS Monograph 140, USGPO, May 1974.

R.A. Day, Use of Loran-C navigational system as a frequency reference, *Signal*, Nov. 1973.

E. Ehrlich, The role of time/frequency in satellite position determination systems, *Proc. IEEE* 60, no. 5, May 1972.

F.F. Mazda, *Discrete Electronic Components*, Cambridge University Press, 1981.

Tom Wilkie, Time to re-measure the metre, *New Scientist*, 27 Oct. 1983.

B. Hagga, Essentials of calibration standards, *Test Electronics*, Jan. 1985.

Accuracy enhancement in d.c. calibration systems, *Electronics Industry*, Jun. 1985.

P. Harold, Solid state d.c. voltage standards rival standard cell performance, *EDN*, 4 Apr. 1985.

Hal Chenhall, Designing a multimeter calibration system, *New Electronics*, 1 Oct. 1985.

Ed Patterson, Calibration becomes an industry, *Test Electronics*, Jul./Aug. 1985.

Calibration systems and equipment, *Test Electronics*, Sep./Oct. 1985.

Frank Capell, New developments in automated calibration systems, *Test & Measurement World*, Oct. 1985.

Ed Patterson, Choosing a calibration service, *Test Electronics*, Sep./Oct. 1985.

Chapter 3

J.R. Pearce, Electronic measuring instruments: results precise, but wrong, *Electronics & Power*, Feb. 1983.

H.D. Young, *Statistical Treatment of Experimental Data*, McGraw–Hill, New York, 1962.

D. Bartholomew, *Electrical Measurements and Instrumentation*, Allyn & Bacon, Boston, 1963.

M.B. Stout, *Basic Electrical Measurements*, Prentice–Hall, N.J., 1960.

F. Mazada (Ed.), *Electronics Engineers' Reference Book*, Butterworths, London, 1983.

G.O. Chalk & A.W. Stick, *Statistics for the Engineer*, Butterworths, London, 1975.

W.C. Hahn, *Modern Statistical Methods*, Butterworths, London, 1979.

S. Lyons, *Handbook of Industrial Mathematics*, Cambridge University Press, 1978.

Chapter 4

S. Gage *et al.*, *Optoelectronics Applications Manual*, McGraw–Hill, New York, 1977.

F.F. Mazda, *Discrete Electronic Components*, Cambridge University Press, 1981.

R. Allan, New applications open up for silicon sensors, *Electronics*, Nov. 6, 1980.

R.J. Corruccini, Interpolation of platinum resistance thermometers, *Rev. Sci. Instrum.* **31**, 1960.

M. Dean (Ed.), *Semiconductor and Conventional Strain Gages*, Academic, New York, 1962.

C.M. Harris & C.E. Cred (Eds), *Shock and Vibration Handbook*, 3 vols., McGraw–Hill, New York, 1961.

T.R. Harrison, *Radiation Pyrometry and its Underlying Principles of Radiant Heat Transfer*, Wiley, New York 1960.

J.J. Merill, *Light and Heat Sensing*, Macmillan, New York, 1963.

H.N. Norton, Specification characteristics of pressure transducers, *Instrum. Control Syst.*, Dec. 1963.

F. Oliver, *Practical Instrumentation Transducers*, Hayden, New York, 1971.

J.M. Ruskin, Thermistors as temperature transducers, *Data Syst. Eng.*, **19**, no. 2, Feb. 1964.

H.N. Norton, *Sensor and Analyzer Handbook*, Prentice–Hall, Englewood Cliffs, N.J., 1982.

C.C. Perry & H.R. Lissner, *The Strain Gauge Primer*, McGraw–Hill, New York, 1967.

R.P. Benedict, *Fundamentals of Temperature, Pressure and Flow Measurements*, Wiley, New York, 1977.

J.A. Hall, *The Measurement of Temperature*, Chapman & Hall, London, 1966.

F.J. Oliver, *Practical Instrumentation Transducers*, Pitman, London, 1972.

I.E. Shepherd, Temperature measurement with low cost thermistors, *Elect. Eng.* Sep. 1974.

K.C. Bhatt, *Thermistors–Development, Manufacture and Applications*, IEE–IERE Proceedings, India, Sep.–Oct. 1973.

K.W. Stanley, Non-linear resistors, *The Radio and Electronic Engineer*, **43**, no. 10, Oct. 1973.

Jim McDermott, Focus on piezoelectric crystals and devices, *Electronic Design*, **17**, Aug. 16, 1973.

Hans-Gunther Steidle, Semiconductor magnetoresistors, *Siemens Review* No. 4, 1973.

Yoshimi Makino, New breed of magneto-resistance element solves magnetic sensing problem, *JEE*, Aug. 1975.

Dave Bursky, Sensors in five areas are getting tinier, cheaper and more precise, *Electronic Design*, **15**, Jul. 19, 1974.

David Tompkins, The selection of pressure transducers, *Electron*, 8 May, 1975.

E.A. Lyons, Force transducers, *Electron*, 12 Sep. 1974.

A.R. Zias & W.F.J. Hare, Integration brings a generation of low-cost transducers, *Electronics*, Dec. 4, 1972.

C. Budd, Thermistor and thermocouple action, *Wireless World*, Feb. 1976.

M. Renneberg, Trends in thermometry, *Electron*, 20 Mar. 1979.

J. Nicol, Dimension measurement using photodiode arrays, *Electron Ind.* Oct. 1979.

R.O. Cook & C.W. Hamm, Fibre optic lever displacement transducer, *Appl. Opt.* 1 Oct. 1979.

E. Bose, Fluid pressure transducers, *Electronic Engineering*, Nov. 1981.

M.L. Sanderson, Electromagnetic and ultrasonic flowmeters, *Electronics and Power*, Feb. 1982.

W. Henning, Microelectronic sensors in semiconductor technology, *Automation*, Oct. 1981.

A. Bead, Measurements using the Hall effect, *New Electronics*, 12 Apr. 1981.

R.M. Langdon, Vibratory process control transducers, *Electronic Engineering*, Nov. 1981.

T. Wearden. Fibre Optic Sensors and Transducers, *Electronic Product Design*, Jun. 1982.

R.S. Medlock, Transducers, Instruments and measurements in the process industries, *Electronics and Power*, Feb. 1982.

P.H. Sydenham, *Measuring Instruments: Tools of Knowledge and Control*, Peter Peregrinus, 1980.

L.R. Wollmann, The thermopile: the commercial infrared detector, *Electro. Opt. Syst. Des.* 9 Sept. 1979.

Jim McDermott, Sensors and transducers, *EDN*, 20 Mar. 1980.

T. Ivall, Developments in optical transducers, *Automation*, Jun. 1983.

Sensors becoming more diversified and lower price, *JEE* Mar. 1983.

H. Gutgesell, Basis of pressure transducers, *Electronics and Power*, Feb. 1983.

R.M. Whittier, Silicon strain gauge technology in aerospace applications, *Electronic Engineering*, Nov. 1981.

E.N. Sharples, Liquid crystals for temperature measurement, *Test & Measurement World*, Jun. 1983.

T. Ivall, Transducer development is slow but promising, *Automation*, May 1983.

W.S. Jaroszynski, Temperature measurement, *Test & Measurement World*, Apr. 1983.

G. Erh, Specifying temperature measuring systems, *Test & Measurement World*, Jun. 1983.

J. McDermott, Sensors and transducers, *EDN*, 4 Aug. 1983.

G. Boyes, High resolution angular measurement, *Electronic Product Design*, Oct. 1983.

H.L. Berman, Infrared temperature measurement, *Test & Measurement World*, Oct. 1983.

A. Petersen, Silicon temperature sensors, *Electronic Components and Applications*, **5**, no. 4, Sep. 1983.

John Proctor, Temperature transducer IC is linear over wide range, *Electronic Design*, 5 Apr. 1984.

P.R. Barabash, Basic limitations of ISFET and silicon pressure transducers, *Sensors Actuators*, **4**(3), Nov. 1983.

Frank Goodenough, Sensor ICs: processing, materials open factory doors, *Electronic Design*, 18 Apr. 1985.

Paul Brokaw, Versatile transmitter chip links strain gauges and RTDs to current loop, *Electronic Design*, 18 Apr. 1985.

Roger Allen, Non vision sensors, *Electronic Design*, 27 June, 1985.

R.E. Newnham, Composite piezoelectric sensors, *Ind. Qual. Cond.*, **60** (1–4), 1984.

J. Williams, Digitise transducer outputs directly at the source, *EDN*, 10 Jan. 1985.

R. Frank, One element transducer brings uniform traits to IC pressure sensors, *Electronic Design*, 18 Apr. 1985.

J. McDermott, Silicon fabrication techniques extend sensor ranges and designs, *EDN* 27 Dec. 1984.

W. Thomas, What the future holds for sensors and transducers, *New Electronics*, 13 Apr. 1985.

Chapter 5

Jim Farrell, Mating micros to the IEEE 488 bus doesn't take many connections, *Electronic Design*, 22 Nov. 1978.

Leon Smith, Join 488 bus instruments and efficient software for fast, automatic tests, *Electronic Design*, 22 Nov. 1978.

Dennis Moralee, Towards intelligent instruments, *Electronics and Power*, Jan. 1979.

Dan Hosage, The office supercontroller, *Telecommunications*, Dec. 1980.

W.A. Levy, WangNet: a bold step forward?, *Mini–Micro Systems*, Nov. 1981.

C. Warren, Understanding bus basics helps resolve design conflicts, *EDN*, 27 May 1981.

A. Santoni, IEEE 488 instruments, *EDN*, 28 Oct. 1981.

R.E. Metzler, IEEE 488 for the eighties, *New Electronics*, 10 Nov. 1981.

W.B. Riley, Local area networks move beyond the planning stage, *Electronic Design*, 11 Nov. 1982.

M. Beishon, Local area networks, *New Electronics*, 23 Mar. 1982.

W.A. Levy & H.F. Mehl, Searching for the 'right' approach, *Mini–Micro Systems*, Feb. 1982.

F.A. Wang, Office automation, *Mini–Micro Systems*, Dec. 1982.

J.S. Mayo, Communications at a distance, *Mini–Micro Systems*, Dec. 1982.

A. Goldberger & C. Kaplinsky, Small area networks fit jobs too small for local nets, *Electronics*, 3 Nov. 1982.

W.D. Livingston, Local area network improves real time intelligence systems, *Defense Electronics*, Dec. 1982.

V. Coleman *et al.*, Controlling local area networks, *Electronic Product Design*, Oct. 1982.

Eric Lundquist, ATE vendors embrace networking, *Mini–Micro Systems*, Dec. 1982.

D. Bailey, The standard (STD) bus, *New Electronics*, 14 Dec. 1982.

M. Tilden & B. Ramirez, Understanding IEEE 488 basics simplifies system integration, *EDN*, 9 Jun. 1982.

P. Snigier, Designers' guide to the IEEE 488 bus, *Digital Design*, Jun. 1982 and Aug. 1982.

R. Connel, IEEE 488 bus test systems, *New Electronics*, 20 Apr. 1982.

G. Kotelly, Local area networks, *EDN*, 17 Feb. 1982.

Bus systems and networks, *Electronic Engineering*, Mar. 1983.

G. Cruzan, Distributed intelligence in GPIB systems, *Test & Measurement World*, Jan. 1983.

R. Wiggins, Intelligent networking, *Telecommunications*, Jan. 1983.

D.B. Davis, Pioneering vendors attempt to develop infant broadband local net markets, *Mini–Micro Systems*, Jan. 1983.

B. Metcalfe, Controller–transceiver board drives Ethernet into PC domain, *Mini–Micro Systems*, Jan. 1983.

W. Twaddell, Varied technologies and approaches contend for network interconnect dominance, *EDN*, 3 Feb. 1983.

E.R. Teja, Powerful local area network controllers make networking more accessible than ever, *EDN*, 3 Mar. 1983.

J.R. Jones, Consider fibre optics for local-network designs, *EDN*, 3 Mar. 1983.

Howard Frank, Broadband versus baseband local area networks, *Telecommunications*, Mar. 1983.

G. Kotelly, Personal computer networks, *EDN*, 3 Mar. 1983.

J.M. Stratford, Networked automatic test equipment improves manufacturing efficiency, *EDN*, 3 Mar. 1983.

D. Ledamun & M. Goodwin, Exploring the possibilities of the 1553 B data bus, *Electronic Engineering*, Mar. 1983.

H. Wilson, Using the IEEE 488 instrument bus, *Electronic Engineering*, Mar. 1983.

B. Donnelly, Board level design – VME bus allows future expansion, *Electronics Industry*, Apr. 1983.

B. Green, STD-bus is both cost effective and flexible. *New Electronics*, 3 May 1983.

D. Casey, Buses – a thoroughly modern method of transport, *Automation*, Jun. 1983.

V. Iyer & S.P. Joshi, Hardware considerations in LANs, *Electronic Product Design*, Oct. 1983.

G. Trudgen, Understanding the IEEE bus, *New Electronics*, 15 Nov. 1983.

V. Iyer & S.P. Joshi, Hardware considerations in LANs, *Electronic Product Design*, Nov. 1983.

M.D. Tilden, Programming techniques speed IEEE 488 system execution, *EDN 27*, Oct. 1983.

C.H. Small, New features, abilities and prices expand IEEE 488 controller choices, *EDN*, 27 Oct. 1983.

R.R. Russ, Getting the best of both buses, *Computer Design*, Oct. 1983.

M.J. Relis, Military avionic LANs point toward fiber optics, *Defence Electronics*, Oct. 1983.

Paul Hensel, Local area networks and the impact of optical fibers, *Electronics & Power*, Nov./Dec. 1983.

M.V. Wilkes & D.J. Wheeler, *The Cambridge Digital Communication Ring*, Proceedings of LACN Symposium, May 1979.

G.W. Litchfield, D. Hunkin & P. Hensel, *Application of Optical Fibres to the Cambridge Ring System*, Proceedings of ICCC, 1982.

Edwin A. Bertress, Baseband LAN fine tunes token passing technique, *Computer Design*, Autumn, 1983.

A.M. Dahod 10 M–bps LAN combines benefits of CSMA/CD and token passing, *Mini–Micro Systems*, Nov. 1983.

Paul Snigier, Designer's guide to the GPIB/HPIL, *Digital Design*, Jun. 1982.

R. Parker & S.F. Shapiro, Untangling local area networks, *Computer Design*, Mar. 1983.

Joseph D. Baugh PC, local net binds elements of Arcnet, Ethernet and Cluster/One, *Mini–Micro Systems*, May 1983.

S. Ohr, Ethernet chips hold the lead in VLSI scramble for local networks, *Electronic Design*, 23 Jun. 1983.

R. Ott, A serial bus for inter-chip data transfer, *Electronic Product Design*, Dec. 1983.

F. Costa, The evolution of Multibus 11, *New Electronics*, 13 Dec. 1983.

L. Zidek, The I²C bus – a small area network, *New Electronics*, 13 Dec. 1983.

Tony Danbury & Geoff Alan, Small area networks using serial data transfer, part 1, *Electronic Engineering*, Jan. 1984. Part 11, *Electronic Engineering*, Feb. 1984.

A. Moelands, *I²C bus in consumer applications, Electronic Components and Applications*, **5**, no. 4 Sep. 1983.

S. Ohr, Three 32 bit wide buses will give 32 bit μCs mainframe performance, *Electronic Design*, 12 Jan. 1984.

J. Richard Jones, Emerging trends in local area networks, *Telecommunications*, Dec. 1983.

Jack E. Hemenway, Powerful VME bus features ease high-level microcomputer applications, *EDN*, 12 Jan. 1984.

The 1553 bus, *Electronic Product Design*, Feb. 1984.

Barrie Nicholson, 32 bit futurebus nears IEEE approval, *EDN*, 23 Feb. 1984.

S.J. Packer & N. Bhasker, Message passing in Multibus II, *Electronic Product Design*, Mar. 1984.

J.V. St. Amand, Local area networks: a matter of choice, *IEEE Trans. (NS)*, **31**(1), Feb. 1984.

Nicholas Mokhoff, Networks expand as PBXs get smarter, *Computer Design*, Feb. 1984.

H. Wurzburg & S. Kelley, PBX-based LANs: lower cost per terminal connection, *Computer Design*, Feb. 1984.

W.E. Burr & R.J. Carpenter, Wideband local nets enter the computer arena, *Electronics*, 3 May 1984.

G.S. Gardiner, A multiprocessor data communications link, *Electronic Product Design*, May 1984.

John Theus *et al.*, Futurebus anticipates coming needs, *Electronics*, 12 Jul. 1984.

Roger Allan, LANs stake their claims and opt for co-existance, *Electronic Design*, 26 Jul. 1984.

Alan V. Flatman, Low-cost local network for small systems grows from LEEE 80213 standard, *Electronic Design*, 26 Jul. 1984.

Haw-Ming Haug & Gerald Moseley, Manchester chip eases the design of Ethernet systems, *Electronic Design*, 26 Jul. 1984.

Walt Sapronov, Gateways link long-haul and local networks, *Data Communications*, Jul. 1984.

S. Joshi & V. Iyer, New standards for local networks push upper limits for lightwave data, *Data Communications*, Jul. 1984.

A.M. Dahod, Local network responds to changing system needs, *Computer Design*, 1 Jun. 1984.

R. Rosenberg & T.E. Feldt, Local nets arrive in force, *Electronics Week*, 6 Aug. 1984.

F.M. Burg & C.T. Chen, Of local networks, protocols and the OSI Reference Model, *Data Communications*, Nov. 1984.

C.D. Tsao, A local area network architecture overview, *IEEE Communications Magazine*. Aug. 1984.

G. Moore, Local area networks, *Electronics and Power*, Oct. 1984.

W. Stallings, Local networks, *Computing Surveys*, Mar. 1984.

D.B. Gustavson, Computer buses – a tutorial, *IEEE Micro*, Aug. 1984.

C. Pabouctsidis & J.L. Ebener, G-64 bus suits mid-range industrial μC applications, *EDN*, 18 Apr. 1985.

P. Harold, Powerful local buses join VME bus, *EDN*, 18 Apr. 1985.

J. Hicks, Still a market for the STD bus, *New Electronics*, 13 Aug. 1985.

W. Neudecker, Why choose Multibus for your next design?, *New Electronics*, 13 Aug. 1985.

T. Balph & D. Artusi, VME – a system architecture for industrial control, *New Electronics*, 13 Aug. 1985.

J. Victor, Multibus II, VME bus clash in 32-bit arena, *Mini–Micro Systems*, Aug. 1985.

Jon Titus, Two buses vie for 32-bit system supremacy, *EDN*, 31 Oct. 1985.

R. Rosenberg, Battle of the buses: and the winner is . . . , *Electronics*, 25 Nov. 1985.

Ed Jacks, Bridging the gap between Multibus I and II, *New Electronics*, 7 Jan. 1986.

M. Tucker, Network software rejuvenates LAN market, *Mini–Micro Systems*, Sep. 1985.

R.V. Balakrishnan, The proposed IEEE 896 Futurebus – a solution to the bus driving problem, *IEEE Micro*, Aug. 1984.

Chapter 6

D. Bartholomew, *Electrical Measurements and Instrumentation*, Allyn & Bacon, 1963.

H.E. Thomas & C.A. Clarke, *Handbook of Electronic Instruments and Measurement Techniques*, Prentice–Hall, 1967.

A.P. Malvino, *Electronic Instrumentation Fundamentals*, McGraw–Hill, New York, 1967.

John D. Lenk, *Handbook of Electronic Test Equipment*, Prentice–Hall, 1971.

B.M. Oliver & J.M. Cage, *Electronic Measurements and Instrumentation*, McGraw–Hill, 1971.

S.D. Prensky, *Electronic Instrumentation*, Prentice–Hall, 1971.

C.N. Herrick, *Instruments and Measurements for Electronics*, McGraw–Hill, 1972.

J. Douglas-Young, *Complete Guide to Electronic Test Equipment and Troubleshooting Techniques*, Parker Publishing Co. 1975.

A.F. Arbel, *Analog Signal Processing and Instrumentation*, Cambridge University Press, 1980.

J.D. Lenk, *Handbook of Electronic Meters*, Prentice–Hall, 1981.

Sol. D. Prensky & R.L. Castellucis, *Electronic Instrumentation*, Prentice–Hall, 1982.

W. Freeman & W. Ritmanich, Cut a/d conversion costs by using d/a converters and software, *Electronic Design*, 26 Apr. 1977.

A. Muto & M. Neil, ADC dynamic performance testing, *Electronic Product Design*, Jun. 1982.

F. Shoreys, New approaches to high speed high resolution analogue to digital conversion, *Electronics & Power*, Feb. 1982.

J. Tsantes, Data converters, *EDN*, Aug. 1982.

H. Blatch, The increasing capabilities of modern DMMs, *New Electronics*, Apr. 1981.

F.F. Mazda, *Integrated Circuits*, Cambridge University Press, 1978.

S. Runyon, DMMs, counters, sources, scopes give answers–not just readings, *Electronic Design*, 22 Nov. 1978.

K. Jessen, Digital voltmeter gets speedy, accurate results, *Electronic Design*, 7 Jun. 1980.

Rick Nelson, $4\frac{1}{2}$ digit hand-held models head the list of the latest innovative low-cost DMMs, *EDN*, 5 Aug. 1981.

Gary F. Chesnutis, Intelligent instruments, *EDN*, 3 Mar. 1982.

Michael Chester, Handheld DMMs: good things keep coming in smaller packages, *Electronic Design*, 25 Nov. 1982.

G. McGlinchey, A high-speed 12 bit a/d converter, *Electronic Product Design*, Aug. 1982.

P.J. Hart, Measurement techniques in r.m.s. multimeters, *New Electronics*, 19 Apr. 1983.

H.J. Lit, The microprocessors' impact on precision measuring equipment, *New Electronics*, 19 Apr. 1983.

M. Dance, Benchtop t & m review – DMMs break price structures, add features, *Electronics Industry*, Nov. 1983.

A.M. Rudkin, Advances in microprocessor controlled instrumentation, *New Electronics*, 19 Apr. 1983.

B. Nicholson, Novel a/d converters, built-in facilities extend DMM accuracies and capabilities, *EDN*, 5 Apr. 1984.

C. Pointer, The development of a new DMM, *Electronic Product Design*, May 1984.

David A. Bell, *Electronic Instrumentation and Measurements*, Reston Publishing Co. 1983.

John A. Allocca & Allen Stuart, *Electronic Instrumentation*, Reston Publishing Co. 1983.

Bill Travis, Data converters, *EDN*, 14 Jun. 1984.

Gary Davies, Multimeters - understanding the differences, *Electrical Equipment*, Aug. 1984.

Glenis Moore, Digital multimeters, *Electronics and Power* Aug. 1984.

B. Brodie, A 160 ppm digital voltmeter for use in a.c. calibration, *Electronic Engineering*, Sep. 1984.

Jim Hicks, Developments in DPM design, *New Electronics*, 5 Feb. 1985.

Bob Milne, DMMs bring laboratory precision to the production floor, *Electronic Design*, 16 May 1985.

R. Sommers, Hand-held instruments, *Electronic Design*, 14 Mar. 1985.

K. Salz & F. Smith, Smart a.c. voltmeter takes giant step forward in analysis and comparison, *Electronic Design*, 8 Aug. 1985.

T & M review – increasing sophistication in DMMs, *Electronics Industry*, Sep. 1985.

Chapter 7

E.C. Crawford, Impedance measurements and the *in situ* component bridge, *Marconi Instrumentation*, **9**, no. 2, 1963.

M.P. MacMartin & N.L. Kusters, A d.c. comparator ratio bridge for four-terminal resistance measurements, *IEEE Trans. Instr.* **IM-15**, no. 4, Dec. 1966.

S. Wolf, *Guide to Electronic Measurements and Laboratory Practice*, Prentice–Hall, 1973.

Common *V & I* measurement cuts *LCR* bridge costs, *Electronics Industry*, Jul. 1981.

D. Tait, Inductance measurement – a new bridge configuration, *Electronics Industry*, Nov. 1980.

David A. Bell, *Electronic Instrumentation and Measurements*, Reston Publishing Co., 1983.

Larry D. Jones & A. Foster Chin, *Electronic Instruments and Measurements*, John Wiley & Sons, 1983.

Chapter 8

Davis Bartholomew, *Electrical Measurements and Instrumentation*, Allyn and Bacon, Boston, 1963.

Fundamentals of R. F. and Microwave Power Measurements, Application Note 64-1, Hewlett Packard, Aug. 1977.

R.E. Henning, Peak power measurement technique, *Sperry Engineering Review*, May–Jun. 1955.

IEEE Standard Application Guide for Bolometric Power Meters. IEEE Std. 470-1972.

W.H. Jackson, A thin-film semiconductor thermocouple

for microwave power measurements, *Hewlett Packard Journal*, **26**, no. 1, Sep. 1974.

Power meter – new designs add accuracy and convenience, *Microwaves*, **13**, no. 11, Nov. 1974.

R.E. Pratt, Very low level microwave power measurements. *Hewlett Packard Journal*, **27**, no. 2, Oct. 1975.

P.A. Szente *et al.*, Low-barrier-Schottky diode detectors, *Microwave Journal*, **19**, no. 2, Feb. 1976.

Thomas S. Laverghetta, *Handbook of Microwave Testing*, Artech House, 1981.

W. Jung, *V–f* converter doubles as clock and input of stable sine-wave source, *Electronic Design*, 15 Nov. 1984.

John H. Mayer, Signal/sweep generators, *Test & Measurement World*, Jun. 1985.

Charles H. Small, Benchtop pulse generators address high-speed applications, *EDN*, 25 Jul. 1985.

C. Palmer & A. Reid, Signal generators – enhanced microprocessor control, *Electronics Industry*, Sep. 1985.

Jim Lewis, Synthesizer uses maths input for signal generation, *New Electronics*, 15 Oct. 1985.

Chapter 9

F.F. Mazda, *Integrated Circuits*, Cambridge University Press, 1978.

Application and Performance of the 8671A and 8672A Microwave Synthesisers, Application Note 218-1, Hewlett–Packard Inc.

Rick Nelson, High frequency instruments, *EDN*, 18 Feb. 1981.

Mike Dance, Developments in signal generators *Electronics Industry*, Sep. 1981.

Michael Chester, Function generators do more, *Electronic Design*, 3 Feb. 1983.

Michaeal Fleischer *et al.*, A new family of pulse and pulse/function generators, *Hewlett Packard Journal*, Jun. 1983.

Robert Baetke, What is a function generator?, *Test and Measurement World*, Nov. 1983.

Peter Connell, Synthesiser techniques, *New Electronics*, 1 Nov. 1983.

Mike Dance, Signal generators – more activity up-market, *Electronics Industry*, Nov. 1983.

M. Sizmur, Applications of modern signal sources, *Electronic Engineering*. Apr. 1984.

Brian Dance, Monolithic function generator devices – a review, *New Electronics*, 1 May 1984.

David A. Bell, *Electronic Instrumentation and Measurements*, Reston Publishing Co., 1983.

John A. Allocca & Allen Stuart, *Electronic Instrumentation*, Reston Publishing Co., 1983.

Cori Hoberg, The signal generator – an introduction to the selection process, *Test & Measurement World*, Sep. 1984.

A.M. Rudkin, Signal generator for maintenance, *Electronic Engineering*, Dec. 1984.

C. Everett, Consider cost and output purity when choosing signal generators, *EDN*, 27 Dec. 1984.

Larry D. Jones & A. Foster Chin, *Electronic Instruments and Measurements*, John Wiley & Sons, 1983.

John H. Mayer, Function and pulse generators, *Test & Measurement World*, Mar. 1985.

Lester Brodeur, Waveform synthesizer relies on equations to define complex signals, *Electronic Design*, 16 May 1985.

Chapter 10

F.F. Mazda, *Integrated Circuits*, Cambridge University Press, 1978.

Fundamentals of Electronic Counters, Hewlett Packard Application Note 200, Jul. 1978.

Fundamentals of Microwave Frequency Counters, Hewlett Packard Application Note 200-1, Oct. 1977.

Fundamentals of Time Interval Measurements, Hewlett Packard Application Note 200-3.

Understanding Frequency Counter Specifications, Hewlett Packard Application Note 200-4.

G.W. Maton *et al.*, Complete counting capability, *Marconi Instrumentation*, **16**, no. 5, Summer 1979.

M.J. Bowman & D.G. Whitehead, A picosecond timing system, *IEEE Trans.* **1M-26**, no. 2, 1977.

J.W. Driscoll, Avoiding compromise in counters and timers, *Electronic Engineering*, Mid-Oct. 1979.

Barrie Nicholoson & Bill Donnelly, DVMs and counter timers – trends and availability, *Electronics Industry*, Jul. 1981.

Counter-timers and frequency meters, *Electronics Industry*, Nov. 1983.

Graham Prophet, Optimising processor-based instruments, *Electronic Product Design*, Dec. 1983.

Brian Hull, Counter-timers aim at higher frequencies, better accuracy, *Test and Measurement World*, Mar. 1984.

Dick Page. Developments in microwave counter design, *New Electronics*, 17 Apr. 1984.

John A. Allocca & Allen Stuart, *Electronic Instrumentation*, Reston Publishing Company, 1983.

L. Wakeman, CMOS counter-timer *IC* watches the clock for machine and user, *Electronic Design*, 18 Apr. 1985.

D. Draper, CMOS timer eases battery demands, *Computer Systems*, 15 Nov. 1984.

B. Neidorff, Dual-mode timer serves many applications, *EDN*, 29 Nov. 1984.

Counter timers and frequency meters–GHz capability for satcomms., *Electronics Industry*, Sep. 1985.

John H. Mayer, Counter/timers compete for higher frequency measurement, broader capabilities, *Test & Measurement World*, Sep. 1985.

Chapter 11

Spectrum Analyser Basics, Hewlett Packard Application Note 150, Apr. 1974.

Spectrum Analysis – Signal Enhancement, Hewlett Packard Application Note 150-7, Jun. 1975.

Spectrum Analysis – Accuracy Improvement, Hewlett Packard Application Note 150-8, Mar. 1976.

Spectrum Analysis – Noise Measurement, Hewlett Packard Application Note 150-4, Apr. 1974.

Network and Spectrum Analysis Primer, Hewlett Packard Application Note 216.

Gene Heftman, Focus on FFT spectrum analysers: know what makes them measuring computers, *Electronic Design*, 22 Nov. 1978.

Peter Osborne & Dick Widenka, Transient recorders, *Electron*, Nov. 1978.

Don Stoddart, Analyser pulse generator extends range, *New Electronics*, 27 Nov. 1979.

Phil Feinberg, Spectrum analysers adapt to diverse tasks, *Electronic Design*, 11 Oct. 1980.

Paul Colwill, The role of the transient recorder, *New Electronics*, 16 Nov. 1982.

Vic Fairchild, Zooming in the frequency domain, *Electronics Industry*, Dec. 1982.

Kenji Nakatsugawa, Double-duty instrument serves networks, spectrum analysis, *Electronics Design*, 3 Feb. 1983.

A.J.R. Lord, Signal processing using FFT techniques, *New Electronics*, 19 Apr. 1983.

H.L. Swain & R.M. Cox, Noise figure meter sets records for accuracy, repeatability and convenience, *Hewlett Packard Journal*, Apr. 1983.

D.R. Glancy, Waveform analysis of simple and complex signals, *Test and Measurement World*, Apr. 1983.

Waveform recorders, *Electrical Equipment*, Dec. 1983.

Mike Dance, Acquiring fast signals with waveform transient recorders, *Electronics Industry*, Feb. 1984.

Ray Ganderton, Spectrum analysers with counter accuracy to 325 GHz, *New Electronics*, 7 Feb. 1984.

Larry D. Jones & A. Foster Chin, *Electronic Instruments and Measurements*, John Wiley & Sons, New York 1983.

Graham Camplin, Digital waveform analysers, *Electronic Product Design*, Apr. 1985.

M. Van den Bergh, 2 MHz phase synthesizer brings NBS accuracy to engineer's test bench, *Electronic Design*, 18 Apr. 1985.

C. Erskine, Chip hardware, software speeds up processing in laboratory instrumentation, *Electronic Design*, 4 Apr. 1985.

J. O'Donnell, Looking through the right window improves spectral analysis, *EDN*, 15 Nov. 1984.

C. Everett, Spectrum analyzers meet measurement needs with a wider price range and model selection, *EDN*, 21 Mar. 1985.

D. Greenwood, Spectrum analysers, *Electronics Test*, May 1985.

R. Pope & R. Irwin, Enhancing the accuracy of spectrum analysis, *Communications International*, Aug. 1985.

John H. Mayer, Spectrum analyzers, *Test & Measurement World*, Oct. 1985.

Chapter 12

F. Mazda, *Discrete Electronic Components*, Cambridge University Press, 1981.

Jerald B. Murphy, Eliminating time-base errors from oscilloscope measurements, *Electronics*, 22 Jun. 1978.

D. Morgan, Digital storage oscilloscopes, *New Electronics*, 17 Apr. 1979.

S. Kennedy, Advances in oscilloscope photography, *New Electronics*, 17 Apr. 1979.

P. Jansen, A digital testing oscilloscope, *New Electronics*, 17 Apr. 1979.

A. Santoni, Laboratory oscilloscopes, *EDN*, 20 Jan. 1980.

D. Hoare & D. Parish, Analogue oscilloscopes for digital measurements, *New Electronics*, 29 Apr. 1980.

C. Gilder, Advances in digital oscilloscopes, *New Electronics*, 29 Apr. 1980.

D. Morgan, Choosing and using digital storage oscilloscopes, *New Electronics*, 29 Apr. 1980.

D. Parish, New developments in digital storage oscilloscopes, *Electronic Engineering*, Nov. 1980.

N. Vaughan, Logic analysers – the third generation, *New Electronics*, 25 Nov. 1980.

L. Farndale, Developments in digital storage oscilloscopes, *New Electronics*, 21 Apr. 1981.

M. Dance, Developments in oscilloscopes – going digital, *Electronics Industry*, May 1981.

M. Gasparian, Variable persistence aids signal display, *EDN*, 10 Jun. 1981.

R. Nelson, Storage oscilloscopes, *EDN*, 10 Jun. 1981.

Using the 468 digital oscilloscope in envelope mode, *Tek News*, Autumn 1981.

H. Blatch, The increasing capabilities of modern DMMs, *New Electronics*, 21 Apr. 1981.

C.M. Boardman & J.P. Michel, Future flat displays – the liquid crystal answer, *Electronic Engineering*, Feb. 1982.

D. Parish & M. Connah, Digital developments in oscilloscopes, *New Electronics*, 20 Apr. 1982.

J. Reed, Direct-view storage-tube displays, *Electronics & Power*, Apr. 1982.

A. Tegen & J. Wright, Oscilloscopes: the digital alternative, *New Electronics*, 20 Apr. 1982.

R. Nelson, Digital storage scope innovations make signal processing easier, *EDN*, 9 Jun. 1982.

Phosphor selection as a design consideration, *Electronic Engineering*, Jun. 1982.

M. Krans, Intelligent oscilloscopes, *New Electronics*, 1 Jun. 1982.

M. Riezenman, Even in a digital age scopes remain the instrument, *Electronic Design*, 2 Sep. 1982.

An Introduction to Automating Measurements with the

1980A/B Trigger Flag, Hewlett Packard Product Note 1980A/B-4, Sep. 1982.

S. Tanaka, Make a good choice of oscilloscopes, *JEE*, Oct. 1982.

I. Yamada, Programmable oscilloscopes play a vital role in labour saving, *JEE*, Oct. 1982.

S. Tanaka, Digital memory scopes offer more varied functions, *JEE*, Feb. 1983.

M. Ota, Integration and growing computer applications enhance demand for wideband oscilloscopes, *JEE*, Feb. 1983.

M.C. Gasparian, Oscilloscope, counter merge to measure time precisely, *Electronic Design*, 3 Feb. 1983.

W. Van Groningen, Digital storage oscilloscopes as spectrum analysers, *New Electronics*, 29 Apr. 1983.

C.H. Small, Oscilloscopes, *EDN*, May 1983.

C. Foley, Digital storage scopes in a test strategy, *Test & Measurement World*, May 1983.

C. Crook, Oscilloscope storage goes digital, *New Electronics*, 14 Jun. 1983.

Liquid crystals filter displays, *Electronic Product Design*, Sep. 1983.

C. Steward, Envelope-mode scopes exorcise random ghosts, *Electronic Design*, 15 Sep. 1983.

T. Engibons & G. Draper, Flat displays – an alternative to crts?, *Computer Design*, Sep. 1983.

R.E. Peterson Jr., Flat-panel displays, *EDN*, 24 Nov. 1983.

Dan Denham *et al.*, Using 'auto-convergence' in a colour display, *Mini–Micro Systems*, Nov. 1983.

Miniature flat screen crt, *Electronic Product Design*, Dec. 1983.

J. Driscoll, Peak monitoring–improving digital storage scope performance, *Electronic Engineering*, Feb. 1984.

Oscilloscopes, *Electronic Engineering*, Apr. 1984.

C. Everett, Storage oscilloscopes, *EDN*, 5 Apr. 1984.

O. Carrado, Measuring crt resolution, *New Electronics*, 15 May 1984.

R. Bristol, Oscilloscopes: the true general-purpose test and measurement instrument, *Test & Measurement World*, May 1984.

R. Peterson, High-capacity, high-contrast lcds become viable crt alternatives, *EDN*, 26 Jul. 1984.

N. Mokhoff, Flat-panel technologies vie to displace crt in terminals, *Computer Design*, Sep. 1984.

C. Crooke & D. Parish, Adding alphanumerics to digital storage, *New Electronics*, 4 Sep. 1984.

E. Evel, Digital oscilloscope is quick on the trigger to nab elusive glitches, *Electronic Design*, 31 Oct. 1984.

F.D. Rampey & M. Karin, 1-GHz digitals scope keeps a close watch on subnanosecond logic, *Electronic Design*, 18 Oct. 1984.

D.J. Oldfield, Testing in-circuit ECL is just routine for digital oscilloscope, *Electronic Design*, 15 Nov. 1984.

R. Bristol, Believable time measurements with oscilloscopes, *New Electronics*, 11 Dec. 1984.

J. Wright & S. Funge, Random sampling speeds up digital scopes, *New Electronics*, 8 Jan. 1985.

B. Furlow, CAE, scope makers jockey for position, *Computer Design*, Dec. 1984.

Ray Kushnir *et al.*, Speed and flexibility are equal partners in digitizing scope, *Electronic Design*, 24 Jan. 1985.

Rod Schlater, Digital scopes gain persistence, *Electronics Week*, 11 Feb. 1985.

Chapter 13

W.S. Richardson, Diagnostic testing of MOS random access memories, *Solid State Technology*, Mar. 1975.

W. Luciw, Can a user test LSI microprocessors effectively? *IEEE Trans. Manufact. Tech.*, MFT-**5**, no. 1, Mar. 1976.

Andy Santoni, Automatic testers can characterize as well as inspect, *Electronic Design*, 22 Nov. 1978.

Phil Nutburn, Economics of IC testing, *Electron*, 23 Oct. 1979.

Bob Botos, Designer's guide to *RCL* measurements, *EDN*, 5 Jun. 1979.

Gene Heftman, *IC* testers turn complex semis to good account, *Electronic Design*, 11 Oct. 1980.

Jonah McLeod, ATE packs new weapons as LSI invades components and boards, *Electronic Design*, 1 Feb. 1980.

Ed Belt & Roy Kole, Testing microprocessors, *New Electronics*, 10 Nov. 1981.

Mike Portsmouth, LSI testing tomorrow, *New Electronics*, 10 Nov. 1981.

B. Nicholson, Real time microprocessor testing requires speed and technique, *Electronic Industry*, Dec. 1981.

A. Santoni, Increased capability, wider range characterize today's benchtop testers, *EDN*, 18 Feb. 1981.

A. Santoni, Semiconductor test systems, *EDN*, 16 Apr. 1981.

A. Muto & M. Neil, ADC dynamic performance testing, *Electronic Product Design*, Jun. 1982.

C. Chrones, Capable analog testers handle variety of devices, *EDN*, 20 Jan. 1982.

T. Masson, Semiconductor parameter analysis, *New Electronics*, 16 Nov. 1982.

T.N. Thompson, Temperature testing electronic components, *Electronic Production*, Jan. 1982.

Y. Kikuchi, Digital *LCR* measurement instruments, *Electronics Industry*, Jul. 1982.

Donald R. Glancy, Analog/linear IC testing – flexible test systems solve complex challenges, *Test & Measurement World*, Sep. 1983.

H.L. Mason, A practical approach to the testing of a high-volume custom linear IC, *Electronics & Power*, Jun. 1983.

Tim Higgins, How digital signal processing can aid production testing of analogue LSI devices, *New Electronics*, 3 May 1983.

Chris Chrones, Software – a critical dimension in testing LSI/VLSI chips, *Semiconductor International*, Mar. 1983.

J.D. Lawrence, Parallel testing of memory devices, *Test & Measurement World*, Oct. 1983.

P.C. Maywell, Testing integrated circuits, *J. Elect. & Electron. Eng. Australia*, **3**, no. 4, Dec. 1983.

Peter H. Singer, Memory testing: as dynamic as ever, *Semiconductor International*, May 1983.

Richard Meredith, Enhanced testing of VLSI devices, *Electronics & Power*, Jun. 1983.

Stephan Ohr, VLSI/LSI testers: speed is primary, *Electronic Design*, 3 Feb. 1983.

K. Krauss & P.D. Via, In-circuit tester puts VLSI through its paces, *Electronic Design*, 3 Feb. 1983.

B. Donnelly, Component ATE – full performance testing or goods inward verification, *Electronics Industry*, Jul. 1983.

F. Hardaway & N. Kelley, Current issues in discrete semiconductor testing, *Test & Measurement World*, May 1983.

R.S. Gibbons, Measuring true contact resistance, *New Electronics*, 31 May 1983.

G. Hill, Component failure analysis with curve tracers, *Test & Measurement World*, Feb. 1983.

D.G. Glancy, IC testers meet the challenges of complex analog/linear circuits, *Test & Measurement World*, Sep. 1984.

D. Hutcheson, Linear IC test equipment, *Test & Measurement World*, Sep. 1984.

Stephen F. Scheiber, Emerging alternatives in VLSI testing, *Test & Measurement World*, Mar. 1984.

Scott Kline, VLSI testers help guarantee chip quality, *Electronics Week*, 29 Oct. 1984.

P. Buckley, Matching performance test demands, *Electronics Manufacture & Test*, Mar. 1984.

P.H. Singer, Testing ultra high speed devices, *Semiconductor International*, Sep. 1984.

Roddy Beat, The memory tester: its architecture and micro-code, *Test Electronics*, Feb. 1985.

C. McGinley, The semiconductor test problem – an analysis, *Test Electronics*, Jan. 1985.

John Coghlan, The design of an advanced VLSI tester, *Electronic Product Design*, Feb. 1985.

J.W. Driscoll, Automated voltage/current technique for cost-effective impedance measurement, *Electronics & Power*, Mar. 1985.

Terence Lee, In-circuit discrete analogue component testing at higher frequencies, *Test & Measurement World*, Apr. 1985.

Patrick Zicollelo, Testing video RAMs, *Test and Measurement World*, Oct. 1985.

Craig Foster, Optimising VLSI test accuracy, *Electronics Manufacture & Test*, Dec. 1985.

Chapter 14

E.G. Fubini & D.C. Johnson, Signal-to-noise ratio in a.m. receivers, *Proc. IRE*, **36**, Dec. 1984.

R.T. Myers & T.A. McKee, Receiver spurious responses – computer improves receiver design, *IEEE Trans. Vehic. Comm.*, Mar. 1966.

J. Linsley Hood, A direct coupled high quality stereo amplifier, *Hi-Fi News*, Nov. 1972 and Dec. 1972.

P.J. Baxandall, Low distortion amplifiers, *J. Brit. Sound Recording Assn.* Aug. 1961 and Nov. 1961.

B.A. Blesser, Digitization of audio, *JAES*, Oct. 1979.

K.J. Wood & M.J. Hawksford, High fidelity digital audio conversion using low-cost components, 65th AES Conv. London, 1980.

Using a Narrow Band Analyser for Characterizing Audio Products, Hewlett Packward, Application Note 192, Oct 1975.

Spectrum Analysis – Distortion Measurement, Hewlett Packard, Application Note 150-11.

Accurate and Automatic Noise Figure Measurements, Hewlett Packard, Application Note 64-3, Jun. 1980.

Wayne Jones, Measuring audio distortion, *Test & Measurement World*, Nov. 1982.

H. Pickler & F. Pavuza, Testing digital audio systems, *Test & Measurement Word*, Apr. 1983.

Chapter 15

R. Gannaway, Signal to noise ratio in receivers using linear or square-law envelope detectors, *Proc. IEEE Letters*, Oct. 1965.

A.L. Lance. *et al.*, Automated phase noise measurements. *Microwave Journal*, Jun. 1977.

C.J. Kikkert, A.M./F.M. Modulation standard, *IEE Proc. F., Commun., Radar & Signal Proc.* **128**(**6**), 1981.

R.H.T. Cartwright, Automated telephone-subscriber line-testing system, *Electronics & Power*, **27**(**9**), 1981.

Spectrum Analysis Amplitude and Frequency Modulation, Hewlett Packard Application Note 150-1, Nov. 1971.

Spectrum Analysis. Field Strength Measurement, Hewlett Packard Application Note 150-10, Sep. 1976.

Digital Phase Modulation (PSK) and Wideband F.M., Hewlett Packard Application Note 164-4, Aug. 1975.

Measuring F.M. Peak-to-Peak Deviation, Hewlett Packard application Note 174-8.

Measuring Electrical Length (delay) of Cables, Hewlett Packard Application Note 174-10, Nov. 1971.

Transmission Line Matching and Length Measurements Using Dual-Delayed Sweep in the Microprocessor Controlled Oscilloscope (Model 1722A), Hewlett Packard Application Note 185-2.

Percent Amplitude Modulation Measurements in the Time Domain, Hewlett Packard Application Note 185-3.

Precise Cable Length and Matching Measurements Using the 5370A Universal Time Interval Counter and 5363B Time Interval Probes, Hewlett Packard Application Note 191-6, Mar. 1980.

Applications and Measurements of Low Phase Noise Signals Using the 8862A Synthesized Signal Generator, Hewlett Packard Application Note 283-1, Nov. 1981.

Successful Buried Cable Fault Locating, Hewlett Packard Application Note 285.

Applications and Operation of the 8901A Modulation Analyser, Hewlett Packard Application Note 286-1, 1980.

Richard Adams, Tester checks telecom circuits with high accuracy and automatic speed, *Electronics*, 29 Dec. 1982.

S.C. Coupe, Automatic testing in the local telephone network, *Electronics & Power*, Jun. 1983.

Roger Allan, Data-comm testers take to the field, *Electronic Design*, 3 Feb. 1983.

Mark Johnston, Data communications line testing, *Test & Measurement World*, Dec. 1983.

A.M. Rudkin, Speed and Accuracy in modulation measurement, *Electronics & Power*, Feb. 1984.

Gabe Kasperek, What to look for in today's analogue test equipment, *Data Communications*, Nov. 1984.

J. Oxenboll & S. Neilson, Future design of telecom measuring instruments, *Telecommunications*, Mar. 1985.

Walter A. Fischer, Taking the mystery out of protocol analysis, *Telecommunications*, Mar. 1985.

David Welch, CATV equipment tests local area networks, *New Electronics*, 16 Apr. 1985.

E. Chapman, Build a quick and easy cable checker, *EDN*, 10 Jan. 1985.

E.S. Gillespie, A review of antenna measurement techniques, *Proc, 4th Intern. Conf. on Antennas and Propagation (ICAP 85)*, IEE, Coventry, 16–19 Apr. 1985.

Robert Panther, Subscriber line testing, *Telecommunications*, Sep. 1985.

Chapter 16

A Designer's Guide to Signature Analysis, Hewlett Packard Application Note 222, Apr. 1977.

Monitoring the IEEE-488 Bus with the 1602A Logic Analyser, Hewlett Packard Application Note 280-2, Jul. 1978.

The 1602A Logic State Analyzer as an Automatic Test Instrument, Hewlett Packard Application Note 280-3, Jul. 1978.

S.E. Scrupski, New digital tools evolve for bus-oriented microprocessor systems, *Electronic Design*, 22 Nov. 1978.

John Marshall, Now choosing the right logic analyzer requires a logical approach, *Electronic Design*, 22 Nov. 1978.

Rick Muething, Flagging down the logic analyzer express, *Digital Design*, Nov. 1978.

Lawrence Lowe, Digital troubleshooting, *Electron*, 19 Jun. 1979.

A. Santoni, Instruments, *EDN*, 20 Jul. 1979.

Minicomputer Analysis Techniques Using Logic Analyzers, Hewlett Packard Application Note 292, Aug. 1979.

A Manager's Guide to Signature Analysis, Hewlett Packard Application Note 222-3, Oct. 1980.

A Signature Analysis Case Study of a Z80-Based Personal Computer, Hewlett Packard Application Note 222-10, Oct. 1980.

Application Articles on Signature Analysis, Hewlett Packard Application Note 222-2, Oct. 1980.

Bruce Farley, Logic analyzers, *Digital Design*, Jan. 1980.

Jonah McLeod, Logic analyzers – sharp fault finding getting sharper, *Electronic Design*, 29 Mar. 1980.

A. Santoni, Logic analyzers, *EDN*, 5 Oct. 1980.

H. Reiney, *et al.*, Logic analyzer responds to user programming language, *Electronic Design*, 5 Jul. 1980.

K. Barnes, Present and future trends in logic analyzers, *Digital Design*, Apr. 1980.

I.H. Spector & R. Muething, Under designer's guidance, logic analyzer deploys its full strength against crashes, *Electronic Design*, 29 Mar. 1980.

Martin J. Weisberg, Designer's guide to testing and troubleshooting microprocessor based products, *EDN*, 20 Mar. 1980.

RS-232-C Communications with HP 64000 Logic Development System, Hewlett Packard Application Note 298-1, Dec. 1980.

Paul Wintz, Fundamentals of microprocessor development systems, *Digital Design*, Nov. 1980.

Guidelines for Signature Analysis, Hewlett Packard Application Note 222-4, Jan. 1981.

D. Peacock, Signature analysis aids production testing, 27 Nov. 1979.

A. Santoni, Newest logic and signature analyzers fulfil more digital testing needs, *EDN*, 25 Nov. 1981.

B. Nicholson, Extending signature analysis, *Electronics Industry*, Mar. 1981.

Functional Analysis of the IEEE 488 Interface Bus, Hewlett Packard Application Note 292-1, Feb. 1981.

John Mills, Micro logic analysers, *New Electronics*, 19 May 1981.

Dick Parrish, Speed considerations in logic analysis, *New Electronics*, 21 Apr. 1981.

A. Santoni, Instruments, *EDN*, 22 Jul. 1981.

Fred E. Warren, Understand the tradeoffs in development system selection, *EDN*, 10 Jun. 1981.

G. Kotelly, Operating systems cost more – but they also do more, *EDN*, 16 Sep. 1981.

Software Project Management with HP 64000 Logic Development System, Hewlett Packard Application Note 298-2, Jan. 1981.

HP 64000 Logic Development System Microassemblers for Bit–Slice Processors, Hewlett Packard Application Note 298-4, Jan. 1981.

Mike Mihalik & Bob Francis, Understand emulator use to increase prototyping skills, *EDN*, 10 Jun. 1981.

K. Rothmuller, Signature analysis software guides circuit troubleshooting, *Electronic Design*, 11 Nov. 1982.

M. Riezenman, Logic analyzers, spurred by microprocessors, move onto faster track, *Electronic Design*, 25 Nov. 1982.

G.M. Murch & J. Huber, Colour display clears up analysis of digital logic data, *Electronic Design*, 25 Nov. 1982.

G.F. Chesnutis, Logic analyzers, *EDN*, 9 Jun. 1982.

J. Mills *et al.*, Aids for the digital designer, *New Electronics*, 20 Apr. 1982.

T. Saitoh, Logic analyzers are essential for microprocessor troubleshooting, *JEE*, Aug. 1982.

M. Connah, Trace control in logic analyzers, *New Electronics*, 16 Nov. 1982.

Chris Bailey, Software strengthens development tools, *Electronic Design*, 4 Feb. 1982.

A. Santoni, Microprocessor development systems, *EDN*, 28 Apr. 1982.

Bobo Wang, Dedicated emulators untangle multiprocessor development, *Electronic Design*, 25 Nov. 1982.

Barrie Nicholson, Troubleshooting microprocessor based equipment, *Electronics Industry*, Jul. 1980.

Ken Lowe, Clock trigger versatility bolsters logic analysis, *Electronic Design*, 2 Sep. 1982.

Larry Palley, E² PROMs bring flexibility to in-system signature analysis, *Electronic Product Design*, Feb. 1983.

Lawrence Lowe, A signature analysis based test philosophy, *Electronics Industry*, Oct. 1983.

Charles H. Small, Enhanced signature analyzers ease microprocessor system diagnosis, *EDN*, 22 Dec. 1983.

Tom Williams, Logic analyzers rise to the challenge of microprocessors, *Computer Design*, 5 Apr. 1983.

John Brampton, Logic analyzers – a production tool?, *Test & Measurement World*, Oct. 1983.

Gail Hamilton, Logic analyzer gives programmers real-time view of software performance, *Electronics*, 5 May 1983.

Randy Steyer, Choose the logic analyzer that best meets your needs, *EDN*, 26 May 1983.

Bill Donnelly, Logic test instruments – analysers more powerful as software evolves, *Electronics Industry*, Jun. 1983.

C. Nobles *et al.*, A new approach to instrumentation, *Electronic Engineering*, Jun. 1983.

Donald R. Glancy, Logic analyzers – a tool for the digital world, *Test & Measurement World*, Oct. 1983.

Bob Milne, Logic analyzers mount a three-pronged attack on hardware and software debugging, *Electronic Design*, 15 Sep. 1983.

Dick Woods, Logic analyzer triggering gives the total picture, *Electronic Design*, 15 Sep. 1983.

Development systems and logic analysers – towards the universal digital workbench, *Electronic Engineering*, May 1983.

Sandra Jumonville, Time after time, logic analyzers got the job done, *Computer Design*, 5 Apr. 1983.

Charles Bilbe, State analyser treats hardware and software, *New Electronics*, 8 Feb. 1983.

S.D. Beste, Crack tough system problems with a dual-time base analyzer, *EDN*, 28 Apr. 1983.

B. Ableidinger, Timing boards complement state analyzer in PC add-on, *Electronic Design*, 24 Nov. 1983.

George Kotelly, Logic analyzer computer combination heralds personal instrument expansion, *EDN*, 31 Mar. 1983.

John W. Hyde, Software development system keeps up with advanced microprocessors, *Electronic Design*, 23 Jun. 1983.

R.C. Houghton, Software development tools: a profile, *Computer*, May 1983.

Charles Malkiel, Microprocessor designers adopt in-circuit emulation, *Test & Measurement World*, Dec. 1983.

S. Ohr, Focus on development systems: sophisticated tools abound, *Electronic Design*, 24 Nov. 1983.

S. Pease, Long and short term factors affect development system choice, *EDN*, 10 Nov. 1983.

Bill Donnelly, Microprocessor development systems –from hardware aid to full computer, *Electronics Industry*, Dec. 1983.

R. Freund & H. Gillette, Instruments for developing microprocessor based systems, *Electronics Industry*, Jun. 1983.

D.B. Richey & J.P. Romano, Emulator for 16 bit microprocessor, *Hewlett Packard Journal*, Mar. 1983.

M.W. Davis *et al.*, Extensive logic development and support capability in one convenient system, *Hewlett Packard Journal*, Mar. 1983.

Charles H. Small, Emulators, software transform personal computers into development systems, *EDN*, 10 Nov. 1983.

S. Vannerson, The engineering laboratory of the 1980s, *New Electronics*, 1 Nov. 1983.

James W. Nash, RX for 16 bit emulation: help is just a menu away, *Electronic Design*, 23 Jun. 1983.

R.A. Nygaard *et al.*, A modular analyzer for software analysis in the 64000 System, *Hewlett Packard Journal*, Mar. 1983.

J.E.M. Tyler, Extending microprocessor emulators, *Electronic Engineering*, Feb. 1983.

Paul Maritz, Software development, *Mini–Micro Systems*, Dec. 1983.

Gary F. Chesnutis, Multifunction development tools, *EDN*, 3 Feb. 1983.

R. Drohan *et al.*, Work station merges hardware software design, *Electronic Design*, 15 Sep. 1983.

Adrian Bishop, Software applies signature analysis to microprocessor chips, *Electronic Design*, 18 Oct. 1984.

P. Kleindienst & B. Greenlish, Microprocessor specific logic analysis, *New Electronics*, 17 Apr. 1984.

John Branton, The logic analyser becomes an essential tool, *New Electronics*, 17 Apr. 1984.

Kenneth Lowe & Mark Van Hook, Hardware/software problems yield to today's logic analyzers, *EDN*, 6 Sep. 1984.

Robert Delp, New directions in logic analysis, *Test & Measurement World*, Nov. 1984.

Malcolm Connah, Trends in logic analysers, *New Electronics*, 17 Apr. 1984.

Peter Heinrich, Multiprocessor development and testing made simple, *New Electronics*, 7 Feb. 1984.

Charles H. Small, Integrated development tools, *EDN*, 29 Nov. 1984.

Microprocessor development systems: a review, *Electronics Industry*, Dec. 1984.

Michael Sykes, Hardware/software integration in microprocessor based systems, *Electronic Engineering*, Nov. 1984.

R. Ercole, Simulation of microprocessor based systems, *Electronic Engineering*, Oct. 1984.

Bob Milne, Emulators, backed by strong debugging, mimic latest microprocessors, *Electronic Design*, 18 Oct. 1984.

B. Hordos. What is emulation and where is it going?, *Test & Measurement World*, Jun. 1984.

Douglas Lundin & Michael Crovitz, Triple-threat instrument debugs microprocessor based systems, *Electronic Design*, 9 Feb. 1984.

Peter Kleindienst, Teach yourself logic analysis, *Test Electronics*, Mar. 1985.

David Fynn, Trace control improves logic analyser performance, *New Electronics*, 5 Feb. 1985.

Trevor Powell, Post processing of logic analyser data, *Electronic Product Design*, Feb. 1985.

Ken Rush, Probing high speed logic, *Test Electronics*, Mar. 1985.

Glenis Moore, Logic analysers – wheedling out the bugs, *Electronics & Power*, Mar. 1985.

John Nichols, Making logic analysis affordable, *New Electronics*, 16 Apr. 1985.

Colin Dawney, The future of logic analysis, *New Electronics*, 16 Apr. 1985.

Stan Lang *et al.*, Analyzer plus scope displays timing, analogue waveforms at same time, *Electronic Design*, 24 Jan. 1985.

H. Van Eijkelenburg, With transitional timing analyzers boost resolution, open wider windows, *Electronic Design*, 15 Aug. 1985.

John Marshall, Integrated tools in software design, *Electronic Product Design*, Feb. 1985.

W. Damm, Analyzer module debugs 500 MHz logic with a scope's resolution, *Electronic Design*, 22 Aug. 1985.

G. Hirst, A portable microprocessor development system, *Electronic Engineering*, Sep. 1985.

P. Kleindienst, Systematic testing using word generators and logic analysers, *New Electronics*, 1 Oct. 1985.

Richard Goering, Logic analyzers offer wide choice of performance and price, *Computer Design*, 1 Sep. 1985.

G. Gohsman, PC-based analyzer links verification with software tools, *Computer Design*, 15 Aug. 1985.

B. Ableidinger, Real-time analyzer furnishes high-level look at software operation, *Electronic Design*, 19 Sep. 1985.

Logic analysis – complex functions, lower cost, *Electronics Industry*, Dec. 1985.

Charles H. Small, Software debuggers struggle to meet engineer's needs, *EDN*, 12 Dec. 1985.

C. Palmer, MDS – PCs to become dominant, *Electronics Industry*, Oct. 1985.

V. Schricker, Emulation for a complete 32-bit microprocessor family, *Electronics Industry*, Oct. 1985.

Tony Stevens, Operating to different principles, *Computer Systems*, Oct. 1985.

M. Wright, Basic skills speed hardware development, *EDN*, 17 Oct. 1985.

John Mayer, Logic analyzers, *Test & Measurement World*, Nov. 1985.

Charles H. Small, Logic analyzers evolve in response to high-level languages, *EDN*, 6 Feb. 1986.

Chapter 17

Jurgen R. Meyer-Arendt, *Introduction to Classical and Modern Optics*, Prentice–Hall, 1972.

Joe Horwath, Shedding light on LED luminance, *Electronic Design*, 12 Apr. 1973.

Brian Putland, Optical properties of LEDs, *Electron*, 18 Jul. 1974.

F. Rose, Measurements, *Electron*, 14 Mar. 1974.

M.L. Alekseeva *et al.*, The FOU general purpose objective photometer, *J. Opt. Technol.* **45**, no. 6. Jun. 1978.

J. Chrostowski *et al.*, Asymmetrical Fabry–Perot interferometer with feedback, *Appl. Opt.* **18**, no. 14, 15 Jul. 1979.

D.G. Taylor & J.N. Demas, Light intensity measurements I: large area bolometers with microwatt sensitivities and absolute calibration of the rhodamine B quantum counter, *Anal. Chem.*, May 1979.

D.G. Taylor & J.N. Demas, Light intensity measurements II: luminescent quantum counter comparator and evaluation of some luminescent quantum counters, *Anal. Chem.*, May 1979.

E.S. Avdoshin & V.B. Nikulin, Low-temperature radiometer with conical light pipe, *Instrum. & Exp. Tech.*, Jan.–Feb. 1979.

J.M. Palmer, Electrically calibrated pyroelectric radiometer without optical chopping, *J. Opt. Soc. Amer.*, Oct. 1979.

I. Lewin *et al.*, Developments in high speed photometry and spectroradiometry, *J. Illum. Eng. Soc.*, Jul. 1979.

T. Wearden, Test equipment for optical fibre installations, *Electronic Engineering*, Nov. 1980.

Tom Wearden, Technology and applications of fibre optics, *Electronic Product Design*, Oct. 1981.

J.D. Archer, Fibre optic communications, *Electronic Product Design*, Nov. 1982.

Robert W. Zimmerer, Laser power and energy measurement, *Test & Measurement World*, May 1983.

Optical time domain reflectometry – pinpointing fibre faults, *Test & Measurement World*, Jan. 1983.

Robert Landon, Fibre optic time domain reflectometers meet expanded system demands, *EDN*, 26 May, 1983.

G.M. Glashauser, Testing fibre optics transmitters and receivers on ATE, *Test & Measurement World*, Feb. 1983.

R. Rickenbach & P. Wendland, Fibre optics cable fault location, *Test & Measurement World*, Nov. 1983.

James E. Hayes, The fibre optics test and measurement market, *Test & Measurement World*, Nov. 1983.

P.H. Wendland *et al.*, Fibre-optic measurement are traceable to NBS standards, *Electronic Design*, 3 Feb. 1983.

W.L. Schumacher, Fibre optic test methods and measurements, *IFOC*, Sep. 1983.

Peter Keller, Optical sensors and sources: measuring infrared, visible and ultraviolet light – a primer, *Test & Measurement World*, May 1984.

Yoshiro Nagaki, The development of optical measuring instruments, *Telecommunications*, Jul. 1984.

Duane A. Burchick, Operational considerations for the design of fibre optics test equipment, *Test & Measurement World*, Jan. 1984.

J.D. Chipman & K. Prescott, Optical fibre bandwidth measurements in the field, *Test & Measurement World*, Mar. 1984.

U. Deserno & D. Schicketanz, Measurement methods for future fibre optic applications, *Test & Measurement World*, Jun. 1984.

E. Klement & K. Rossner, Principles of optical measurements for communication systems, *Test & Measurement World*, Jun. 1984.

Harold Winard, Focus on fibre optic cables: steadily forging the link, *Electronic Design*, 8 Mar. 1984.

A. Eckert & W. Schmid, Optical stimulus and receivers for parametric testing in fibre optics, *Hewlett Packard Journal*, Jan. 1985.

W. Berkel *et al.*, A versatile, programmable optical pulse power meter, *Hewlett Packard Journal*, Jan. 1985.

W.E. Schneider, Automated spectroradiometric systems: components & applications, *Test & Measurement World*, Jun. 1985.

David Peri & Israel Fainaro, Direct display of optical fibre perform deflection functions, *Test & Measurement World*, Sep. 1985.

J.M. Wiesonfeld & J. Stone, New methods for measuring dispersion and light loss in optical fibres, *Test & Measurement World*, Mar. 1985.

S. Barber, Optical fibre testing, *Telecommunications*, Jul. 1985.

Alan Wiltshire, Fault location in optical fibres, *Electronics & Power*, Feb. 1986.

J. Van der Donk, Beam propagation method, *Electrical Communications*, **59**, no. 4, 1985.

Index